Readings from
American Scientist

Earth's Energy and Mineral Resources

Edited by

Brian J. Skinner

Yale University

Member, Board of Editors
American Scientist

WILLIAM KAUFMANN, INC.   LOS ALTOS, CALIFORNIA

Library of Congress Cataloging in Publication Data

Main entry under title:

Earth's energy and mineral resources.

   (Earth and its inhabitants)
   1. Power resources—Addresses, essays, lectures.
2. Mines and mineral resources—Addresses, essays,
lectures.   I. Skinner, Brian J., 1928–
II. American scientist.   III. Series: Earth and its
inhabitants.
TJ163.24.E18   333.79   80-23495

ISBN 0-913232-90-4

Contents

Introducing

Earth and Its Inhabitants

A new series of books containing readings originally published in *American Scientist*.

The 20th century has been a period of extraordinary activity for all of the sciences. During the first third of the century the greatest advances tended to be in physics; the second third was a period during which biology, and particularly molecular biology, seized the limelight; the closing third of the century is increasingly focused on the earth sciences. A sense of challenge and a growing excitement is everywhere evident in the earth sciences—especially in the papers published in *American Scientist*. With dramatic discoveries in space and the chance to compare Earth to other rocky planets, with the excitement of plate tectonics, of drifting continents and new discoveries about the evolution of environments, with a growing human population and ever increasing pressures on resources and living space, the problems facing earth sciences are growing in complexity rather than declining. We can be sure the current surge of exciting discoveries and challenges will continue to swell.

Written as a means of communicating with colleagues and students in the scientific community at large, papers in *American Scientist* are authoritative statements by specialists. Because they are meant to be read beyond the bounds of the author's special discipline, the papers do not assume a detailed knowledge on the part of the reader, are relatively free from jargon, and are generously illustrated. The papers can be read and enjoyed by any educated person. For these reasons the editors of *American Scientist* have selected a number of especially interesting papers published in recent years and grouped them into this series of topical books for use by nonspecialists and beginning students.

Each book contains ten or more articles bearing on a general theme and, though each book stands alone, it is related to and can be read in conjunction with others in the series. Traditionally the physical world has been considered to be different and separate from the biological. Physical geology, climatology, and mineral resources seemed remote from anthropology and paleontology. But a growing world population is producing anthropogenic effects that are starting to rival nature's effects in their magnitude, and as we study these phenomena it becomes increasingly apparent that the environment we now influence has shaped us to be what we are. There is no clear boundary between the physical and the biological realms where the Earth is concerned and so the volumes in this series range from geology and geophysics to paleontology and environmental studies; taken together, they offer an authoritative modern introduction to the earth sciences.

Volumes in this Series

The Solar System and Its Strange Objects Papers discussing the origin of chemical elements, the development of planets, comets, and other objects in space; the Earth viewed from space and the place of man in the Universe.

Earth's History, Structure and Materials Readings about Earth's evolution and the way geological time is measured; also papers on plate tectonics, drifting continents and special features such as chains of volcanoes.

Climates Past and Present The record of climatic variations as read from the geological record and the factors that control the climate today, including human influence.

Earth's Energy and Mineral Resources The varieties, magnitudes, distributions, origins and exploitation of mineral and energy resources.

Paleontology and Paleoenvironments Vertebrate and invertebrate paleontology, including papers on evolutionary changes as deduced from paleontological evidence.

Evolution of Man and His Communities Hominid paleontology, paleoanthropology, and archaeology of resources in the Old World and New.

Use and Misuse of Earth's Surface Readings about the way we change and influence the environment in which we live.

Introduction: Earth's Energy and Mineral Resources

Earth is finite and so are its nonrenewable resources. That realization has finally been impressed on the populace at large and has thereby challenged our complacent assumptions about abundant supplies and the good life forever. But resource limitations should not be considered in overly simplistic terms. Though locally some resources may already be exhausted and even global exhaustion may lie but a short time ahead, substitution, alternate sources, and changing use patterns make long-term predictions difficult. The papers in this volume have been selected because they explore some of these difficult resource issues.

No resource is so obvious nor so necessary for the well-being of a technological society as energy. Growing populations and increasing per capita consumption of energy have kept the global use of energy rising steeply. We must question whether any source of energy is capable of meeting this double-barreled challenge. The first paper, by Eric S. Cheney, examines the economic, environmental, political and military dangers inherent in a reliance on each of the major energy resources. But the problems are not solely sociological, they are technological too. As Ralph Roberts points out, energy sources are not equivalent in ease of access, in methods of storage, or in feasibility of conversion from one form to another.

Few questions in recent years have raised such uncertainty as the magnitude of petroleum resources. Will more drilling discover new, large fields, or have we largely found what is to be found? Meyerhoff points out that 79% of the known petroleum is present in giant fields that aggregate only 2% of all fields discovered. When this fact is combined with the analysis of petroleum discovery successes, by Root and Drew, who use data drawn from the intensively drilled Permian Basin in Texas and New Mexico, an uncomfortable conclusion is reached. The initial drillings in a new area soon find the giant fields, and thereafter the amount of petroleum discovered per unit of drilling declines steadily. Because increased drilling is unlikely to find new giants in previously explored areas, the long-term trend in a well-drilled country such as the United States must inevitably be down.

The knowledge that petroleum production proceeds on a downward trend in the United States is not new. Imports continue to grow and the OPEC countries seem able to raise prices at will. Because domestic petroleum sources are insufficient to stop the rising tide of prices, alternate sources of energy are being closely examined. High among the possibilities are sources for liquid fuels—the so-called synthetic fuels—that can be used in existing engines. Ogden Hammond and Robert Baron discuss the various sources of carbonaceous material such as coal, oil shale, and tar sand and how they can be used to produce synthetic fuels. In the light of events since 1976, when their paper was published, their estimated cost figures probably need revision, but their technical data and opinions are as fresh and pertinent as the day they were written.

Nuclear energy sources have attracted more vociferous supporters and detractors than any other. Because the issues are extraordinarily complex in both technological and sociological terms, the situation is highly polarized. Kulcinski et al. discuss the possibilities of the two nuclear energy sources—energy released when a very heavy atom fissions into two intermediate weight atoms and energy released when two very light atoms are fused to make a heavier atom. For each system there are hazards to be faced in disposing the radioactive waste, and because the most likely fission fuel that can be used for the long-run, U^{238}, requires the still-to-be-perfected breeder reactor, technological and development costs could be enormous for both fission and fusion.

But can society afford to abandon nuclear power? Some countries, such as Japan, clearly think not, because they continue to build electrical power generators drawing energy from nuclear reactors. Others, such as the United States, apparently remain ambivalent. Alvin Weinberg, a long-time student of energy problems in general and nuclear sources in particular, discusses the historical development of nuclear power and the problems facing its acceptance. His paper and the two subsequent papers, by David J. Rose et al. and R. Phillip Hammond, were published as *Views*, or personal perspectives on the problems. Because all of the authors have played and continue to play influential roles in the development of nuclear energy, their *Views* are most informative. Rose and his associates ask what nuclear power is to be compared with and consider the costs and drawbacks of turning to other energy sources such as coal. Hammond looks at some of the hazards involved in the operation of reactors and the huge problem of managing and disposing of the radioactive wastes that lies ahead. "The hazard is real" says Hammond, and "somewhere along the line in a nuclear economy there will be some lives lost, some injuries, and some nasty messes to clean up and decontaminate." But he concludes that in all probability "there will be no catastrophe," and "other paths will have higher costs in lives, in dollars, and in damage to the environment."

Not everyone is prepared to accept the conclusions of those who examine the nuclear power question. There are other sources of energy, they argue, and solar energy in particular is cheap and environmentally acceptable. Long before most people were aware of pending energy problems, Farrington Daniels turned his attention to direct use of solar energy. His paper is an elegant discussion of the many ways it can be used. But the big question concerns electricity, of course, because electricity is one way in which energy can be readily transported. W.D. Johnston, Jr., discusses solar cells that convert the sun's rays directly to electricity. In discussing the design, construction, and future prospects, he is led to conclude that promising though the method may be, "photovoltaic conversion is further in the future than other solar technologies." The other technologies, as outlined by Daniels, are based principally on the use of solar heat. Duguay examines the possibility of solar energy for heat, light, and electricity, showing how a building could be designed to be essentially self-sufficient in all respects. Perhaps the small-scale solar installations discussed by Duguay will turn out to be successful, but Pollard doubts that large-scale electrical systems will be competitive. He calculates that base-load electricity (meaning that generated steadily 24 hours a day) from solar sources would be at least four times more expensive than power from present-day coal-fired plants.

There are other ways solar energy can potentially be captured. The most obvious is in green plants that obtain energy from sunlight by photosynthesis. Pollard argues that to some extent replacement of oil and gas by fuels derived from animal and agricultural wastes is promising. Calvin describes the photosynthetic process itself and argues persuasively that the photosynthetic system in certain plants can provide a design for a synthetic system that will provide a renewable resource for both materials and fuels. But if growing plants are to be used as energy sources, one of the first questions to be asked is whether increased yields will supply both food and fuel. Increased

yields mean higher energy consumption, and this aspect of alternate energy sources is discussed by G.H. Heichel.

Geothermal power, drawn from Earth's internal heat, is a local source of energy that has been used extensively and effectively in certain countries. A.J. Ellis, a New Zealand scientist closely associated with development of one of the most advanced geothermal systems in the world, discusses the problems and prospects. As with most energy sources, geothermal systems have their problems, but most can be overcome; the development of geothermal energy can be expected to expand rapidly.

Lest we be lulled into thinking that all resource problems are focused in energy, the final four papers discuss aspects of other mineral resources. Skinner argues that the geological occurrence of many metals is such that accessible and recoverable supplies of silver, tin, copper, lead, and many others will soon be exhausted, while supplies of iron, aluminum, manganese, magnesium, and titanium are large and practically inexhaustible. For the very long run—beyond the end of the present century—he argues that technology will increasingly have to do without many specialty metals and increasingly use iron and aluminum. The intervening period will be one of intensive prospecting as still undiscovered rich deposits are sought. Pressures will be tremendous on those who wish to preserve some areas from changes wrought by mining, but pressures will be equally great on those who wish to mine to designate areas where ores are likely to be found and to estimate how much. The difficult task of assessing undiscovered resources is in its infancy, and so far only one large-scale attempt has been made for a specific region. Singer and Ovenshine discuss how their analysis of potential mineral resources in Alaska was made. Zumberge addresses another problem facing mineral hunters. Certain parts of the world—the deep seafloor and Antarctica—are international property. Geopolitical conflicts, he argues, will be fierce when mineral resources are discovered in Antarctica.

Eventually any mineral deposit is worked out and depleted. When all the world's deposits of a certain kind are gone, or are controlled by OPEC-like cartels, what do we do? Bauxite, the present-day ore of aluminum is a prime example, and Patterson discusses its future use and the alternatives to follow. For aluminum, as for iron and a few other metals, there are, fortunately, alternatives with slightly higher costs. But most metals are not so fortunate. As with nonrenewable resources, energy or mineral, answers to questions of future supplies are still ambiguous. But on one thing all students of the subject seem to agree—the time when major changes will appear is not far ahead.

Suggestions for Further Reading

Brobst, D.A. and W.P. Pratt, editors, *United States Mineral Resources* (U.S. Geological Survey Professional Paper 820, 1973).

Cloud, P.E. Jr., editor, *Resources and Man* (San Francisco: W.H. Freeman and Co., 1969).

Cook, Earl, *Man, Energy, Society* (San Francisco: W.H. Freeman and Co., 1976).

Daniels, Farrington, *Direct Use of the Sun's Energy* (New Haven, Connecticut: Yale University Press, 1964).

Landsberg, H.H., editor, *Energy: The Next Twenty Years; Report of a Study Group Sponsored by The Ford Foundation* (Cambridge, Massachusetts: Ballinger, 1979).

Metz, W.D. and A.L. Hammond, *Solar Energy in America* (Washington, D.C.: American Association for the Advancement of Science, 1978).

Nuclear Energy Policy Study Group, *Nuclear Power Issues and Choices* (Cambridge, Massachusetts: Ballinger, 1977).

Skinner, Brian J., *Earth Resources*, 2nd ed. (Englewood Cliffs, New Jersey: Prentice-Hall, 1976).

Steinhart, Carol and John Steinhart, *Energy: Sources, Use and Role in Human Affairs* (North Scituate, Massachusetts: Duxbury Press, 1974).

Stobaugh, Robert and Daniel Yergin, editors, *Energy Future; Report of the Energy Project at the Harvard Business School* (New York: Random House, 1979).

Authoritative and up-to-date reviews and summaries of important issues regarding energy resources and energy use can be found in the volumes of *Annual Review of Energy*, published by Annual Reviews, Inc., Palo Alto, California, 94306. The following articles, listed by volume, should prove especially interesting:

Volume 1, 1976

Brown, Harrison, "Energy in our Future," p. 1–36.

Volume 1, 1976 (continued)

Morse, Frederick H. and Melvin K. Simmons, "Solar Energy," p. 131–158.

Post, R.F., "Nuclear Fusion," p. 213–256.

Rattien, Stephen and David Eaton, "Oil Shale: The Prospects and Problems of an Emerging Energy Industry," p. 183–212.

Schmidt, Richard A. and George R. Hill, "Coal: Energy Keystone," p. 37–64.

Zebroski, E. and M. Levenson, "The Nuclear Fuel Cycle," p. 101–130.

Volume 2, 1977

Bolin, Bert, "The Impact of Production and Use of Energy on the Global Climate," p. 197–226.

Grenon, Michael, "Global Energy Resources," p. 67–94.

Häfele, W. and W. Sassin, "The Global Energy System," p. 1–30.

Volume 3, 1978

Lovins, Armory B., "Soft Energy Technologies," p. 417–518.

Merriam, Marshal F., "Wind, Waves and Tides," p. 29–56.

Spinrad, Bernard I., "Alternative Breeder Reactor Technologies," p. 147–180.

Volume 4, 1979

Brooks, Harvey and Jack M. Hollander, "United States Energy Alternatives to 2010 and Beyond: The CONAES Study," p. 1–70.

Harris, DeVerle P., "World Uranium Resources," p. 403–432.

PART 1 *The Energy Problem*

Eric S. Cheney

U.S. Energy Resources: Limits and Future Outlook

The environmental, economic, political, and military dangers inherent in each of the major energy resources could be decreased by zero per capita power growth

There are three critical concepts of worldwide significance that have emerged from previous discussions of energy crises: (1) the conventional energy sources are not inexhaustible; (2) the world's consumption of energy is doubling every decade (which is equivalent to a growth rate of 7% per year); and (3) this exponential exploitation of finite natural resources may be one of the major limits to growth of the world's population and industrial society within the next century (Meadows et al. 1972). Table 1 summarizes Hubbert's (1971) estimated lifetimes of the world's important energy sources. Virtually all of the world's coal and petroleum will be burned in the present millennium of the entire life span of civilized man on earth (Hubbert 1969). Because of the consequent environmental problems, the burning of fossil fuels has been termed man's greatest geochemical experiment (Holland 1972).

The lifetimes for the fuels shown in Table 1 depend somewhat upon the estimated reserves. Theobald et al. (1972) show that considerable dis-

Eric S. Cheney is an associate professor of geological sciences at the University of Washington. As a result of his teaching and research on the geology, geochemistry, and origin of nonrenewable resources and his experience exploring for ore deposits, he has become deeply interested in the consequences of the exploitation of natural resources. This article has its origins in illustrated lectures presented to geological and nongeological audiences during the past two years. The author thanks his colleagues J. D. Barksdale, H. A. Coombs, R. L. Gresens, and D. P. Dethier for many fruitful discussions and for reviews of earlier drafts of his manuscript. Address: Department of Geological Sciences, University of Washington, Seattle, WA 98195.

agreement exists on ultimate oil and gas reserves of the United States; the U.S. Geological Survey's estimates are 6–7 times greater than Hubbert's. However, the reserves are still finite, and with demand growing exponentially, an arithmetic increase in the reserves does not appreciably extend their lifetimes. For example, a doubling period of 10 years would consume an 8-fold increase in reserves in only 30 years. Hubbert (1971) notes that half of the world's coal mined up to 1969 had been burned after 1940 and half of the petroleum had been burned since 1959.

U.S. energy resources

Uses of energy in the United States. To understand the magnitude of the energy crises, one has to consider our energy sources and the problems of using them in substantially larger amounts. Natural gas and petroleum produce about three quarters of America's energy, coal about a fifth, and hydropower about 4%. All of the other energy sources, including nuclear power, produce about 1% (Beall 1972). A quarter of the energy produced is used by electrical utilities. Half of the total energy produced ultimately is discarded as waste heat (Cook 1971).

Since World War II, America's total annual energy consumption has doubled every 16 years, and electrical consumption has doubled every 10 years (Hubbert 1971). If these conditions remain unchanged, the capacity of all existing electrical power plants—hydroelectric, nuclear, coal, oil, and natural gas—will have to be duplicated in 10 years. This would mean that, in ad-

dition to arguing about the environmental impact of various power plants, we would actually have to build increasing numbers of them each year and increase our imports of petroleum (mostly via tankers) every year. The per capita consumption of electricity in the United States is about six times the world average. The per capita consumption of electricity in the Pacific Northwest is twice the national average (Bonneville Power Admin. 1970), largely because of aluminum smelting, low cost electricity that is the envy of the rest of the nation, and the extensive use of hydropower instead of fossil fuels.

Electricity. In 1970 in the United States the energy sources for electricity were coal 46%, gas 24%, hydropower 16%, oil 13%, and nuclear only 1% (Beall 1972) but nuclear power has now grown to 4%. Nearly 70% of all this energy is lost in the generation and transmission of electricity (Cook 1971).

Because so much electricity is generated from coal, which adds sulfur dioxide and particulate matter to the atmosphere, it currently makes little sense to combat air pollution problems by replacing gasoline-powered cars with electrically powered cars. However, Mencher and Ellis (1971) suggest that when nuclear power generates about 30% of the electricity, electrically powered automobiles would produce less sulfur dioxide and respirable particulate matter (less than 5 microns) than gasoline-powered cars. Because the efficiency of the internal combustion engine and the generation, distribution, and utilization of electricity are about the same, energy would not be conserved by electric cars. We would have to increase electric-generating capacity by about 75% to operate the 100 million cars that are now on the road (Summers 1971).

Natural Gas. Natural gas is in great demand (Bureau Natural Gas 1972) because it is easy to handle and burns cleanly. However, it is the scarcest of all the major fuels in the United States (Table 2). Reserves in known gas fields in 1971 were equivalent to 11 years and are declining. The major reason for this shortage is that until August 1972 the Federal Power Commission encouraged consumption and discouraged exploration for new reserves by prohibiting producers from charging distributors more than 20¢ per thousand cubic feet of gas. Until proven reserves increase dramatically (it will take 7 to 10 years to find, develop, and market any major new discoveries), it is folly to encourage automotive, industrial, or residential consumption of this fuel.

Petroleum. Despite declining reserves (now equal to 12 years of current production if Alaska's North Slope reserves are included) and declining production, the United States is still the world's leading oil producer (Table 3). However, domestic production meets fewer of our requirements each year and is now less than 70% of our consumption (Fig. 1). About 40% of the petroleum we use is in the form of motor gasoline. Ironically, at the same time that domestic reserves are declining, the efficiency of automobiles, as defined by miles per gallon, has dropped drastically since 1971 owing to the installation of antipollution devices and air conditioning (Table 4). As a result, oil imports in 1973 exceeded the pessimist's estimates made in 1972. If automotive use continues at such a rate, the demand for oil will nearly double by 1985, and declining domestic production, including Alaska, will provide less than half of our consumption (Beall 1972).

Because the Alaskan pipeline is much in the news, three popular misconceptions about Alaska's North Slope oil need to be dispelled: the first is that the present liquid fuel crisis has been caused by delays in building the trans-Alas-

Table 1. Lifetimes of the world's energy resources (from Hubbert 1971).

Coal and lignite	300 years
Oil, gas, and tar	64 years
Hydro and tidal, total equals 60% of world's 1970 energy consumption	Renewable
Solar, total equals 120% of world's 1970 energy consumption	Renewable
Geothermal	50 years
Fission	
U^{235}	90 years
Breeder	Unlimited
Fusion	
$Li^6 \rightarrow H^3$	300 years
$H^2 + H^2$	Unlimited

kan pipeline. The gasoline shortage in the summer of 1973 may have been caused by earlier price controls that disrupted corporate planning and investment and, perhaps, by the industry itself. Most important, however, was a lack of refining capacity to meet the demands of an increased number of cars, many of them with reduced mileage efficiencies. More simply stated the 4% per year rate of growth of gasoline consumption jumped to 6–7% in 1972 and 1973. This has been termed a "consumption crisis"

(Hartley 1973). Because the hard-pressed refining companies must make a seasonal change from maximum gasoline production for the summer to maximum heating oil production for the winter, the "consumption crisis" and any abnormally cold weather can be expected to generate a heating oil shortage during the winter of 1973–74.

Secondly, it is instructive to speculate where Alaskan oil will be marketed. Because supertankers are too large to pass through the Panama Canal, and no major pipeline now exists from the West Coast eastward, the major American market for this petroleum must be the West Coast. In 1970, the states in Petroleum Administration District V (Arizona, California, Oregon, Washington, Alaska, and Hawaii) consumed almost exactly 2 million barrels of oil a day (bpd) (Kirby and Moore 1972). Two million bpd is the planned output of the Alaskan pipeline (Special Interagency Task Force 1972). Assuming a growth rate of 4% a year—which was about the average projected national growth rate for petroleum consumption (*World Oil,* 15 Feb. 1973) before the liquid fuel crisis was publicized in 1973—assuming an adequate refining capacity, and further assuming that the combined volume from present domestic and imported sources of petroleum into PAD V remain substantially unchanged, it will take 18

Table 2. Fuel reserves of the United States. References: (*1*) National Coal Association 1972; (*2*) Total recoverable, paramarginal, and submarginal discovered and undiscovered resources calculated by Theobald et al., 1972, divided by present production; (*3*) Total recoverable production from Theobald et al., 1972; (*4*) *World Oil,* 15 Aug. 1972; (*5*) Bureau of Natural Gas, 1972; (*6*) *Mining Engineering,* May 1973.

	Percent of total energy (1)		*Years: reserves/production*	
	Present technology and reserves	*Ultimate recovery*	*Present technology and reserves*	*Maximum ultimate recovery (2)*
Coal	88	74	700 (*3*)	6,000
Petroleum	2	5	11 (*4*)	130
Natural gas	3	5	11 (*5*)	340
Natural gas liquids	—	1	—	—
Oil shale	5	13	0	1,500
Uranium	2	2	20 (*6*)	670

years for the PAD V market to double and to thereby absorb Alaskan oil. In other words, until about 1988 a very significant portion of Alaskan oil might be sold to another major industrial nation that borders the Pacific Ocean, has huge tankers, and needs oil. In exchange, that country might dispatch Middle Eastern oil to the eastern United States.

According to SOHIO (1973), the North Slope petroleum reserves presently are 9.6 billion bbl. Assuming that production is 2 million bpd, these reserves will last 13 years. The earliest date North Slope oil could reach the West Coast is 1978. Thus, whether exported or not, the North Slope oil, if produced at a rate of 2.0 million bpd, would be depleted soon after West Coast consumption had grown large enough to use the daily production. The specters of exportation and early depletion of North Slope oil could be avoided if the output of the pipeline were less than the officially announced 2 million bpd or if reserves were greater than 9.6 billion bbl. Spahr (1973) states that the maximum efficient output of the presently known reserves on the North Slope would be 1.5 million bpd; initial production, SOHIO estimates, probably would be 1.2 million bpd.

Now that the trans-Alaskan pipeline has been approved, exploration will begin anew, and the reserves on the entire North Slope ultimately might prove to be greater than 9.6 billion bbl. SOHIO's merger with British Petroleum is dependent upon the refining, sale or exchange, and use of up to 0.6 million bbl per day in the United States or Canada (Moller 1973). Thus, at least this much Alaskan oil seems destined for domestic use. Nonetheless, if additional reserves are discovered on the North Slope, the following incentives to produce 2.0 million bpd and to export the surplus will be overwhelming: (1) the companies would recapture their investments sooner; (2) according to SOHIO the total costs per barrel for outputs of 1.2 and 2.0 million bpd are $1.435 and $1.134 respectively; (3) the oil could be transported to Japan in foreign tankers cheaper ($0.38/bbl) than to Los Angeles in U.S. tankers

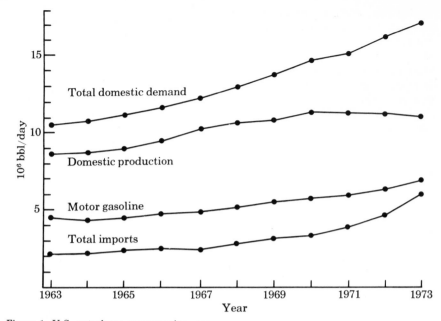

Figure 1. U.S. petroleum consumption, production, imports, and gasoline consumption. Source, *World Oil,* February 15th issues, 1964–1973.

Table 3. Nations with the greatest proven petroleum reserves, 1972. References: (*1*) *World Oil,* 15 August 1973; (*2*) Albers et al. 1973; (*3*) U.S. Bureau of the Census 1973

Country	Reserves (*1*)	Production (*1*)	Net exports, 1970 or 1971 (*2*)	U.S. imports (*3*)
	(All figures in 10⁶ bbl)			
Saudi Arabia*	137,100	2,201	1,378	159
Kuwait*	73,937	1,097	1,276	19
Iran*	62,202	1,849	1,573	57
USSR	42,000	2,896	233	3
United States	36,339	3,457	(−1,325)	—
Iraq*	33,000	536	546	2
Libya*	24,100	822	999	38
Abu Dhabi*	18,234	384	383	12?
Venezuela*	13,872	1,178	1,219	448
Neutral Zone*	13,500	208	(with Kuwait and Saudi Arabia)	
Nigeria*	12,600	665	533	89
China	12,500	192	(−1)	0?
Indonesia*	10,700	395	263	60
Algeria*	9,750	398	351	36
Canada	8,020	564	9	371
Ecuador*	5,964	29	(−8)	6
Qatar*	5,832	177	131	2
Total	519,686	17,050	8,894	1,372
OPEC* total	420,827	9,939	8,644	998
World total	562,295	18,638	—	1,651

* Member, Organization of Petroleum Exporting Countries (OPEC)

($0.47/bbl); and (4) the revenue of the state of Alaska will be largely dependent upon oil production.

After being subjected to an energy shortage, the American public may demonstrate a certain amount of economic nationalism about exporting Alaskan oil and could decide to limit its production or to require (and possibly subsidize) a trans-Canadian pipeline to deliver the oil to the markets in the midwestern United States. The same public also might question whether it should assume such grave environmental risks for the exportation of Alaskan oil to another nation. If the reserves prove greater than 15 billion bbl, a second pipeline, perhaps across northern Alaska and Canada, might be needed.

A third misconception is that the national energy problem will be solved by the importation of Alaskan oil. This would be true only if the demand for oil decreases, which is unlikely, and if the oil also could be delivered to the eastern United States. If the demand for oil does increase from 17 million bpd in 1973 to the estimated 26 million bbl in 1985, 2 million bbl of Alaskan oil would only meet 7.7% of the demand (Special Interagency Task Force 1972); and an additional 14.8 million bbl will have to be imported each day from other sources.

Where will these 14.8 million bbl come from? Major new oil discoveries in the North Sea and western Siberia are larger than those in Alaska (Weeks 1973), but these oils almost certainly will be consumed by Europe, which heretofore has had to import most of its oil. Because exploration, development, production, and distribution of oil generally take 7 to 10 years lead time, only countries with presently producing oil wells and huge proven reserves can meet our increased demand throughout the next decade. Most of the nations with large reserves and exporting capacities, the majority of which are Arab (Table 3), are members of the Organization of Petroleum Exporting Countries. The OPEC was formed in 1960, and it now seeks to bargain as one with the international oil companies. The Middle East already supplies most of the requirements of Western Europe and

Table 4. Variables in automotive efficiency in terms of gasoline mileage. References: (1) Lincoln 1973; (2) Cook 1973; (3) Sansom 1973; (4) Hartley 1973.

2,500–5,000 lb car	65–100% decrease in efficiency (3, 4)
Rotary engine	Up to 35% decrease (3)
Emission controls	15–35% decrease (1, 2, 4)
Air conditioning	9–20% decrease (3, 4)
Lead-free gasoline	*12% decrease (1)
55 mph to 65 mph	*11% decrease (4)
Automatic transmission	5–6% decrease (3)
Power steering	3–4% decrease (4)
Steel belted radial tires	Up to 10% increase (3)

* This gasoline consumption is approximately equivalent to the 1973 gasoline shortage (4).

Japan, and if the United States also becomes highly dependent on this source (for our automobiles), the political and economic strength of the Arab nations will increase accordingly. In November 1973, following the fourth Arab-Israeli war, the Arab countries decreased production and sale of petroleum to the importing countries in order to reverse the fortunes of Israel. The future of Israel, despite four brilliant military victories in the past twenty-five years, clearly is in doubt (unless the United States and Saudi Arabia agree to maintain Israel and Jordan as buffer states between Saudi Arabia and the radical governments of Egypt, Syria, and Iraq).

For politicians of the near future, it is equally significant that the world's greatest known natural-gas reserves (30% of the total) appear to be in Siberia (Albers et al. 1973). Furthermore, owing to recent discoveries in Siberia, the oil reserves of the USSR may be considerably greater than indicated in Table 3; they may be about equal to Kuwait's (Albers et al. 1973).

Some Arab nations, especially Saudi Arabia, have received so much money from petroleum that increased production and sales may be of little importance to them. For example, in 1974 Saudi Arabia may earn $15 billion but can only spend $3 billion at home. Furthermore, cash on hand is a detriment during monetary devaluations; Arab dollars and Eurodollars were a major cause of the dramatic increase in the price of gold and the devalua-

tion of the American dollar in February 1973. In addition to obtaining in June 1973 an 11.9% price increase for their oil, to offset the February 1973 devaluation of the dollar, the OPEC countries are demanding still higher prices and are looking for other ways to protect their black gold. Some countries may curtail production because oil in the ground is their best hedge against inflation in an energy-hungry world. By 1980, Saudi Arabia may have accumulated $100 billion in reserves—more than the United States, Europe, and Japan combined (*Christian Science Monitor*, 11 Oct. 1973). To protect this fortune and to prolong the economic benefits of petroleum after her resources become depleted, Saudi Arabia is likely to become the world's banker and is already contemplating investing in new American oil refineries.

Other economic, environmental, and military problems may be imported with the oil. The balance of payments may worsen. The United States, which paid $3 billion for oil imports in 1970, will be paying (in terms of 1972 undevalued dollars) $20 to $30 billion by the early 1980s (McLean 1972); by comparison, in 1972 the United States spent about $30 billion on all kinds of imports. Oil imports in 1973 are estimated to be $9 billion (Abelson 1973). The potential for pollution problems also will multiply enormously as off-shore drilling, supertankers, and offshore oil terminals for these tankers outside shallow or crowded harbors (like New York) become more commonplace. The northernmost deep water "port" on the West Coast of the United States is Puget Sound, which is within the major population centers of the state of Washington. Because of narrow rocky channels, frequent fog, and swift tidal currents, Puget Sound is particularly susceptible to oil spills and the damage they inflict. The military wisdom of becoming dependent upon foreign oil sources can also be questioned, especially when it is remembered that the Soviet navy has 408 submarines, compared to Hitler's 57 at the beginning of World War II (Blackman 1972; Morison 1963). In this sense, Alaskan oil is "foreign" oil.

To maintain their profits and corporate existence beyond the 70-year lifetime of oil, several of the major oil companies have aggressively acquired coal and uranium companies since 1968. Because the government chose not to contest these acquisitions as monopolistic, the country now is run by a handful of energy companies whose financial and political power would be difficult to challenge.

Nuclear power. Nuclear power, especially breeder reactors that produce more fuel as plutonium than the rather scarce uranium they consume, is considered by many to be the solution to our difficulties. Owing to technological problems, it is generally conceded that commercial breeder reactors will not be available for another 15 to 30 years. Meanwhile environmental conflicts and technological considerations have significantly postponed the mass construction of conventional nuclear plants. Nuclear power may not capture 25% of the energy market until A.D. 2000 (Lincoln 1972). In the interim, the nation's energy demands may have doubled 3 times.

Geologically and seismologically safe nuclear power plant sites with abundant coolant water are still available, but many are not environmentally acceptable. The integrity of some sites can be greatly enhanced; for example, a massive flood wall was built around the Trojan plant downstream from Portland, Oregon, to protect the plant in case any of the dams above it on the Columbia River fail.

Nonetheless, despite the best geological and nuclear engineering, some risk exists at any site.

The fail-safe disposal of highly radioactive waste products is a critical problem that grows more serious each year. As yet, no disposal system exists. Because the half-life of plutonium-239 is 24,360 years, waste material must be isolated from the biosphere for hundreds of thousands of years, that is, for periods of time during which enormous geological changes can take place. Perhaps we should consider putting the wastes in Saturn rockets and expending the tremendous amount of energy needed to fire them into the sun; however, such a solution would require totally fail-safe launchings! The best earth-bound disposal sites may be horizontal, impervious, bedded salt formations 100 to 300 m thick at depths between 400 and 700 m in geologically stable areas on land far from major population areas (National Academy of Sciences–National Research Council 1970; Gera 1972). The number of such sites is extremely limited: Lyons, Kansas, and Carlsbad, New Mexico, are under investigation; other areas with bedded salt that might be geologically suitable are western New York, southern Michigan, central Saskatchewan, and Nova Scotia. Another type of disposal site might be burial at sea in the deltas of those major rivers where sedimentation and subsidence rates are abnormally high. Alternatively, because radioactive wastes generate temperatures in excess of 900°C, we might explore the possibility of using them as a source of thermal energy, but once again, an exceedingly safe site would be needed.

The limiting factor to high levels of energy consumption on earth probably will be the disposal of vast amounts of hot water into rivers, oceans, and the atmosphere (Cook 1971; Summers 1971). Because nuclear reactors produce 50 percent more of this thermal pollution as coal or oil-fired plants, the long-range solution to dissipating this heat and to conserving other forms of energy would be to locate reactors within cities so that the hot water can be utilized to provide space heating and hot water for commercial and residential build-

ings. At present, the relatively low temperature of the outlet water (15°F warmer than the intake water) makes it difficult to use efficiently (Summers 1971).

Solid fuels. Until the commercial breeder reactor is developed, the only way the United States can begin to meet projected energy demands domestically is to mine its abundant solid fuel reserves. This country is liberally endowed with coal. Table 2 indicates that, on an energy basis at current prices and technology, coal represents 88% of the nation's available energy. Oil shales in Colorado, Wyoming, and Utah and tar sands in Alberta both contain more fuel than the original oil and gas reserves of the U.S. Unfortunately three-fourths of the coal reserves and all of the oil shale are west of the Mississippi, whereas three-fourths of the population are east of it. Wherever these solid fuels are within a few hundred feet of the surface, they will be mined by huge open-pit methods because surface mining, although environmentally more damaging, is cheaper, more efficient, and, above all, safer than underground mining. Nonetheless, in the autumn of 1973 the U.S. Congress was acting on bills to prohibit surface mining on federal lands. Underground mining and retorting of the oil shales may be accomplished by multiple detonations of nuclear devices (Perry 1968).

Ironically, the country's largest coal mine is currently being developed in Centralia, Washington, within the very heartland of hydropower. Because even this coal mine will not meet anticipated demands, northwestern and midwestern power companies are developing the huge coal reserves of southeastern Montana and southern Wyoming (Davenport and Duell 1972). The lignite that underlies the western half of North Dakota at shallow depths is another obvious source of energy, especially as it contains uranium that could be recovered as a by-product.

Plans also are well developed to strip the vast coal reserves of northeastern Wyoming and adjacent Montana in order to produce synthetic "natural" gas (*Mining Engineering* 1973). Several types of pilot plants currently are being tested (Sisselman 1972). The first commercial gasification plant probably will not be in operation before 1978 (National Coal Association 1972). Once hydrogen is inexpensively available in large amounts (see below), coal would have a tremendous advantage over oil shales as a source of synthetic fuels for the following reasons: (1) the greater reservoir of coal-chemistry knowledge; (2) yields of 100 to 150 gal of oil per ton of coal compared to 25 to 40 gal per ton of oil shale; (3) smaller waste disposal problem, and (4) proximity of eastern coal deposits to large markets (Perry 1968).

The shales in Colorado, Utah, and Wyoming could yield 25 to 40 gal of synthetic oil per ton of shale from solid organic materials in the shales (Perry 1968). Oil recovered from these shales will become economic when the price of petroleum increases sufficiently. In 1970 the estimated cost of oil from shale yielding 30–35 gal per ton was $4.30 to $5.30 per 42 gal barrel (Culbertson and Pitman 1973), whereas the price of crude oil was about $3/bbl. In early 1973 the price of crude oil was $3.30/bbl. In mid-1973, Libya nationalized the operations of Occidental Petroleum and posted a price of $4.90/bbl (*World Oil*, Sept. 1973). In late 1973, the OPEC was bargaining for a price of $6/bbl. As the price of oil rises, the politics of oil shale will become fascinating because the government owns most of the best oil shale lands, and it alone, not the private energy companies, has access to the huge amount of capital that will be needed to develop the oil shales on a scale that would really solve a liquid fuel crisis.

The availability of water may limit the mining and processing of oil shale. Depending upon various assumptions, enough water may still exist in each of the three states to permit the mining and partial refining of 0.6 to 5.0 million bpd in each state (Perry 1968). Assuming the worst conditions of water availability, the three states might produce about 2.9 million bpd, which would be about 11% of the nation's projected oil consumption in 1985. Greater productive capacity presumably could be achieved only by

discovering abundant supplies of groundwater or by importing water, probably from the Columbia/Snake drainage basin. The waters of the Columbia and Snake rivers are vitally needed for irrigation and the generation of hydroelectric power elsewhere.

Unless nuclear or other in situ methods of breaking and retorting the oil shale are utilized, production of 2.9 million bpd would require underground mining on a scale 48 times greater than the country's largest underground copper mine, at San Manuel, Arizona. Obviously a comparison with San Manuel is far from perfect, but it does illustrate the environmental and capital costs involved. The recent expansion of mining, milling, and smelting capacities at San Manuel cost $250 million (Beall 1973). This would suggest a capital investment of $4,000/bpd producing rate for the oil shales compared to $4,000 to $5,000/bpd required for plants to process the tar sands of Alberta described below. If, because of these problems, oil from shale does not exceed 100,000 bpd in 1985 (Beall 1972), it cannot appreciably solve a fuel crisis.

Canadian energy resources

Some of the same problems that beset U.S. oil shales exist with the Alberta tar sands, but by 1976 two companies will be producing a total of 170,000 barrels of synthetic crude oil per day (Spragins 1972). Canada also has huge uranium reserves, and the oil and gas discoveries in the McKenzie River delta and Arctic islands ultimately may rival the Alaskan North Slope.

In 1970, Canada became a net exporter of energy for the first time. However, the Canadian government is becoming increasingly reluctant to commit its large reserves to any proposed "North American" energy program. A few examples should suffice. In 1970 the Canadian government prohibited the sale of the largest Canadian uranium company, Denison Mines, Ltd., to an American energy corporation and agreed to subsidize Denison. In 1971 it blocked the sale of the second-largest domestically owned oil

company, Home Oil Co., Ltd., to another American oil company. On 1 March 1973, the exportation of crude oil was added to energy sources regulated under the National Energy Board Act of 1959. According to its provisions, the Energy Board must now satisfy itself that the quantity of oil, gas, or power to be exported does not exceed the surplus remaining after allowance has been made for the foreseeable requirements of Canada (Frazer 1973). In November 1973 the export tax on petroleum was raised from 40¢ to $1.90/bbl. D. M. Frazer, Vice-Chairman of the National Energy Board, concludes that "there is no supply crisis in Canada, and there is no need for Canadians to panic. In energy matters at least we don't have to develop a high fever when the U.S. starts to sneeze" (1973).

Other energy sources

Geothermal power is extremely limited and, contrary to popular belief, depletable (see Table 1). Full development of the world's hydropower would not be equivalent to the world's 1970 energy consumption (Table 1). Only an infinitesimal amount of hydropower could be generated by the oceans' tides because most tides are not high enough to provide a significant hydrostatic head. Seattle has calculated that its municipal sewage system produces enough methane to power 20% of the city's buses. Similarly, water added to the manure produced by 100,000 cattle in a feedlot would produce enough methane to supply 30,000 people, and the urban and agricultural wastes of the United States could theoretically produce 1.5 times the methane currently used (Cook 1973).

Earth-bound solar power would be four times as expensive as nuclear power, and an orbiting solar plant that transmits energy back to earth by microwaves would be at least twice as expensive as nuclear power (Summers 1971). It may be decades before nuclear fusion, used in the hydrogen bomb, can do useful work. Exothermic chemical reactions within electrochemical batteries do not produce as much energy as is expended in recovering the metals from their ores (Roberts

1973). Hydrogen has been suggested as the synthetic fuel of the future and as a method of converting and storing electricity for use as peaking power, but it is costly and difficult to store (Maugh 1972) and cannot solve an energy shortage because electricity is needed to manufacture it. However, windmills might be used to provide the electricity to decompose the water into hydrogen and oxygen.

The above discussion demonstrates that none of the conventional or innovative energy resources available to the United States are without financial, political, environmental, or military risks if used in much greater amounts than at present. In the long term, the least objectionable resources may be those which currently are too expensive to consider. The following section discusses ways in which energy can be conserved so that some of the above energy resource problems might be significantly postponed or avoided. Ultimately, however, the United States will have to adopt zero per capita power growth and the steady state (instead of growing) economy that it implies.

Energy conservation

The best way to postpone some of the energy crises would be to conserve energy. Reactions to the shortage of hydroelectric power in the Pacific Northwest in the autumn of 1973 are good examples of the magnitude of energy reductions that can be made voluntarily. Despite growth in their facilities since 1972, the University of Washington, the Boeing aerospace company, and the Seattle public utility system (serving populations of approximately 50,000, 51,000, and 600,000, respectively) achieved 10%, 16 to 20%, and 5% voluntary reductions in their consumption of electricity by mid-November 1973.

We might be able to reduce permanently about a quarter of our consumption of energy without significantly affecting our life-styles. The greatest saving would result from the proper insulation of commercial and residential buildings, especially glass-clad, high-rise office buildings (Hammond 1972); such a move would reduce heating

in the winter and air conditioning in the summer. Air conditioning is the fastest growing end-use of electricity, increasing 20% per year from 1967 to 1970 (Cook 1971). More stringent insulation specifications could be written into Federal Housing Authority regulations and into all government building contracts. Hirst and Moyers (1973) calculate that with optimum insulation the average home would use 42% less energy for space heating, at a considerable economic savings to the owner. In 1970, this would have reduced the nation's energy consumption by 4.6%. Domestic furnaces are about 75% efficient when sold but, owing to the homeowner's neglect, are less than 50% efficient in practice. Mobile homes, which constitute a quarter of all new dwelling units in the United States, are notoriously difficult to heat. Where fossil fuels must be burned to generate the electricity, electrical heating is less efficient (30%) than heating by fossil fuels (60–90%) (Summers 1971; Hammond 1972). Setting all residential thermostats 2° lower in the winter and 2° higher in the summer could reduce the energy needed for household heating by 10% (Lincoln 1973).

The energy efficiency of everything from air conditioners to individual factories producing the same manufactured item varies widely. Frost-free refrigerators and freezers use almost twice the energy of manual defrosting units (Hammond 1972). The illumination levels of most commercial buildings have tripled in the past 15 years and could be easily reduced (Hammond 1972). Although fluorescent lights use a quarter as much energy as incandescent light bulbs, they already supply 70% of the country's illumination (Summers 1971); thus, phasing out incandescent bulbs from residences will not significantly conserve energy on a national scale.

The automobile is responsible for many of the crises we face. The United States had 46% of the world's passenger vehicles in 1970 (Motor Vehicle Manufactr. Assn. 1972), and they consumed 21% of the nation's energy (Hammond 1972). Yet most people think that the only serious automotive prob-

lem is air pollution. For example, since 1970, in attempting to clean up the air, we have adopted a number of pollution control strategies that have drastically reduced the mileage per gallon of automobiles (see Table 4) and have thereby accelerated the liquid fuel crisis. Balgord (1973) suggests that the abundant submicron-size particles that will be produced by automotive emission-control catalysts may also pose a major health problem. Because automobiles consume about one-half of our crude oil requirements, the true extent of the automotive problem is all of the aforementioned environmental, financial, and political problems associated with petroleum and synthetic petroleum.

In addition, compared to trains, buses, bicycles, and walking, the automobile, while admittedly very convenient, is extremely inefficient in energy consumed per passenger mile; and trucks are inefficient in energy consumed per ton mile (Hirst in Hammond 1972). Airplanes are far less efficient than cars and trucks. Yet since the 1950s the government has lavishly subsidized automotive and air transport at the expense of bicycles, railroads, buses, and mass transit systems. Clearly the only way to simultaneously solve the air-pollution and liquid-fuel problems is to relax the air-pollution standards and to drive smaller and fewer cars at slower speeds.

Contrary to present practices, one way to conserve or ration energy is to increase the price of exorbitant consumption. Presently, the more electricity a factory or home uses, the lower the unit price of electricity—witness the all-electric rate. Charging higher rates during the peak periods of use might discourage consumption (this is a policy that telephone users, for example, understand). By the same token, automobiles could be heavily taxed on a formula of horsepower times tonnage, or of gasoline consumption per mile. Several such bills have been introduced in Congress (see Hartley 1973). The beauty of these methods of pricing and taxation is that they avoid governmental dictation; those who elect to consume extravagantly our nonrenewable resources pay the most.

Recycling of materials is generally an attractive form of conservation of energy and materials. For example, electricity is the most costly item in producing aluminum from bauxite, and the cost of recycling an aluminum can is about one-third the cost of producing aluminum from bauxite (Cook 1973). However, the efficiency of the collection system determines whether recycling is true conservation. A housewife who drives around town collecting aluminum cans and glass bottles from the neighbors could easily expend more energy in gasoline alone than would be saved in reprocessing the aluminum or the glass!

Steady state economy

Ultimately, to solve our energy problems we will have to practice not only zero population growth but also zero per capita power growth as well. Although such a policy is decidedly foreign to the present American way of life, it may not be synonymous with a stagnant economy, as many critics suppose. For example, early computers built with vacuum tubes consumed more power than their more sophisticated transistorized descendants. Similarly, four passengers can be transported in a Pinto at about one-eighth the consumption of metal and fuel—but admittedly not with the same convenience—as one passenger in a Cadillac.

After World War II, the real (uninflated) price of electricity, petroleum, and most other forms of energy decreased. Then, in 1971, electrical rates began to rise, largely because of increased environmental costs (Chapman et al. 1972). Costs probably will continue to rise because of fuel shortages and the growing tendency to impose the formerly indirect, or hidden, environmental and societal costs of pollution on the individual power plant. For coal these indirect costs may be approximately equal to the direct costs of producing electricity (Morgan et al. 1973).

Furthermore, the birth rate in the United States began to decline in the 1960s and in 1972 fell below the replacement level of two children per family. For the first time in this century, therefore, the United States may be simultaneously experiencing increased energy costs and decreasing population rates, both of which should significantly reduce the rate of increase of energy consumption. As a result, the energy demands of the United States in A.D. 2000 may not be much greater than at present (Chapman et al. 1972). These trends must be encouraged if we are to solve our problems. So far, however, the government has favored a cheap energy policy by increasing oil imports, imposing ceiling prices on gasoline, and promoting offshore drilling and the installation of additional generators in powerhouses in federal hydroelectric dams.

The United States is becoming progressively less self-sufficient in meeting its energy requirements. The economic, social, and military dangers of not being self-sufficient in resources in this industrial era are well known (Cheney 1967). The United States probably does have the technological and financial resources both to solve its energy crises and have its expanding economy as well, although to do so probably would invite unacceptable environmental and political risks. The only way our national crises can be solved without these grave risks is to initiate a very radical conservation of energy resulting in the permanent lowering of per capita consumption of energy. This probably cannot be done voluntarily but will be accomplished by the effective rationing mechanisms of increased prices and taxes. Every energy consumer should, therefore, be prepared to see fuel costs increase by a third in the next 10 years and double before the 21st century.

The international picture is much bleaker. Although the per capita consumption of agricultural and forest products in industrial countries eventually levels off, the per capita consumption of mineral resources, especially fuels, has always increased over the long term (Potter & Christy 1962). Most of the rest of the world is growing at a faster rate than the United States and its industrial allies, both in terms of population and in industrial output. The world's biggest political problem is that the pres-

ent industrial nations (with large per capita consumptions of energy) can hardly say to the billions in Asia, Africa, and Latin America that, for environmental and economic reasons, we will not tolerate your appetite for mineral resources and the fruits of industrialization. For this reason alone, the United States would do well to practice zero per capita power growth.

References

Abelson, P. H. 1973. Importation of petroleum. *Science* 180:1127.

Albers, J. P., D. M. Carter, A. L. Clark, A. B. Coury, and S. P. Schweinfurth. 1973. Summary petroleum and selected mineral statistics for 120 countries, including offshore areas. *U.S. Geol. Surv. Prof. Paper 817*, 149 p.

Balgord, W. D. 1973. Fine particles produced from automotive emissions-control catalysts. *Science* 180:1168–69.

Beall, J. V. 1972. Muddling through the energy crisis. *Mining Engineering* 24(10):41–48, esp. Figs. 2, 5, and 9.

Beall, J. V. 1973. Copper in the U.S.—a position survey. *Mining Engineering* 25(4):35–47.

Blackman, R. V. B., ed. 1972. *Janes Fighting Ships, 1972–1973.* N.Y.: McGraw Hill, 745 p.

Bonneville Power Admin. 1970. *Pacific Northwest Economic Base Study for Power Markets,* Vol. 1 (Summary). Portland, Ore.: Bonneville Power Admin., 223 p., esp. Fig. 44.

Bureau of Natural Gas. 1972. *National Gas Supply and Demand 1971–1990.* Washington: U.S. Fed. Power Commission, Staff Rept. No. 2, 166 p., esp. Fig. 1.

Chapman, D., T. Tyrrell, and T. Mount. 1972. Electricity demand growth and the energy crisis. *Science* 178:703–08, esp. Figs. 3, 4.

Cheney, E. S. 1967. Mineral resources in national and international affairs. *Mining Engineering* 19(12):47–49, 51, 54, esp. Fig. 3.

Christian Science Monitor, 11 Oct. 1973.

Cook, C. S. 1973. Energy: Planning for the future. *American Scientist* 61(1):61–65.

Cook, E. 1971. The flow of energy in an industrial society. *Scientific American* 224(3):134–44, esp. Figs. 3, 5.

Culbertson, W. C., and J. K. Pitman. 1973. Oil shale. In D. A. Brobst and W. P. Pratt, eds., United States Mineral Resources. *U.S. Geol. Surv. Prof. Paper 820,* pp. 497–503.

Davenport, C. P., and G. Duell. 1972. Western coals look good to Northwest power companies. *Mining Engineering* 24(10):80–81.

Frazer, D. M. 1973. Some Canadian energy prospects: An address to the Canadian Institute of Steel Construction, Le Château Montebello, 19 May 1973. Available from the National Energy Board, 24 p.

Gera, F. 1972. Review of salt tectonics in relation to the disposal of radioactive wastes in salt formations. *Geol. Soc. Amer. Bull.* 83:3551–74.

Hammond, A. L. 1972. Conservation of energy: The potential for more efficient use. *Science* 178:1079–81.

Hartley, F. L. 1973. *On S. 1055, Automotive Transport Research and Development Act of 1973, and S. 1903, Motor Vehicle Fuel Economy Act: Testimony before the Senate Commerce Committee.* Available from the Union Oil Company, Box 7600, Los Angeles, CA, 32 p.

Holland, H. D. 1972. The geologic history of sea water—an attempt to solve the problem. *Geochim. Cosmochim. Acta* 36:637–51.

Hubbert, M. K. 1969. Energy resources. In *Resources and Man.* San Francisco: W. H. Freeman, pp. 157–242.

Hubbert, M. K. 1971. The energy resources of the earth. *Scientific American* 224(3):60–70, esp. Figs. 6, 10.

Kirby, I. G., and B. M. Moore. 1972. Crude petroleum and petroleum products. *U.S. Bur. Mines Minerals Yearbook,* Vol. 1., pp. 870–71.

Lincoln, G. A. 1973. Energy conservation. *Science* 180:155–62.

Meadows, D. H., D. L. Meadows, J. Randers, and W. W. Behrens III. 1972. *The Limits to Growth.* N.Y.: Signet, 207 p., esp. Figs. 35 and 36.

Maugh, T. H., II. 1972. Hydrogen: Synthetic fuel of the future. *Science* 178:849–52.

McLean, J. G. 1972. *The United States Energy Outlook and its Implications for National Policy.* Stamford, Conn.: Continental Oil Co., 12 p.

Mencher, S. K., and H. M. Ellis. 1971. *The Comparative Environmental Impact in 1980 of Gasoline-Powered Motor Vehicles versus Electric-Powered Motor Vehicles.* N.Y.: Gordon Associates, Inc., 104 p.

Mining Engineering. 1973. 25(5):10.

Moller, H. S., Jr. 1973. Proxy statement. Cleveland: The Standard Oil Co. (Ohio), 23 p.

Morgan, M. G., B. R. Barkovich, and A. K. Meier. 1973. Social costs of producing electrical power from coal: A first order calculation. *Proc. Inst. Elect. Electronic Eng.* 61:1431–42.

Morison, S. E. 1963. *The Two-Ocean War.* Boston: Little, Brown, 611 p.

Motor Vehicle Manufacturers Association of the U.S., Inc. 1972. *1971 Automobile Facts and Figures.* N.Y.: Motor Vehicle Mfg. Assn. U.S., 88 p., esp. p. 29.

National Academy of Sciences–National Research Council, Committee on Radioactive Waste Management. 1970. *Disposal of Solid Radioactive Wastes in Bedded Salt Deposits.* Washington, D.C.: Nat. Acad. Sci.–Nat. Res. Council, 28 p.

National Coal Association. 1972. *Bituminous Coal Facts 1972.* Washington, D.C.: Nat. Coal Assoc., 95 p.

Perry, H. 1968. *Prospects for Oil Shale Development, Colorado, Utah, and Wyoming.* Washington, D.C.: U.S. Dept. Interior, 134 p., esp. Figs. 1, 7, 9.

Potter, N., and F. T. Christy, Jr. 1962. *Trends in National Resource Commodities.* Baltimore: Johns Hopkins, 586 p., esp. Chart 23.

Roberts, Ralph. 1973. Energy sources and conversion techniques. *American Scientist* 61(1):66–75.

Sisselman, R. 1972. Coal gasification a partial solution to the energy crisis. *Mining Engineering* 24(10):71–78, esp. Fig. 2.

Sansom, R. L. 1973. *The Automobile as a Social Machine.* Washington, D.C.; U.S. Environmental Protection Agency, 14 p.

SOHIO. 1973. *Detailed Report on Matters Related to a Trans-Alaska or Trans-Canada Pipeline.* Report to Subcommittee Public Lands, House Committee on Interior and Insular Affairs, U.S. House of Representatives. Available from the Standard Oil Co. (Ohio), Cleveland, Ohio. 24 p.

Spahr, C. E. 1973. *A Trans-Alaska or a Trans-Canada Pipeline.* Statement before Subcommittee Public Lands, House Committee on Interior and Insular Affairs, U.S. House of Representatives. Available from the Standard Oil Co. (Ohio), Cleveland, Ohio. 5 p.

Special Interagency Task Force for the Federal Task Force on Alaskan Oil Development. 1972. *Final Environmental Impact Statement, Proposed Trans-Alaskan Pipeline: Introduction and Summary,* Vol. 1. Washington, D.C.: U.S. Dept. of the Interior, 322 p.

Spragins, F. K. 1972. The new look in the Syncrude Canada Tar Sands Project. *Mining Engineering* 24(10):90–92.

Summers, C. M. 1971. The conversion of energy. *Scientific American* 224(3):149–60.

Theobald, P. K., S. P. Schweinfurth, and D. C. Duncan. 1972. Energy resources of the United States. *U.S. Geological Survey Circ. 650.* Washington, D.C., 27 p.

U.S. Bureau of the Census. 1973. U.S. imports for consumption and general imports. *TSUSA Commodity and Country Report FT 246, 1972 Annual,* 646 p.

Weeks, L. G. 1973. Accelerating demand challenges petroleum industry performance. *World Oil* 175(1):42–44.

World Oil. Issues of 15 February 1964–73, 15 August 1973, and September 1973.

"Remember — it's better to light just one little thermonuclear power station than to curse the darkness."

Ralph Roberts

Energy Sources and Conversion Techniques

What is our capability of meeting the energy needs of the future within the limitations of known energy resources and energy conversion technology?

Energy sources and their conversion into useful forms of electricity or mechanical motion have become of increasing concern during the past decade, as testified to by the number of recent studies and reviews of the problems of energy requirements, resources, and the environmental impact of energy production. In forecasts, it is frequently assumed that currently noncommercialized methods will become available, yet there has been little assessment of the barriers that must be overcome to make them viable and economically competitive. Several of these methods were once described by H. E. Liebhafsky, during a discussion of fuel cells, as systems where "the publicity is ahead of the technology." In this article I will stress the present status of these technologies and some of the problems still to be solved in order to assess their potential in meeting our energy needs.

The need for these alternatives was emphasized in *Energy Research and Development and National Progress*, issued in 1964 by the Office of Science and Technology:

> As national needs become more varied, the country must be given additional alternatives and options for obtaining energy. Viewed in this context, current research and development efforts, while basically

sound, ought to be improved so as to provide the nation with options for energy resources, and to do so more economically [1].

The continuing need for energy for this and other countries was emphasized by J. F. O'Leary at the December 1971 meeting of the American Association for the Advancement of Science:

> Unless industrial systems change or there is a serious interruption of one of the geometric progressions [world population growth and per capita energy consumption], we can anticipate that before the end of the century energy supplies will have become so restricted as to halt economic developments around the world [2].

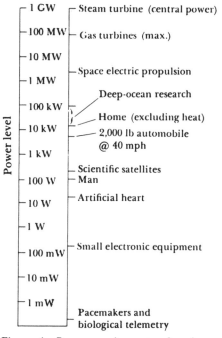

Figure 1. Power requirements of various energy applications. GW = gigawatt; MW = megawatt; kW = kilowatt; W = watt; mW = milliwatt.

The validity of studies of overall needs for energy and their extrapolation to future requirements is dependent on the accuracy of the projected growth rate. Table 1 summarizes the differences in estimated future energy requirements for the United States based on two different projections (3). Such growth rates are often expressed in doubling times, a 3.2 percent growth rate yielding a doubling time of 22 years. For the world energy needs, in view of the greater rate of increase in energy requirements for the developing and less industrialized countries, this doubling time is estimated as being of the order of 15 years.

Currently, the electrical energy produced in the United States is consumed by various sectors of the economy, distributed as follows (4):

Transportation	24.6%
Industrial	37.2%
Residential and commercial	22.4%
Conversion and transmission losses	15.8%

A look at the various energy applications in Figure 1 shows that energy is used at power levels ranging from fractions of a watt to multimegawatts. Not shown in Figure 1 is the power requirement for the Saturn launch vehicle, estimated as 10 gigawatts—greater than the power requirement for a large American city. However, the energy requirement must be considered in relation to the period of operation. For example, the launch vehicle operating for approximately one minute utilizes 160 MW/hrs of energy; a 2,000 lb compact automobile, operating two hours a day in a five-year period, will use 36.5 MW/hrs—over 20 percent of the energy of the Saturn. Thus, in determining energy requirements, both the power and time of use must be considered.

Dr. Ralph Roberts, Director of the Power Program of the Office of Naval Research, has been concerned with a number of aspects of the energy conversion field, including chemical propulsion, electrochemical power sources, and advanced methods of electrical generation. He received his Ph.D. degree from Catholic University in 1942 and has been associated with the Office of Naval Research since 1946. During the academic year 1971–72 he was a Visiting Fellow at Yale University. Address: Office of Naval Research, Arlington, VA 22217.

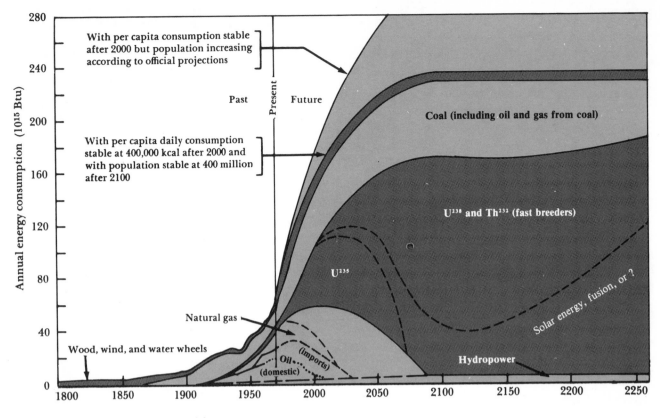

Figure 2. Energy sources, past, present, and future.

Energy sources

Figure 2 (5) shows the rate of utilization and the projected use of various energy resources. Based on these and other extrapolations, various estimates have been made of the time of their depletion. As in the case of the projections of energy needs, there are a number of uncertainties because the figures involve not only estimates of reserves of energy resources but also estimates of the development of the technologies related to fuel utilization and energy production. For purposes of summary, energy sources have been subdivided into four categories: environmental, chemical, nuclear, and mechanical.

Environmental energy sources. These include solar energy, energy resulting from thermal effects or gradients in the earth and ocean, and gravitational

forces of tides and water flow. The power available from these sources is given in Table 2 (6).

The sun is the basic source of energy for the earth and the only essentially nonconsumable one. Of the total energy it emits, only that incident on the surface of the earth can be utilized, unless a space system, such as is discussed in the section on photovoltaics, can be developed. In spite of the large total energy arriving on the earth from the sun, its density is a low 1.39 KW/m². Thus the use of solar energy as a heat source has been limited primarily to home water heaters and some exploratory development for home heating, by converting the absorbed radiant energy into hot water or storing the heat in a solid substance with a high specific heat or high heat of fusion.

Tybout and Löf (7), in their evaluation of solar heating of homes, have shown that it is currently competitive with electrical and fossil fuels in climatic conditions such as those existing in Santa Maria, California. The estimate is based on collector costs of $2 and $4 per square foot, other capital costs, depreciation, and interest on the investment. The use of solar heat to decrease the use of fossil fuel or electricity for home heating and cooling has definite possibilities, although the technology has not been commercialized (8).

Historically there have been many attempts to use solar energy as a heat source for various heat engines or other thermal energy conversion systems. These, although feasible, have not been economic. Recently, Meinel and Meinel (9) suggested the use of specially designed collector materials to maximize high solar energy absorbability and low reflectivity. Two different approaches to achieve this, one based on the principle of the interference filter and the other on the use of modern semiconductor technology, were considered.

The thermal energy is used to heat a liquid metal, which transfers the heat to a thermal storage medium. This

Table 1. Energy utilization in the United States.

Year	3.2% growth 10¹² kWh (t)	4.35%*–3.5% growth 10¹² kWh (t)
1970	20.2	20.2
1980	27.8	30.8
2000	51.9	61.5

* 4.35% growth 1970–1980
kWh (t) = kilowatt hours of thermal energy

Table 2. Environmental power sources.

		Watts
Solar		1.73×10^{17}
Incident	30%	
Atmospheric	47%	
Hydrologic cycle	23%	
Earth		
Conducted to surface via rocks		32×10^{12}
Convection (hot springs and volcanoes)		0.3×10^{12}
Tides, tidal currents		3×10^{12}
Winds, waves, convection, currents		370×10^{12}
Water power		3×10^{12}

heat is used to generate steam for a turbine-driven electric power plant. On the assumption of an efficiency of solar energy conversion of 30 percent, Meinel and Meinel estimate an allowable cost of $60/m² for the entire collection and storage system, with a plant life of forty years. Based on the density of solar energy, this is a cost of $144/kW of installed power, an estimate highly competitive with current nuclear technology. However, in an earlier evaluation of flat-plate solar collectors, Hottel and Howard (*10*) concluded that, until better and cheaper collectors are available, further studies of large-scale power from the sun are "a waste of time and money." It will, therefore, be necessary to evaluate experimentally the proposed collectors and to determine their economics.

Power plants have been operated on other environmental sources. Water power, used directly for small mechanical systems, dates back to ancient times and, with the advent of the electrical generator, has become an important power source. However, its growth potential is limited, and it can never supply more than a small part of national needs.

Geothermal energy sources have become of increasing interest for the production of power without chemical pollution or the hazards of radiation (*11*). The heat is derived primarily from radioactive decay within the earth which is transported to the surface in small quantities. Large deposits of heat in various locations are attributed to recent intrusions of molten rock from the earth's mantle. Where water comes into contact with hot rock, hot springs and steam can form. Where moisture-free steam

occurs, such as in the Geysers area of California, it can be piped to a steam power generation plant. This is being done by the Pacific Gas and Electric Company, and it is producing electricity at a cost less than that for a fossil or nuclear-fueled power plant. These sources are geographically limited, and the life of the steam wells is uncertain. In spite of this, where available, they constitute a valuable energy source.

Two other geothermal sources—lower-temperature wet steam and hot rock—are currently being studied to determine their potential. The Russians have developed a power plant in which the heat of wet steam is utilized to produce a working fluid of freon for operating a closed-cycle organic vapor turbine. Within the United States such hot-water sources have been located in the southwest, Oregon, and Idaho, but have not been commercially utilized.

Hot rock, at temperatures of about 300°C, also is being considered as an energy source for future exploitation. One suggestion, made by the Los Alamos Scientific Laboratory, is to fissure such rock beds and use the heat to obtain high-temperature water to operate a power plant similar to that described in the preceding paragraph. Success in using such energy sources will decrease the dependence of these areas on fossil and nuclear energy sources.

Thermal gradients in the ocean also have been investigated as potential sources of power. Studies have concluded that a minimum temperature differential of 20°C is required between the warm and cool water zones. A power plant to operate on ocean

thermal gradients was designed for Abidjan on the Ivory Coast, and although a number of the components were built and tested, the power plant was not completed. It also has been suggested that in polar regions the temperature difference between the atmosphere and the open ocean, which exceeds 20°C, could be the basis for this type of power plant. There also has been a revival of attempts to use wind power, and claims have been made that this may be economical in regions where the wind velocity exceeds 20 mi/hr.

Tidal power has been considered in the United States, and a project to achieve this was initiated around 1940 but not completed. Such a power plant is operating on the Channel Island coast of France, and other tidal and wave-motion machines have been proposed from time to time. In summary, although environmental sources have and can be utilized, the total energy available, excluding solar energy, can at best fulfill only a small part of the world's projected energy needs and only in specific regions.

Chemical energy sources. Since man first learned to control fire, fossil fuels—wood, coal, petroleum, and natural gas—have been the primary energy source. In addition, large untapped sources of related fuels are locked up in shale. The finite nature of these resources and the pollution resulting from their utilization are problems of much concern. Fossil fuels, the removal of their primary contaminating pollutant, sulfur, and processes for converting solid fuels to liquid and gaseous forms are discussed in considerable detail in the recent publication by Hottel and Howard (*12*), and several processes have been developed for coal gasification and liquefaction (i.e. preparation of petroleum-like hydrocarbons) which offer promise toward meeting future needs for these fuels as the natural supplies decrease. While these processes have not proved competitive for large commercial production, they have been tested in pilot plants and are ready for production evaluation. Moreover, we now have the basic technology for using coal as a source of low-ash and low-sulfur fuel—thus a low-polluting fuel.

It should be noted that the time at which these processes become competitive will depend on the cost of petroleum fuels. For example, during World War II coal hydrogenation was used in Germany to produce products normally derived from petroleum, which was in short supply. The use of hydrogen to replace fossil fuels, although suggested from time to time, recently has been revived (13). Its primary drawbacks, other than economic, are its low density, more difficult handling characteristics, and probably greater hazards. Present methods for producing hydrogen make it more expensive for heating than fossil fuels. However, this situation could change with the availability of lower cost electricity—as projected for nuclear sources.

In addition to fossil fuels any exothermic chemical system can, in principle, be utilized as an energy source. Although some of the light metals show greater heats of reaction than the hydrocarbon fuels, their use is limited to specialty applications. For example, aluminum has been broadly used to increase the energy available from solid propellants, and the oxidation of metals is utilized in electrochemical primary and secondary batteries. Yet it should be noted that it takes more energy to recover these metals from their ores than can be obtained when they are used as a heat source.

Nuclear energy

The use of nuclear energy as a power source has been increasing. A recent press release by the International Atomic Energy Agency (14) reported that, at the beginning of 1972, 51 nuclear power plants were in operation in 12 countries, an additional 123 reactors were under construction, and 108 others were being planned in 24 countries. Within the United States nuclear power plants are primarily based on thermal-neutron light-water reactors using enriched uranium. Most of these are pressurized- or boiling-water reactors. In addition, one gas-cooled reactor is in operation and another is under construction. Some of the characteristics of these installations are shown in Table 3 (15). The lower efficiency and absence of stack gases require greater local heat dissipation in the coolant water of nuclear reactors compared with fossil-fuel power plants. This leads to the problem of localized thermal pollution.

Table 3. Thermal neutron nuclear reactor temperatures.

Type	Pressurized water	Boiling water	Gas cooled
Reactant coolant			
Inlet temp., °F	545	376	760
Outlet temp., °F	610	546	1,430
Pressure, lb/in²	2,250		
Working fluid			
Inlet temp., °F	506	546	1,000
Inlet pressure, lb/in²	720		
Efficiency	32–33%	32–33%	39%

Breeder reactors. Fast neutron breeder reactors are currently under development in which neutrons produced by the induced fission of U^{235} or Pu^{239} are used to convert the more abundant U^{238} to Pu^{239}, which can be utilized as a fissionable nuclear fuel. The other breeding reaction involves converting Th^{232} to U^{233}. Three types of reactors are under investigation, namely, liquid metal, fused salt, and gas cooled, with the first currently being emphasized throughout the world. The fuel is a mixture of uranium and plutonium oxides, and heat is removed by liquid sodium, which is used to heat a secondary sodium loop. The latter is the heat source for generating steam at 1,050°F for the operation of a steam turbine-driven electrical generation system. While the breeder reactor requires the handling of materials at a greater level of radioactivity than the thermal reactor, it has the long-range advantage of breeding additional nuclear fuel, thus making fissionable material available far longer than can be anticipated with current thermal-neutron reactors. Breeder reactors are now under development in various countries, with Great Britain, France, and the USSR closer than the United States to completion of a demonstration plant.

Although the long-range goal of the breeder has the support of those concerned with national energy needs, the current high priority given to the liquid metal fast breeder has been questioned. Cochran, in a recent study reported in *Science* (16), is critical of this approach, noting that according to his assessment (1) capital costs per unit of installed power will be greater than AEC computations, (2) the estimate of uranium reserves is conservative, and (3) the costs of breeder fuel will increase if the "cooling-off" period for the spent fuel must be in-

creased to lower radiation hazards. He has ranked the gas-cooled breeder reactor ahead of the liquid metal one economically and environmentally. There are also strong advocates of the fused-salt breeder reactor.

Fusion power. Next to solar energy, nuclear fusion offers the closest approach to an inexhaustable source of energy. It is based on utilizing the heat released when hydrogen isotopes, or a hydrogen isotope and helium, interact to produce a new nucleus and release energy. Selected fusion reactions and energy distribution of the products are as follows, the numbers in parentheses representing million electron volts:

1. $D + D \rightarrow He^3 (0.8) + n(2.45)$
2. $D + D \rightarrow T (1.0) + H(3.0)$
3. $D + T \rightarrow He^4 (3.6) + n(14.1)$
4. $D + He^3 \rightarrow He^4 (3.6) + H(14.7)$

In order for these reactions to take place, a very large energy barrier to the reaction must be overcome which is equivalent to temperatures of 75 to 100 million degrees and which requires confining the plasma to avoid contact with the walls of the containing vessel and to increase the particle interactions. Thus, an external magnetic field is imposed on the plasma, in addition to the field generated by its motion. J. D. Lawson, a British physicist, has calculated that the product of ion density in the plasma and confinement time must be of an order of 10^{14} ions/sec/cm³ for the third reaction.

The closest approach to meeting fusion conditions has been achieved in the Russian Tokamak experiment. Here, ion densities of 2.5×10^{13}/cm³ and confinement times of 0.02 seconds have been reported. However, the plasma temperatures, 1 keV electrons and 0.5 keV ions, are below the fusion threshold. In addition to experiments using discharge-generated plasmas, there is currently much interest in the possibility of laser-induced plasmas.

Although feasibility has yet to be demonstrated, an evaluation of the promise of various approaches to confined fusion plasmas as the basis for a fusion reactor has been made by Rose (17). In addition to the unproven feasibility of fusion, there are a number of foreseeable engineering problems such as materials, development of cryogenic magnets, shielding, and energy conversion. For shielding, the use of molten lithium (whose availability is limited) is advocated. This element reacts with neutrons producing tritium as one of the products, thus acting as a breeder for additional tritium fuel. Because the temperature of the lithium shield is estimated as 2,000°C, this heat could be utilized with various energy conversion methods, including high-temperature gas turbines and magnetohydrodynamics, as topping stages to standard steam generating plants.

A novel approach to the direct conversion of plasma energy to electricity has been proposed by Post (18). The high-energy particles diffusing out of a mirror machine—one of the experimental approaches to fusion—pass into a magnetic field gradient. Here, electrons are removed and ions follow the magnetic field to a series of collectors where the particle kinetic energy is converted to electrical potential which then yields a high-voltage direct-current power source. It has been estimated that a collection efficiency of 92 percent can be achieved, and laboratory experiments using energetic ions have demonstrated an efficiency of over 85 percent.

Based on past achievements in understanding plasma instabilities and recent experimental progress, confidence in the feasibility of fusion power has increased in the last five years, and estimates of the time necessary to prove this feasibility have been made, varying from two to three years to a decade or more. Once established, it may take another decade or two to overcome the engineering difficulties. The minimum economic size of such a fusion reactor power plant is estimated between 2,000 and 5,000 MW. However, smaller ones cannot be ruled out,

especially if laser-initiated fusion proves to be practical.

Radioactive isotopes. For small power packages, less than 10 kW and primarily below 1 kW, long half-life beta emitters such as Sr^{90} and Pu^{235} are useful as a heat source. With thermoelectric energy conversion, such long-lived power sources have been developed, for example, the system that powered the Transit satellite for over five years. For larger units, greater than 1 kW, consideration has been given to organic vapor closed-cycle turbines. In addition to satellite applications, radioactive-isotope power sources have been used in telemetering systems for remote meteorological and oceanographic data gathering and transmitting systems. Because of the limited availability and separation cost of isotopes, they are relatively expensive per unit of energy. However, for the specialized applications noted, they are frequently the only energy source capable of meeting operational demands.

Mechanical energy storage. Mechanical sources such as springs and flywheels have seen little use for transport or power generation; the latter has had very limited application in buses. Recently, advances in energy storage in flywheels have been reported by Weatherbee (19) and Rabenhorst (20), who anticipate storage of 30 W hrs/lb, an energy density that would make them competitive with present-day storage batteries. The use of a flywheel with a small gas turbine has been suggested as a low-pollution power plant for the automobile.

Energy conversion methods

Given an energy source, it is necessary to convert it into a useful form of power such as electricity or mechanical motion. For convenience, these energy conversion methods will be described in two groups, rotating machinery and nonrotating methods.

Because readers are probably familiar with open-cycle rotating machinery, I will discuss it only briefly. Open cycle means that the working fluid is not recycled. In the turbine, the kinetic energy of gases or liquids is converted into power by converting linear motion into rotation. This principle is the basis of the primitive water wheel as well as the modern aircraft gas turbine. Combined with an

electrical generator, both water and gas turbines are used for the production of electricity. This application for the latter, especially for stand-by and emergency power, is of growing importance. Because of its promise of decreased pollution and less stringent fuel requirements, the gas turbine is under investigation for application to automotive transport, especially large trucks and buses. A major problem, however, is the compact heat exchanger, which transfers heat in the exhaust to the incoming air, thus reducing the temperature of the former.

Although the power stroke in the Diesel and Otto cycles is linear, these are included with the rotating systems, as the piston motion is converted to rotation by means of the piston rods and crankshaft. This conversion step is avoided in the Wankel engine, which has recently been applied to small engines and automotive power and for which an evaluation of performance and prospects was recently published by Cole (21). The major mechanical problem, the rotating seals, has been solved well enough to include the Wankel in production-line automobiles. In spite of its lower efficiency, the Wankel leads to a smaller engine. Internal combustion engines, especially the Diesel, are used with electrical generators for small, local electric power sources.

In addition to these open-cycle systems, there is a parallel set of closed-cycle systems, in which the working fluid is recovered and recycled—for example, the steam turbine used in electric power plants. Organic and metal vapors have also been investigated as working fluids for closed-cycle turbine systems, especially for space applications, and organic vapor closed-cycle power plants of one to several kilowatts operating on the Rankine cycle, in which the working fluid is condensed before recycling, are now available. It is also possible to use the Brayton cycle, a gas-phase closed working cycle, and engines of this type have been considered for space and underwater applications.

In addition to the turbine types, there are also variations of piston engines working on a closed cycle. Of these the best known is the steam engine, in which much of the steam is recovered and returned to the cycle. This principle can also be applied to various organic vapors, further paralleling the closed-turbine cycles. In addition, the Stirling cycle—a Brayton piston cycle in which the working fluid is intermittently passed into the cylinder, expanded, recovered, reheated, and recycled—has been investigated both in the United States and in Europe, especially in the Netherlands. Another important class of rotating machinery for energy conversion is the electric generator. Here, the new major development is in the application of super-conducting magnets to decrease the weight, size, and cost of electric motors and generators. Although this technology is not yet commercially available, experimental systems have been developed.

Nonrotating methods

Thermoelectrics. Nonrotating methods of energy conversion have been the subject of increased research and development in the last two decades, the most advanced of the static devices relying on the solid-state properties of semiconducting materials such as thermoelectrics and photovoltaics. The former was the subject of intensive studies in the first five years of the 1960s. Using a cascade of thermoelectric materials, devices with thermal efficiencies of the order of 8–10 percent have been developed. Power sources using thermoelectrics with radioisotope heating have shown remarkable reliability, some of them having been in operation for periods of five to ten years.

Thermoelectric energy generation is based on the Seebeck effect, in which the generation of a potential is due to the temperature differential across a given material. These materials are characterized by a figure of merit, Z, given by the relationship

$$Z = \alpha^2 \, \sigma / k$$

where α is the Seebeck constant of the material, σ is the electrical conductivity, and k the thermal conductivity. Since these properties are temperature-dependent, the figure of merit varies with temperature, and different materials have their best figure of merit over a limited temperature range, which has led to the use of staged thermoelectrics to optimize performance over a given temperature span. Although high-temperature (1,800–2,000°C) operation of thermoelectrics has been projected, the current technology ceiling, utilizing a silicon-

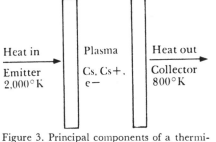

Figure 3. Principal components of a thermionic generator.

germanium semiconductor, is in the order of 980°C. Based on present knowledge, there appears to be little prospect of an increased temperature range or large increase in the figure of merit of thermoelectrics.

Photovoltaics. As mentioned earlier, a second approach to solar energy conversion is by photovoltaics. These are inorganic semiconductors in which the electron is raised from a nonconducting energy level to a conducting one by the absorption of a light quanta. Thus, only that part of the solar spectrum with energies equal to or greater than that required for electron excitation can be used, and this energy threshold is determined by the band-gap of the material. The other important characteristic is the electrical conductivity of the material once the electron has been photo-excited. Silicon solar cells, with an experimental efficiency of 14 percent, are the most promising of those currently available. The limited theoretical efficiency of photovoltaics and the low-energy density of sunlight require large areas of photovoltaic materials in order to generate appreciable blocks of power. A rough estimate is that, at 20 percent efficiency, 1 MW of electric power would require an area of 26,000 m² (approximately 64 acres).

At present, the major application of photovoltaic systems is in the space program, and their cost is of the order of $500/W, making the MW cost $500 million. Even with an estimated reduction of cost of a factor of 100 (*22*), the installed power cost would be $5 million per MW, or twenty times that of a nuclear power plant (*23*). In addition, photovoltaic energy systems require an energy storage system to ensure energy during dark periods. In the space program this is achieved with storage batteries, which add to the cost and decrease the overall efficiency of the system.

An interesting extrapolation of the use of photovoltaic materials for a power source has been made by Glaser (*24*), who has suggested that organic photovoltaic materials with a projected efficiency of 80 percent be used for the conversion of solar energy to electricity. This material would be installed in the collector array of a space station, and the energy generated would be converted to microwaves and transmitted as 10 cm microwaves to the earth, where it would be received and reconverted to electrical power. He estimates that a 25×10^4 MW power source, to meet the requirements of the northeastern United States, would require a collector having a diameter of 5.3 km, a transmitting antenna 2 km in diameter, and a receiver 3 km in diameter. The basis for the projection of an 80-percent efficient organic photovoltaic was not specified; however, recent evaluations indicate lower efficiencies than conventional photovoltaics (*25*).

Thermionic power generation. The direct conversion of heat to electrical energy by thermionics is one of the simplest physical configurations, and, of the several approaches to thermionic power generation, this discussion will center on the cesium plasma diode schematically shown in Figure 3. The emitter, temperature approximately 2,000°K, is the source of electrons and cesium ions. The plasma between the emitter and conductor acts as a low-resistance conductor. The electrons are transported to the lower temperature collector surface, 700 to 900°K. The potential developed depends on the emitter and collector materials, the cesium pressure, and the emitter and collector temperatures. Theoretical efficiencies have been estimated in the order of 27 percent, and present diodes have an efficiency of approximately 10 percent. The output is high current-density, low-voltage direct current, e.g. 10 amps/cm² at 0.4V. In order to convert the output power to the higher voltage range used in most electrical equipment, it is necessary to operate the individual cells in series and to condition the power to meet load requirements. A small, solar-heated thermionic power source has been built in this country and tested for space use, and the Russians have built a 10 kW thermionic nuclear reactor. In the U.S., a development and design effort has been initiated for a 40–50 kW

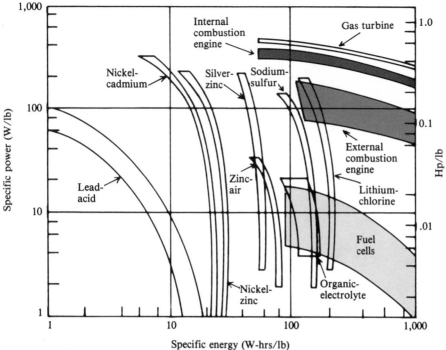

Figure 4. Power and energy density of various batteries. A 2,000 lb vehicle, 500 lb motive power source, and steady driving are assumed. Power and energy are taken at output of conversion device.

Electrochemical sources. Electrochemical power sources, especially primary and secondary batteries, are important contributors to current energy technology. However, they are mostly utilized in small power packages. Secondary batteries have been projected for use in automobiles, and Figure 4 compares a number of battery systems, available and experimental, with the gasoline engine. The adoption of secondary batteries for automotive purposes would increase by 10–20 percent the demands for electrical energy, which is needed to recharge them. Secondary batteries also have been proposed as the energy storage part of solar photovoltaic systems.

In contrast, fuel cells (*27*) that utilize the fossil fuels have been suggested for automobiles and for local and central electrical power generation. The most advanced of the fuel cells is the one based on hydrogen and oxygen which is being used in the Apollo system (Fig. 5). While the preferred electrolyte is a concentrated potassium hydroxide solution, others also have been used. The solution is circulated between the electrodes, and the water formed is removed by the circulating gases. Most hydrogen-oxygen fuel cells operate at elevated temperatures, 60–240°C, although low-temperature performance, to −10°C, has been achieved.

electrical space thermionic reactor (*26*).

In addition to the configuration using thermionic units as an integral part of the reactor, external units heated by a molten metal coolant from the reactor and combustion-heated thermionic generators have been investigated. Currently the thermionic generator appears to be best suited for space or other remote operations requiring 40 to 100 kW. The construction of the projected space thermionic nuclear generator will provide us with information for a better evaluation of this technology.

While the systems developed for space programs are highly reliable, they are not economic as a general power source. One of the costs is that of the catalysts, platinum being preferred for low-temperature and acid electrolyte cells. The related handling properties and economics of hydrogen have been noted in the section on chemical energy sources. Acid electrolyte fuel cells, operating on hydrogen produced from hydrocarbons, are now being investigated as electrical power sources for homes; however, the cost per unit of installed power is presently uneconomic.

Fuel cells which utilize the readily available fossil fuels have been the subject of over three-quarters of a century of investigation. These studies have included systems with acid electrolytes operating at temperatures in the vicinity of 150°C. Because they reject carbon dioxide, they are more compatible with carbon-containing fuels and air than alkaline electro-

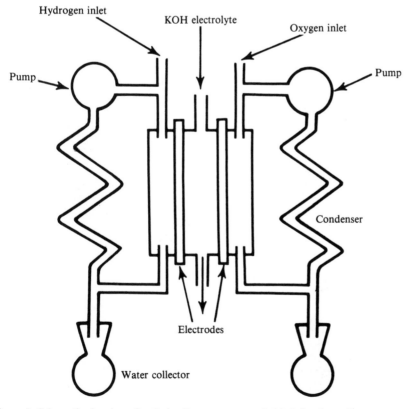

Figure 5. Schematic drawing of a fuel cell. The hydrogen and oxygen are circulated in gas spaces behind the electrodes.

lytes. Direct operation with hydrocarbon fuels is poor, and conversion to hydrogen and carbon dioxide by reforming is necessary to obtain acceptable performance.

Greater success with hydrocarbon fuels has been obtained in cells operating above 500°C using an electrolyte containing molten mixed-alkali carbonates. With experience and the partial solution of difficult material problems, much improved performance has been obtained. Yet in spite of this progress, the start-up difficulties for intermittent application makes their general use unlikely. Such batteries might be used with natural gas or gasified coal for furnishing domestic electric power.

Another approach to the direct utilization of hydrocarbons and coal has been in cells with an oxide ion-conducting solid electrolyte such as zirconium oxide with 10 percent yttrium oxide; the electrodes are platinum. This cell operates efficiently at temperatures greater than 850°C (1,000°C with coal). Recent studies have shown a lifetime of several thousand hours with acceptable deterioration in performance (28). With coal, there is an accumulation of ash which degrades the output. This system is best suited to a continuous load.

All in all, fuel cells, although useful in certain contexts, have restrictions which prevent them from becoming competitive with combustion power sources for all but specialty applications. Future advances may lead to their increased application, but present evidence does not suggest that they will be a near-term major source of local electrical power or replacement for the automobile engine.

Magnetohydrodynamics. The generation of electricity by means of the flow of a conducting fluid through a magnetic field, leading to an induced electrical potential, has been a goal of energy conversion research since the time of Faraday. In the past several years advances have been made that indicate several promising approaches. The principle of magnetohydrodynamics (MHD) is the same as that of conventional electrical generators, except that the conducting wire is replaced by a moving conducting fluid (illustrated for an open-cycle system in Fig. 6).

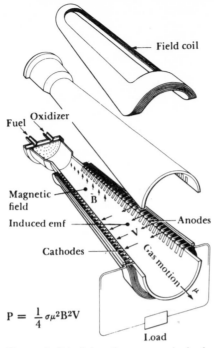

$$P = \frac{1}{4} \sigma \mu^2 B^2 V$$

Figure 6. Principles of a magnetohydrodynamic electrical generator. B = magnetic field; V = induced voltage; μ = flow velocity.

The fuel and oxidizer are selected so that the temperature of the combustion products is of the order of 2,800°K. Not shown in Figure 6 is the introduction into the hot gases of the ionizable seed, a potassium compound, to the extent of 1 percent. The resulting gas has an electrical conductivity σ, and flows through the channel with a velocity u. The power, P, is extracted by applying the generated voltage, V,

to an external load. The exhaust gases, which are at a temperature near 2,300°K, can be used both for heating the incoming oxidizer, formed of air or oxygen-enriched air, and as the heat source for a steam power plant. The seed material and atmospheric pollutants must be removed from the exhaust gas.

This type of generator is presently being evaluated in the USSR in a demonstration plant (Fig. 7). It consists of a natural gas-fueled system utilizing preheated oxygen-enriched air to obtain a generator-inlet temperature of 2,600°C. The seed source is potassium carbonate. The design power output of the MHD generator is 25 MW. This is to be a topping unit to a 75-MW steam generating plant. The net efficiency of this system is expected to be 33 percent. However, for full-scale operation, this type of power plant is projected to have an overall efficiency of greater than 50 percent.

A number of components or processes for the open-cycle system have been developed, including the high-temperature heat exchanger and the seed recovery process. The needed electrode life has not been achieved, although renewable semiconducting oxide surfaces (29) have been developed and an electrode made of eroding potassium hydroxide has been successfully tested for the order of 1,000 hours (30). The

Figure 7. MHD generator of the Russian U-25 experimental power plant.

latter can also be the source of the seed material.

The use of coal or petroleum fuels will duplicate the present problem of oxide of sulfur removal in conventional plants and removal of oxides of nitrogen. In spite of its current status, open-cycle MHD has not yet been established as an economic means of increasing power plant efficiency. Whether or not, as is currently stressed in the United States, open-cycle MHD will be applicable as a peak-load and emergency-overload power source remains to be demonstrated. This type of generator is most competitive with the gas turbine both as a topping cycle and for peak and emergency loads. Which will be the more economic is yet to be definitely established; however, the recent study of Howard and Hottel favors the combination of gas turbine and steam power plant (*31*).

Closed-cycle MHD systems are not as advanced. There are basically two types, one consisting of a noble gas seeded with an alkali metal (cesium is preferred), and the other, a liquid metal (usually an alkali), selected for the compatibility of its melting and boiling points with the MHD cycle. It is the consensus that the first, using cesium-seeded noble gas, will operate with a working fluid which has a temperature in the vicinity of 2,000°K. The conductivity of the fluid depends on nonequilibrium ionization in a magnetic field: i.e. the temperature of the electrons is higher than that of the seeded carrier gas. It has been demonstrated that greater than 10 percent of the energy content of the gas can be converted to electricity, and there is substantive evidence that this figure can be increased to 15–20 percent. It also has been shown that high voltages, i.e. 1,100V, can be generated. While these results indicate that this is potentially a viable system, technological difficulties—a major one being an adequate heat source—remain, even if generator performance fulfills current expectations.

There are several approaches to liquid metal MHD generators, all requiring that the thermal energy in the metal-vapor gas mixture be converted into the kinetic energy of a high-velocity conducting fluid in the generator, which operates at temperatures in the order of 1,000°K. The two-phase system, a mixture of liquid metal and gas (metal-foam), promises to have

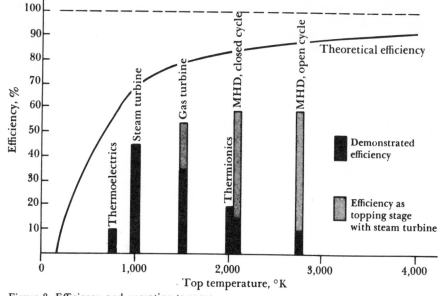

Figure 8. Efficiency and operating temperature of various energy conversion processes.

high efficiency. Assuming that this can be realized, it offers the possibility either of a topping cycle or, if high efficiency is achieved, as the generator of a fixed power plant.

Temperature and efficiency

The selection of a particular energy conversion system must take into account a number of factors. Much has been written on the pollution of the atmosphere, the chemical and thermal pollution of water, and the long-range effects both of carbon dioxide on the thermal balance of the atmosphere and of radioactivity on health. The economics of power production in terms of its use continues to be the dominant consideration within the growing constraints of environmental and health impacts. In spite of these factors, considerations of operating temperature and efficiency are important technological inputs in the selection of thermal energy conversion processes, and these are closely tied together through thermodynamic considerations.

Figure 8 shows the effect of temperature on the theoretical thermal efficiency, assuming that the working fluid is cooled to 300°K (80°F). It shows that higher efficiencies can be achieved by the use of a combination of conversion techniques rather than by a single one. The difference between the bars and 100 percent indicates the heat rejected into the atmosphere or cooling water of the power plant—local thermal pollution.

In principle, by improved heat management, part of the heat loss between the bar top and the theoretical efficiency can be recovered by decreasing heat losses and by lowering the heat rejection temperature. Schemes for achieving the latter have included use of residual thermal energy for desalination, as a source of heat for agricultural use, and, in Sweden, for the prevention of snow accumulation on city streets. It should be noted that all the distributed energy from the power plant is eventually dissipated as heat. However, it is not localized, but becomes a part of the total environmental thermal balance.

Since the conversion of turbine mechanical energy to electricity is of the order of 98 percent, no efficiency improvement in this phase of power generation can be anticipated. The newer developments in superconducting magnetic materials offer the possibility of reduced cost of electrical generators. Superconductivity itself can lead to lower loss in power transmission, thus increasing the usable fraction of the energy produced.

The efficiency of electrochemical power sources is not restricted by the thermodynamic limitations of the heat engine. These sources operate with an efficiency of between 50 and 90 percent, depending on the rate of power production. Unlike mechanical systems, the electrochemical power sources show maximum efficiency at low rates of energy production. Secondary battery efficiency must include the energy required for recharging, thus decreasing the efficiency by the ratio of energy output to recharge energy. In estimating fuel cell efficiency, the use of hydrogen produced from hydrocarbons reduces thermal efficiency because there is a net loss in available heat in this process. When this factor is taken into account, this type of fuel cell does not promise to be more efficient than some of the projected large central power plants. However, the fuel cells would be more efficient if the fossil fuel were utilized directly.

Energy sources and conversion techniques that are within current or projected technology offer alternates to available fossil fuel and nuclear electrical power sources. With the depletion of fossil fuel resources and their use in other aspects of our technology, such as sources of raw materials for the petrochemical and related industries, we will become more dependent on nuclear, solar, and geothermal energy sources. Fossil fuel utilization can be extended through the commercialization of processes for the conversion of coal and shale to gaseous and liquid fuels. Natural gas, whose availability is of great concern, can be replaced by gaseous fuels from coal by processes which are feasible but require investigation in pilot and full-size plants to determine their economics. Inexpensive electrical power or other process leading to cheap hydrogen is a possible alternative.

In spite of efforts to develop fuel cells that can consume carbon-containing gaseous fuels as well as hydrogen, they are not economic when compared with central power plant electricity generation. While the use of such cells for domestic power is being evaluated, initial costs of available units are excessive. In addition, the output would have to be conditioned to the prevailing alternating current equipment, or direct current domestic electrical appliances would have to be manufactured. Studies made on the use of fuel cells for automobiles show that, from an economic point of view, they are unable to replace the internal combustion engine. However, there have been projections that alternate electrochemical systems will begin to replace the current automotive engine as early as 1980.

The use of highly efficient gas turbines or open-cycle MHD as topping cycles to steam turbine power plants offers a means of extending fossil fuel supplies. Large gas turbines for power generation have been demonstrated, and a test facility for open-cycle MHD has been constructed in the USSR. The next five to ten years should demonstrate which of these approaches is the more economic within prevailing antipollution requirements. Successful operation of coal-fired MHD plants will give them a marked long-range advantage over the gas-turbine if cheap sources of liquid and gaseous fuels are not developed.

The breeder reactor offers much promise for extending the availability of uranium and thorium for fission reactors. Of the three approaches, the liquid metal one has received most support, and demonstration plants are being built. However, there are strong advocates of gas-cooled and fused-salt breeder reactors as alternates. One major problem in the fast breeder, which is especially emphasized in its liquid metal form, is the high level of radioactivity that must be handled and stored. Similarly, the radiation hazards of fission reactors are a matter of continuing debate, which, until resolved, may slow their rate of construction. The use of nuclear fusion for power generation has not yet been proved feasible; however, recent research increases the probability of ultimate success. Once this technology is established, much effort will be required to make use of fusion in a practical power plant.

Except for solar energy, the use of environmental energy sources is limited by their geographical distribution and low energy density. Conversion of solar energy using photovoltaic devices is limited by their high cost and low efficiency. Even with projected cost decreases, photovoltaics will be much more expensive than fossil or nuclear power plants. Higher-efficiency forms have been suggested, but present knowledge indicates that they have low probability of success. New approaches to more efficient thermal collectors of solar energy have been proposed, and experimental evaluation is highly desirable. At the present time, however, the most promising use of solar heating is in reducing fossil fuel and electricity requirements for home heating. In the area of electrical generators and electricity transmission, superconductivity and superconducting magnets promise to decrease the cost of these components of power systems.

Overall, the potential for improved power generation which will meet the growing requirements for energy with the needed decrease in atmospheric and water pollution is within our technological capability. Yet much research and development are still required to select the technologies that can most economically achieve these goals.

References

1. U.S., Office of Science and Technology. 1964. *Energy Research and Development and National Progress.* Washington: U.S. Govt. Print. Off., p. xvii.
2. J. F. O'Leary. 1972. *Chemical and Engineering News.* Jan. 3, p. 4.
3. H. C. Hottel and J. B. Howard. 1971. *New Energy Technology and Assessment.* Cambridge, Mass.: MIT Press, p. 13.
4. C. M. Summers. 1971. *Scientific American* 224(3):149.
5. *Chemical and Engineering News.* 1972. Jan. 10, p. 28.
6. M. K. Hubbert. 1971. *Scientific American* 224(3):61.
7. R. A. Tybout and G. O. G. Löf. 1970. *Natural Resources J.* 10:268.
8. W. Clark. 1971. *Smithsonian.* Nov., p. 14.
9. A. B. Meinel and H. P. Meinel. 1972. *Physics Today* Feb., p. 44.
10. H. C. Hottel and J. B. Howard. 1971. *Op. cit.,* p. 340.
11. A. L. Hammond. 1972. *Science* 177:978.
12. H. C. Hottel and J. B. Howard. 1971. *Op. cit.,* Chap. 3.
13. L. W. Jones. 1971. *Science* 174:367.
14. Press release reported in the *New Haven Register,* March 5, 1972.
15. H. C. Hottel and J. B. Howard. 1971. *Op. cit.,* p. 233.
16. A. L. Hammond. 1972. *Science* 176:391.
17. D. J. Rose. 1970. *Science* 172:797.
18. R. W. Moir, W. L. Bass, R. P. Freiss, R. F. Post. 1971. Preprint CN-28/IC-1, Proceedings of the Fourth Conference on Plasma Physics and Controlled Nuclear Fusion Research. University of Wisconsin, June 17–23, 1971.
19. A. Svensson, Jr., A. E. Weatherbee, Jr. 1969. *Feasibility of Flywheel High Energy Storage.* Pratt and Whitney Aircraft, Report No. PWH 3676, April 1969.
20. D. W. Rabenhorst. 1971. *Potential Applications for the Flywheel.* Society of Automotive Engineers, Intersociety Energy Conversion Engineering Conference Proceedings (reprint), Aug.
21. D. E. Cole. 1972. *Scientific American* 227(2):14.
22. Final Report of the Subcommittee on Solar Power. Prepared for the NASA Research and Advisory Committee on Space Power and Electric Propulsion, May 1971.
23. H. C. Hottel and J. B. Howard. 1971. *Op. cit.,* p. 235.
24. P. E. Glaser. 1968. *Science* 162:857.
25. A. Gloickov. *Organic Photovoltaic Devices.* Report of the Space Physics Laboratory, Air Force Cambridge Research Laboratories (no date).
26. A. S. Gietzen et al. 1971. *A 40 kWe Thermionic Power System for a Manned Space Laboratory.* Gulf GA-A 10535, July 1, 1971.
27. R. Roberts. 1971. Fuel and continuous feed-cells. In *Primary Batteries,* G. Heise and N. C. Cahoon, eds. N.Y.: J. Wiley, p. 293.
28. R. Steiner. 1972. *Energy Conversion* 12:31.
29. F. Hals. 1967. MHD power generation. ASME Paper 67-PWR-12 ASME-IEEE Joint Power Conference, Detroit.
30. D. Warszawski. 1972. *Energy Conversion* 12:25.
31. H. C. Hottel and J. B. Howard. 1971. *Op. cit.,* p. 288.

PART 2 *Fossil and Synthetic Fuels*

Arthur A. Meyerhoff

Economic Impact and Geopolitical Implications of Giant Petroleum Fields

Two percent of the world's petroleum fields contain about 79 percent of all known oil and gas reserves, and these giant fields thus assume unusual importance in international relations.

Nearly 22,050 oil and gas fields have been discovered in the world, 16,100 of them in the United States and Canada. This statistic reflects only the mature state of oil and gas development on this continent and is not an indicator of the number of fields that will be discovered in the world. However, it is safe to predict that ultimately between 70,000 and 80,000 oil and gas fields will be discovered, more than 52,000 of them outside the United States and Canada. Most of these (about 36,000) will be found in the Soviet Union. These figures emphasize the fact that natural oil and gas resources in North America—outside of Mexico—are becoming scarce and indicate that the time has arrived for both Canada and the United States to find alternate, economic sources of energy.

Definitions

I begin by defining a few terms that are frequently misused and misunderstood. As employed by the petroleum industry, *petroleum* includes all mobile hydrocarbons which can be recovered from the subsurface by the drilling of wells at the surface—crude oil (oil), oil-associated gas (casinghead gas, or gas dissolved in oil), natural

Arthur A. Meyerhoff is President, Meyerhoff and Cox, Inc., and Director of the Geological Research Corp. of Houston, Texas. He received his B.A. from Yale University in 1947 and his Ph.D. from Stanford University in 1952. He worked for the Standard Oil Company of California in Latin America and as publications manager of the American Association of Petroleum Geologists and was Professor of Geology at Oklahoma State University. His specialties are worldwide petroleum geology, the geopolitics of petroleum, and continental drift and plate tectonics. Address: P.O. Box 4602, Tulsa, OK 74104.

(free) gas which is not associated with oil, and liquid condensates (distillate) which form from natural gas. I use the term *petroleum* in this paper in the sense employed by the petroleum industry.

The term *field* refers to a continuous producing area (as seen in plan or map view) which may contain, in the subsurface, (1) a single pool or reservoir uninterrupted by permeability barriers; (2) an area with multiple reservoirs trapped by a common geologic feature, such as a single dome or anticline; or (3) an area of composite reservoirs or pools, of different geologic ages and rock types but interconnected through permeable rocks within the same trap (Halbouty et al. 1970). The term *reservoir* or *pool* refers to single oil or gas accumulations within a petroleum trap. *Pool* is not synonymous with *field* except where only one reservoir or pool constitutes the field.

The phrase *giant petroleum field* is used here in the sense defined by Holmgren et al. (1975) and by Moody and Esser (1975): a giant oil field contains a minimum of 500 million barrels (68 million metric tons) of recoverable oil; a giant gas field contains a minimum of 3 trillion feet (86 billion cubic meters) of recoverable gas. The gas-oil equivalency is based on the conversion factor of 1 barrel of oil to 6,000 cubic feet of natural dry gas (the U.S. Bureau of Mines 1964 standard). Most oil fields contain associated gas, and some contain strata bearing natural gas as well as strata bearing oil. Thus any field containing both oil and gas in recoverable volumes which, combined, are equivalent to 500 million barrels of oil or 3 trillion cubic feet of gas also qualifies as a giant field.

A *supergiant* field is one containing 10 billion barrels (1.4 billion metric tons) or more of recoverable oil or 60 trillion cubic feet (1.7 trillion cubic meters) or more of recoverable gas.

The world's first giant field was discovered at Talara, coastal Peru, in 1868; two more were discovered in 1871—one at Bibi-Eybat on the western shore of the Caspian Sea in what is now the USSR and one at Bradford, Pennsylvania, in the United States. However, it was not until a rash of giant-field discoveries was made in the Middle East from 1927 through 1948 that the immense importance of such fields became apparent to the petroleum industry (Halbouty et al. 1970). Among these Middle Eastern fields are ten supergiants containing recoverable reserves of oil ranging from 10 to 75 billion barrels. The only supergiant fields outside the Middle East are in the Soviet Union (8 supergiants), the Netherlands (1), Venezuela (1), Mexico (1), and—probably—northern Alaska (1). Giant and supergiant fields are relatively scarce in the Western Hemisphere, totaling only three.

What factors cause giant fields?

The difference between a giant petroleum field and an ordinary one is the size of the accumulation, and size is a direct product of the geologic history. The principal factors involved are the following.

1. Petroleum comes from organic matter. Therefore, sufficient organic matter or source materials must be present for petroleum, either oil or gas, to generate.

2. The organic matter must be preserved. This is accomplished by burial within almost impermeable sediments, such as clay and mud, and by the presence of interstitial salts in the mud and clay. The salt, which comes from seawater and from the waters of interior lakes, acts as a preservative.

3. The source materials must be buried deeply enough to be under sufficient pressure and temperature for the conversion to petroleum to take place.

4. Conversion to petroleum must take place on a large scale for commercial accumulations to form.

5. The newly generated petroleum must be expelled from the clay or mud in which it generated.

6. Carrier beds (permeable sandstone, limestone, or dolomite) must be present to receive the petroleum as it is expelled from adjacent beds of compacting clay and mud.

7. The petroleum must be able to move through the carrier bed, or beds.

8. Traps must be present where the migrating petroleum can accumulate (Fig. 1). Anticlines (or domes, in laymen's language) are the most important type of trap (diagram *a* in Fig. 1).

9. A trap is not a trap unless impervious seals (roof rocks) are present (Fig. 1). Beds of shale are the most common type of seal, although evaporites (gypsum, anhydrite, halite) also form excellent seals in many petroleum fields.

10. The sizes of the reservoir and trap are critical. Reservoirs are usually sandstone or carbonate rocks—limestone and/or dolomite. If the reservoirs are thick and cover a large area, the accumulation may be large, provided that sufficient petroleum was generated and able to move along the carrier beds. The larger the trap, the larger the accumulation. Most petroleum fields associated with salt domes are small, because the total area involved rarely exceeds a few square kilometers. In contrast, anticlinal (domal) arches within stable continental interiors, as in Oklahoma, northern Texas, and Kansas—and particularly in the Middle East, North Africa, and many parts of the USSR—may cover large areas. In these situations, if reservoirs are thick

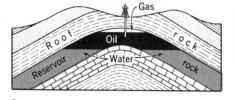

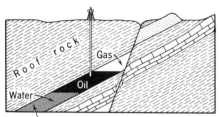

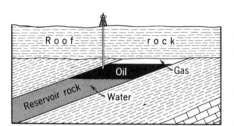

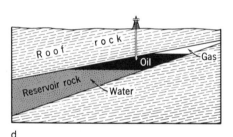

Figure 1. Four of the many kinds of petroleum traps are sketched: *a, b,* and *c* are structural traps; *d* is a stratigraphic trap. Gas (*white*) overlies oil (*black*), which floats on water (*gray*), saturates reservoir rock, and is held down by a seal of claystone (roof rock). Oil and gas fill only the pore spaces in the rock. (From Flint and Skinner 1974, by permission.)

or if numerous reservoirs exist, the resulting accumulation can be gigantic.

11. After the petroleum is accumulated, the area must remain tectonically stable. Mountain-making movements which produce faulting, folding, uplift, and erosion tend to destroy large accumulations. In contrast, within the continental interiors, removed from coastal mountain belts, the accumulations almost everywhere in the world are larger and stand an excellent chance of being preserved.

It was the early recognition by geologists that oil and gas accumulate along the crests of anticlinal uplifts, arches, and domes which led to the rapid development of the modern petroleum industry.

Petroleum's history and major producers

Petroleum has been a commercial commodity at least since 3000 B.C. Good cuneiform records pertaining to petroleum exploitation and sales from various parts of the Middle East and Central Asia discuss contracts for the oil trade, complaints about the shortage of supply, and many related matters (Owen 1975). Almost the only topic not discussed in these records is import regulations! Even the prices for oil approved by King Hammurabi's government trade commission about 1875 B.C. are recorded on cuneiform tablets. Yet methods for recovering large amounts of petroleum did not exist, and supplies were limited by the necessity of collecting oil from surface seeps and from shallow, hand-dug wells.

Confucius, about 600 B.C., reported the first discovery of natural gas in wells drilled with bamboo in China. The first deliberate drilling for commercial gas deposits dates at least to 211 B.C., when Chi-lui-ching field, Szechwan Province, China, was discovered (Meyerhoff 1970). The gas was used as fuel to evaporate salt and the associated condensate was used to construct fire bombs for war. Yet a viable, modern, economic petroleum industry did not begin for another 2,070 years, when the Drake well was completed on 28 August 1859, at Oil Creek, near Titusville, in northwestern Pennsylvania. Petroleum production was slow to be developed, however, and not until the early 1900s, when the internal combustion engine came into its own, were oil and gas widely used. Until the early part of the nineteenth century, man's most important and—in many cases—only source of energy was wood. During the industrial revolution and until World War II, coal gained rapid ascendancy in those nations which had a potential to become industrial powers (i.e. nations which had large commercial coal deposits).

During World War II, the United States accelerated the conversion of its energy base from coal to petroleum. By 1973, the U.S. energy base was

Table 1. World crude oil production in 20 leading nations, 1974

Nation	Total bbl (1,000s)	Bbl/day (1,000s)	% of world production
USSR	3,373,650	9,243	16.4
United States	3,199,328	8,765	15.6
Saudi Arabia	2,996,543	8,210	14.6
Iran	2,210,627	6,057	10.8
Venezuela	1,086,332	2,976	5.3
Kuwait	830,580	2,276	4.0
Nigeria	823,347	2,256	4.0
Iraq	679,803	1,862	3.3
Canada	616,532	1,689	3.0
United Arab Emirates	616,485	1,689	3.0
Libya	555,291	1,521	2.7
Indonesia	501,838	1,375	2.4
China	474,500	1,300	2.3
Algeria	372,753	1,021	1.8
Mexico	238,271	653	1.2
Qatar	189,348	519	0.9
Argentina	151,110	414	0.7
Australia	140,890	386	0.7
Rumania	107,964	296	0.5
Oman	106,046	291	0.5
20-nation total	19,271,238	52,799	93.7
World total	20,518,139	56,214	100.0

SOURCE: American Petroleum Institute 1976.

44% oil, 34% gas, and 18% coal. The remaining 4% came largely from hydroelectric power, with only 0.4% from nuclear sources (Meyerhoff 1973). In addition, during 1973 the United States consumed approximately 36% of all the energy produced in the world while producing less than 17% of it (American Petroleum Institute 1976). Since that year, the United States' share of total world consumption has decreased to about 32% and continues to decline, largely as a result of the increasing industrial output of the Soviet Union, West Germany, Japan, Brazil, China, and other growing industrial powers. Also since 1973, U.S. production of energy has decreased to 16% and continues to drop. This decrease is a result of the declining supply of oil and gas reserves in this country. By mid-1976, the United States was consuming 16,500,000 barrels of oil a day, of which more than half was imported, thus producing a staggering annual imports bill of nearly $30 billion for 1975 and about $33 billion during 1976 (Meyerhoff 1976). This oil-imports bill will continue to rise as U.S. production declines from its peak of 9,637,000 barrels a day in 1970 (American Petroleum Institute 1976).

Other major industrial powers long have been in the same serious energy situation in which the United States now finds itself. Most notable of these is Japan, which must import 7,500,000 barrels a day, a figure which is rising each year. Japan produces only 3% of her petroleum needs (Miyamori 1975). West Germany also is a major oil importer, although she produces more than 20% of her petroleum requirements. Great Britain and France likewise are major importers, but Britain's position is improving steadily as a result of the North Sea discoveries. Brazil, one of the most rapidly growing of the world's industrial nations, imports 76% of her petroleum needs.

It is instructive to compare the statistics for the USSR. In 1950, the USSR energy base was 17.4% oil (versus 39.7% in the U.S.), 2.3% gas (versus 18.1% in the U.S.), and 66.1% coal (38.0% in the U.S.) (Shabad 1969; American Petroleum Institute 1976). Most of the remaining energy source was wood. By 1973, the USSR energy base was 41% oil (44% in the U.S.), 17% gas (34% in the U.S.), and 33.5% coal (18% in the U.S.) (Meyerhoff 1973). The percentage of world energy consumption by the USSR was only 16%. By early 1976, USSR petroleum production rose to 10,100,000 barrels a day, nearly 2,000,000 barrels more than U.S. production. Of this amount, 24% is exported, and less than 1% of USSR needs is imported. Consequently the USSR, whose daily production rate is climbing steadily, has no petroleum imports bill.

The People's Republic of China provides an interesting example of a newly rising industrial power which is developing a petroleum-based economy at a very rapid rate. In 1973, 73% of China's energy base was coal, 15% firewood, 10% oil, and 1% gas (Meyerhoff 1973). Today, oil and gas account for 21% of China's energy base (Connell 1974; Kambara 1974), and her production, which was only 2,450 barrels a day in 1949, now has surpassed 1,600,000 barrels a day. Most of this increase has been since 1970. China is self-sufficient, although this observation should be tempered by the knowledge that China's economy is a controlled one in which the private sector receives almost no petroleum for its use. China, however, faces difficult years ahead, because, with a population of about 850,000,000 persons, and a limited natural-resource base, China cannot expand indefinitely into the major petroleum-exporting nation that many persons predict (Saga Petroleum-Noroil 1975; Auldridge 1975; Harrison 1975; see summary by Fan 1975). Her proved, probable, and potential petroleum resources are no more than 50 or 60% of what the United States has produced and is still capable of producing, with a population less than 30% of China's (Meyerhoff 1970; Willums 1974). Thus the extravagant claims that China will be "another Middle East" are, on the basis of present geologic knowledge, groundless (Meyerhoff 1973; Salmanov 1974).

In contrast, many nations, particularly those of North Africa and the Middle East, have huge petroleum reserves which will be sold to foreign have-not nations. A possible exception is Iran, which, of all of the great Middle Eastern producing nations, is the only country that ultimately may become an important world industrial power. These countries consume less than 1 to 5% of their production and export the rest. The most notable is Saudi Arabia, which produced 8,210,000 barrels a day in 1974 (after deliberately reducing her production from 9,200,000 barrels a day in late 1973). The daily production of the leading 20 petroleum-producing nations during 1974 is shown in Table 1. Of the total, 76% came from giant fields—or 2% of all fields discovered.

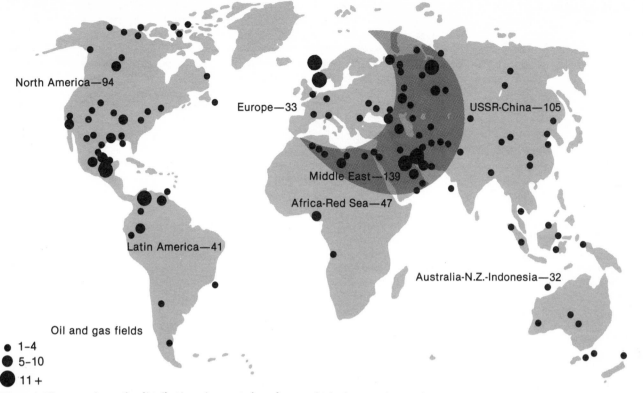

Figure 2. The map shows the distribution of giant petroleum fields of the world. More than 60% of the giant fields are found in the crescent-shaped area, which also contains nearly 68% of the world's known oil and gas reserves. (From Meyerhoff 1973.)

Giant fields in world politics

From the preceding, four broad categories of nations can be defined: (1) *Industrial powers with insufficient petroleum and the need to import large quantities:* these include the United States, Japan, Italy, West Germany, France, and Brazil. To a lesser degree, East Germany, Czechoslovakia, and India belong in this category. Canada, although producing about 80% of its needs, will have to find new and large reserves to keep out of this category, as will Argentina and Rumania. (2) *Industrial powers with sufficient petroleum for their needs:* within this category are the Soviet Union and the People's Republic of China. Venezuela and Mexico soon will join their ranks. The Netherlands and Great Britain also may be classified as this type of nation, although the Netherlands' main petroleum resource is gas, and Great Britain must put additional North Sea fields on stream before imports can be halted. (3) *Underdeveloped to moderately developed countries with a great excess of petroleum for export:* these include Algeria, Libya, Egypt, Iraq, Iran, Kuwait, Saudi Arabia, Qatar, Oman, the United Arab Emirates, Indonesia, Nigeria, Trinidad-Tobago, and Ecuador. (4) *Underdeveloped to moderately developed nations with a small or no petroleum industry:* this category includes the majority of the world's nations, a fact which poses a threat to world stability. A prime example is Bangladesh, with a very large population, few natural resources, and a geologic history that will make it very difficult, if not impossible, for that nation to develop sufficient energy for its population. The great question is this: How will such large numbers of underdeveloped, have-not nations become economically viable in a world of high and rising petroleum prices? The same question may be asked of the industrialized nations which need to import huge volumes of petroleum and petroleum products.

The geographic distribution of giant fields is shown in Figure 2, and some relevant statistics are given in Table 2. It is apparent from the figure and table that 294 (60%) of the world's giant fields are in the USSR, the Middle East, and North Africa. Within these 294 giant fields is 1.035 trillion barrels of oil and gas-equivalent oil (Btu's of gas converted to oil). The sum of 1.035 trillion barrels is 80% of the world's known reserves in giant fields and 64% of all the world's known reserves, whether in small or giant fields. Two Middle Eastern fields alone—Ghawar (Saudi Arabia) and Burgan (Kuwait)—ultimately will produce 141 billion barrels, or 11% of the known world total. Similarly, two USSR gas fields—Urengoy and Orenburg—are capable of producing 14% of all the world's known gas reserves. The gas reserve at Urengoy alone—210 trillion cubic feet— is almost as large as the total remaining gas reserves of the United States—225 trillion cubic feet distributed in several thousand fields.

Figure 2 and Table 2 also show that North America has only 94 giant fields (most of them largely depleted now). They represent 19% of the world's total number of giant fields and will produce only 7% of the world's known reserves. In fact, the 135 giant fields of the entire Western Hemisphere will produce only 11% of the world's total known reserves, or the same amount of oil as the two Middle Eastern supergiants, Ghawar and Burgan. The recent giant discoveries in Mexico (states of Chiapas and Tabasco) will alter this situation somewhat:

Another striking feature of Figure 2

is that 60% of the world's giant fields lie within a crescent-shaped outline covering North Africa, the Middle East, and the west-central part of the Soviet Union. In terms of foreign policy and political maneuvering, the diagram is self-explanatory. The gradual movement of the USSR during the last two decades into almost all parts of the Middle East and North Africa has been governed not only by a desire for foreign ports and bases but also by a deliberate policy aimed at control of the areas of major petroleum reserves. The nation which controls the world's oil and gas spigots controls the economic, military, and—possibly—political futures of those nations having to import huge quantities of oil and gas (e.g. the United States, Japan, West Germany, Italy, France). The discovery of North Sea oil has alleviated the situation in Western Europe but not in the United States and Japan, which are two of the three greatest industrial powers of the world. I believe that the distribution of giant fields shown in Figure 2 (which mirrors the distribution of total petroleum reserves of the world) is a major geopolitical factor that has been ignored by too many persons in positions of political responsibility (see Meyerhoff 1973 for a more complete discussion). The numerous and ramified implications of control of the crescent-shaped area of Figure 2 by any one major power cannot be ignored. The facts (1) of petroleum distribution and (2) of USSR influence throughout the Middle East and North Africa speak for themselves. The question then arises: What can be done to counter the geopolitical implications of Figure 2? A look at the world's future petroleum reserves (those still undiscovered) is not reassuring.

Future reserves

Moody and Esser (1975) are among many, including myself (Meyerhoff 1973), who have computed the world's future reserves. Moody and Esser estimated that 925 billion barrels of oil remains to be discovered in the world. I estimated that 1,275 billion barrels remains to be found. The two figures are not as disparate as they may appear, if one considers the uncertainties involved in our respective calculations. I also estimated that about 4,180 trillion cubic feet of gas remains to be found, in addition to the 3,120 trillion cubic feet already discovered. Of the 4,180 trillion to be found, 3,000 trillion cubic feet will be in the USSR and the Middle East, particularly in Western Siberia and the Persian Gulf. Thus, future supplies will come mainly from the area outlined by the crescent in Figure 2. More than 70% of the future gas will come from giant fields.

These conclusions, based wholly on geologic considerations, are not encouraging. Large reserves will not be found in the United States and Canada, despite (1) some success in northern Alaska, the Mackenzie delta, the Canadian Arctic islands, and offshore Labrador and (2) hopeful expectations, still unproved by drilling, for the areas along the eastern continental shelf of the United States, in the Bering Sea, and in the Baffin Bay–Davis Strait area of northeastern Canada. Even if a reasonable number of giant-field discoveries are made in these "hopeful" areas, they will alleviate only temporarily the U.S. and Canadian petroleum shortage; they cannot cure it. As long as the have-not industrial nations such as the United States and Japan continue to rely on petroleum as their energy base, they will have to import heavily and pay world-market prices. The solution to this huge dilemma is only partly to encourage more exploration, because the geology of the areas remaining to be explored does not favor the discovery of sufficient reserves to reduce significantly the $33-billion annual imports bill. Rather the solution is to encourage both private and public organizations to find and develop alternate sources of energy.

Conclusions

A total of 491 giant fields will produce 79% of the known world reserves of oil and gas. Future discoveries will be increasingly smaller as exploration intensifies, but the relatively few giant fields ultimately will produce

Table 2. Number and reserves of world's giant oil and gas fields, and percentage of world's total proved reserves in giant fields

	Oil		Gas		Total	
	No. of fields	P&P* ultimate 10^6 bbl	No. of fields	P&P ultimate 10^6 BOE[†]	Total no. of fields	P&P ultimate 10^6 BOE
USA	53	64,782	27	37,324	80	102,106
Canada	8	6,866	6	3,600	14	10,466
Total N. America	61	71,648	33	40,924	94	112,572
Latin America	34	59,396	7	4,234	41	63,630
North Africa	36	46,017	11	33,356	47	79,373
Greater Europe	21	20,850	12	22,454	33	43,304
Middle East	88	515,148	51	148,463	139	663,611
Far East	17	16,485	15	15,569	32	32,054
China	5	3,528	2	1,800	7	5,328
USSR	37	103,009	61	188,716	98	291,725
World total	299	836,081	192	455,516	491	1,291,597
World P&P ultimate		1,105,000		520,604		1,625,604
% in giants		76%		87%		79%

*P&P = Proved and prospective. [†] BOE = Barrels of oil-equivalent gas.
SOURCE: Holmgren et al. 1975, updated through 1 January 1975 by Holmgren and Meyerhoff. I thank Dr. Dennis A. Holmgren for permission to use this table in modified form.

more than 50% of all recoverable oil and gas in the world. In the United States and Canada, most of the giant fields already have been discovered, and they are small by Middle Eastern and USSR standards. Moreover, the petroleum in the United States and Canadian giants has been largely depleted. As the great industrialized powers continue to consume energy at ever-increasing rates, they must accept as a fact their gradual decline as preeminent industrial nations, unless their leaders are willing to direct themselves and their nations' efforts *now* to the development of alternate forms of energy.

Since 1973, there has been much political debate in the United States about an "energy crisis." If the problem is adequately defined and circumscribed by facts, there can be no debate. The United States is short of oil and natural gas, but there is no *world* shortage. If the United States persists in maintaining a petroleum-based industrial economy, it can do so only at a dual price: (1) the increasing monetary cost of the petroleum itself and (2) the loss of control over its economic destiny. It has the resources in coal, oil shale, and other raw materials, and the technological competence to regain and retain a large measure of independence, but every delay makes the goal more difficult to achieve. Thus the crisis is not what many politicians and petroleum-company spokesmen would have us believe—a crisis of prices, profits, controls, or divestiture. The crisis is the future of this country. And that future is even now upon us.

References

American Petroleum Institute. 1976. Basic petroleum data book. Washington, DC.

Auldridge, L. 1975. Mainland China striving to boost crude exports. *Oil and Gas J.* 73(1) 26–27.

Connell, H. R. 1974. China's petroleum industry: An enigma. *Am. Assoc. Petrol. Geol. Bull.* 58(10)2157–72.

Fan, P. H. 1975. Chinese oil-industry image changing. *Oil and Gas J.* 73(32) 110–12.

Flint, R. F., and B. J. Skinner. 1974. *Physical Geology.* NY: Wiley.

Halbouty, M. T., A. A. Meyerhoff, R. E. King, R. H. Dott, Sr., H. D. Klemme, and T. Shabad. 1970. World's giant oil and gas fields, geologic factors affecting their formation, and basin classification. *Am. Assoc. Petrol. Geol. Memoir 14*, pp. 502–55.

Harrison, S. E. 1975. China: The next oil giant. *Foreign Policy*, no. 2, pp. 3–27.

Holmgren, D. A., J. D. Moody, and H. H. Em-

merich. 1975. The structural settings for giant oil and gas fields. 9th World Petroleum Cong., Tokyo 1975, Panel Disc. no. 1(4) (preprint).

Kambara, T. 1974. The petroleum industry in China. *China Quarterly*, no. 60, pp. 699–719.

Meyerhoff, A. A. 1970. Developments in mainland China, 1949–1968. *Am. Assoc. Petrol. Geol. Bull.* 54(9)1567–80.

——. 1973. Geopolitical implications of Russian and Chinese petroleum. In *Exploration and Economics of the Petroleum Industry,* vol. 11, pp. 79–127. Southwestern Legal Foundation Proc., Dallas, NY: Matthew Bender and Co., Inc.

——. 1976. Oil and gas. *Geotimes* 21(1)39–31.

Miyamori, K. 1975. The petroleum industry in Japan. Japanese Natl. Comm. of the World Petrol. Cong., Tokyo.

Moody, J. D., and R. W. Esser. 1975. An estimate of the world's recoverable crude oil resource. 9th World Petroleum Cong., Tokyo 1975, Panel Disc. no. 1(4) (preprint).

Owen, E. W. 1975. Trek of the oil finders: A history of exploration for petroleum. *Am. Assoc. Petrol. Geol. Memoir 6.*

Saga Petroleum, Noroil. 1975. The People's Republic of China. pp. 89–94. Oslo.

Salmanov, F. 1974. V paschete na nesvedushchikh, po povodu odnogo "geologicheskogo otkritiya" (Calculations of ignorance, apropos of one "geologic discovery"). *Sotsialisticheskaya Industriya*, 1 Aug., p. 1.

Shabad, T. 1969. *Basic Industrial Resources of the USSR.* Columbia Univ. Press.

Willums, J.-O. 1975. Prospects for offshore oil and gas developments in the People's Republic of China. Dallas, Am. Inst. Min. Metal. Eng., Offshore Technology Conf. Preprint OTC-2086, pp. 541–550.

"We struck flounder."

David H. Root
Lawrence J. Drew

The Pattern of Petroleum Discovery Rates

Why does the amount of petroleum discovered per unit of exploratory drilling drop off so sharply after the initial phases of exploration?

The petroleum industry in the United States is usually said to have begun with the drilling of the first successful exploratory hole in 1859, at Titusville, Pennsylvania. Crude oil was known before that from seeps and other inadvertent discoveries, but 1859 was the beginning of deliberate exploration in the U.S. By 1920, crude oil had been discovered in 25 states; it was supplying 13% of the nation's energy; and 251,000 wells were producing oil (DeGolyer and MacNaughton 1978). In that year, U.S. production was 443 million barrels.

From the beginning, it was plain that only a finite amount of oil was in the ground and that no level of production, however low, could be maintained indefinitely. But as long as oil was being discovered faster than it was being produced, this limitation was a matter of only vague concern. The industry continued to expand almost without interruption for 50 years after 1920, until in 1970 annual

David H. Root is a mathematician at the USGS. He received his Ph.D. in 1968 from the University of Washington. After teaching mathematics and statistics at Purdue University, he joined the USGS in 1974 as an analyst of energy resources, particularly oil- and gas-discovery rates. Lawrence J. Drew received his Ph.D. in petrology and statistics from the Pennsylvania State University in 1966. He was employed from 1967 to 1969 as a research statistician at GEOTECH Inc., from 1969 to 1972 as a mathematical geologist at Cities Service Oil Co., and since 1972 as a research geologist at the USGS. His areas of expertise include discovery-process modeling, analysis of petroleum-discovery rates, construction and manipulation of large petroleum-data files, and petroleum-resource appraisal. Address: USGS, Reston, VA 22092.

U.S. crude-oil production reached 3.52 billion barrels. Since then, U.S. crude-oil production has declined, reaching 2.99 billion barrels in 1977. As of the end of 1977, 118 billion barrels of crude oil had been produced, and approximately 2.5 million holes had been drilled, about 1.4 million of which had at one time produced some crude oil. Oil was being produced by 503,268 wells in the U.S.

The number of barrels of crude oil found per foot of exploratory drilling in the conterminous U.S. declined sharply just after World War II (Fig. 1). Before World War II, the discovery rate was high and fairly stable; after World War II, the discovery rate dropped suddenly; and since 1953, the discovery rate has been low but fairly stable. The decline occurred despite advances in exploration technology and has, so far at least, proved to be irreversible. From the history of the discovery rate, Hubbert concluded in a 1967 paper that only about 30 billion barrels of crude oil remained to be discovered in the conterminous U.S. and that the ultimate total production would probably be between 150 and 170 billion barrels and would be most unlikely to exceed 200 billion barrels. He later updated that analysis and reached essentially the same conclusion (Hubbert 1974).

The use of past discovery rates either to forecast future discovery rates or as evidence that crude oil is physically scarce has prompted much criticism. Barry Commoner (1976, p. 52) said of the sharp decline in the discovery rate, "This suggests that the decline is the result not of some gradual process, such as the progressive depletion of accessible oil deposits, but of some more abrupt event, such as a change in drilling procedures." DeVerle Harris (1977) presented a lengthy discussion of many factors that could affect the discovery rate but did not rank them in order of importance or draw any definite conclusions about the shape of the curve. He wrote, "Data on dQ/dh [Q is cumulative barrels of crude oil discovered, and h is cumulative exploratory footage drilled] which represent all provinces and changes in geologic understanding, economics, and technology severely violate the conditions for a monotonically decreasing function. There is no way of knowing or theorizing about the appropriate functional form for such data" (part 5, chap. 2, p. 34).

An explanation of why the petroleum-discovery-rate curve shown in Figure 1 has a step-function form has been derived from a detailed examination of the exploration history of the Permian Basin, a large petroleum-producing province in Texas and New Mexico (Fig. 2). Two important principles of the petroleum-exploration process become clear: first, most of the oil and gas discovered in a region is contained in a few large fields; and second, most of the large fields are discovered early in the exploration of a region.

Petroleum exploration in the Permian Basin

Discoveries in the Permian Basin as of 31 December 1974 totaled 23.9 billion barrels of recoverable crude oil and 71.0 trillion cubic feet of recoverable wet natural gas. These discoveries were 18% of all the crude oil and 10% of all the natural gas dis-

Figure 1. The amount of oil discovered in relation to the number of feet of exploratory drilling in the conterminous U.S. dropped sharply after about 1945 and has remained at a low but fairly stable level since about 1953. (After Hubbert 1967.)

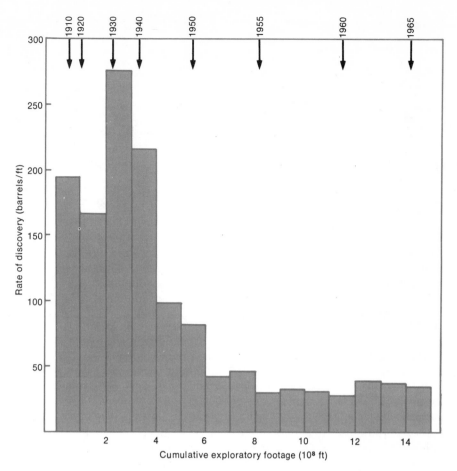

covered (i.e. past production plus proved reserves) in the conterminous U.S. up to that time. In the discussion that follows, information on gas discoveries is combined with information on oil discoveries: all discoveries are reported in barrels of oil equivalent (BOE), and 5,270 cf of wet gas is the equivalent of 1 barrel of crude oil. By the end of 1974, 4,014 oil and gas fields had been discovered in the basin by the drilling of 30,417 exploratory wells. A field is a group of reservoirs whose surface projections overlap. (The data on the sizes of oil and gas fields in the Permian Basin were furnished by the Dallas Field Office of the Department of Energy. The drilling data were purchased from Petroleum Information Inc. of Denver, CO.)

Basing an analysis of exploration in the Permian Basin on a count of the number of field discoveries per unit of exploratory drilling effort through time would be misleading, because oil and gas fields vary widely in size (Fig. 3). A small field in the Permian Basin contains as few as 1,000 BOE, whereas a large field contains as many as 2 billion BOE. This is a difference of six orders of magnitude—a disparity that is brought home by the comparisons of Table 1.

Exploratory drilling can be measured either in feet—i.e. the total depths of the exploratory holes—or simply in number of holes; and thus discovery rates can be measured either in barrels per hole or barrels per foot. Because there is a trend toward deeper exploratory holes in the Permian Basin, when the discovery rate is measured in barrels per foot it declines more rapidly than when it is measured in barrels per hole. The

difference, however, is only of secondary importance, because the trend toward deeper holes has been gradual. The average depth of exploratory holes in the Permian Basin during four successive ten-year periods beginning with 1935–45 and ending with 1965–75 were 4,000 ft, 6,200 ft, 6,300 ft, and 7,500 ft.

Looking at the great diversity of field sizes from another angle, Figure 4 shows the cumulative totals of BOE discovered in fields of 4 size classes through 1974. Most of the oil and gas discovered—23 billion out of 37 billion BOE—was found in 70 fields containing 100 million BOE or more each. More than 60% of the oil and gas came from less than 2% of the fields! In fact, 5.6% of all the petroleum discovered in the basin was contained in the single largest field. On the other hand, the 2,795 fields discovered in the basin that each contained less than 1 million BOE are a large part of the total number of fields discovered through the end of 1974, but all these fields together contributed only 0.49 billion BOE, or 1.3% of the total discovered.

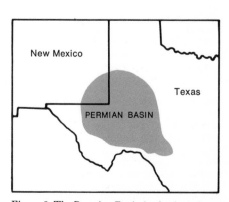

Figure 2. The Permian Basin in the American Southwest holds 18.4% of all the crude oil and 10.0% of all the natural gas discovered in the conterminous U.S. through 1974.

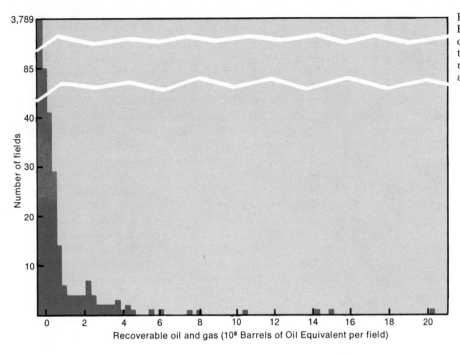

Figure 3. Petroleum fields in the Permian Basin vary tremendously in size—one field contains over 2 billion BOE—barrels of oil or the equivalent in natural gas—but the greatest number of fields contain amounts smaller by as much as six orders of magnitude.

The shape of the discovery-rate curve for the Permian Basin is thus largely determined by the positions of the discoveries of the 70 large fields in the discovery sequence of the basin. Figure 4 shows that most of these large fields had been discovered by 1950. Yet by the end of that year, only 6,286 exploratory holes had been drilled in the basin—20.7% of the 30,417 exploratory holes drilled in the basin through the end of 1974.

The trend over time in the average

size of the oil and gas fields discovered in the basin is shown in Figure 5. The curve was constructed by computing the average size of the fields discovered within each of 14 intervals of exploratory drilling between 1921 and 1974. Each of these intervals covers the drilling of approximately 2,000 exploratory wells.

The average size of the 111 fields discovered by the first 2,015 wells (1921–38) is 121.2 million BOE. During the 1939–46 period, 1,937 exploratory wells were drilled, resulting in 171 discoveries. But the average size of the fields discovered is 27.7 million BOE—a decline of 77.1% from the average size of the fields discovered in the first interval of drilling. During the 1947–50 period, 277 fields were discovered by the drilling of 2,334 exploratory wells. The average size of these fields is 22.7 million BOE— 18.1% smaller than the average size of the fields discovered in the previous period of exploratory drilling. In the fourth drilling interval, covering the 1951–52 period, 2,122 exploratory wells were drilled and 251 fields were discovered. The average size of these 251 fields is 8.4 million BOE—63% smaller than the average size of the fields found in the previous drilling period. By the end of 1952, when a total of 8,408 exploratory wells had been drilled in the basin (27.6% of the 30,417 wells drilled by 1974), the av-

erage size of the fields discovered had declined by more than an order of magnitude.

The average field size continued to decline during each of the next 4 drilling intervals, until it reached an average size of 2.3 million BOE for the fields discovered during the eighth interval, covering the 1959–60 period of exploration in the basin. During the next 6 intervals, 12,251 exploratory wells resulted in the discovery of 1,861 oil and gas fields. The average size of the fields discovered during this period fluctuated between 1.4 and 4.1 million BOE.

We conclude from this analysis that the rates of discovery per unit of exploratory drilling (measured either in number of holes or number of feet of drilling) in the Permian Basin fall into 3 phases. The first was a short phase when the discovery rate was high; this initial phase was over by 1938, when the first 2,015 wells had been drilled in the basin. During this phase, 31 of the 70 largest oil and gas fields were discovered. The second phase spanned the period 1939–50, during which an additional 4,271 exploratory wells were drilled. This was a transition phase, when the number of large fields discovered per hole declined relative to the total number of discoveries per hole, and consequently the volume of petroleum

Table 1. Differences of six orders of magnitude

Category	Small item	Large item
group of people (number)	baseball team	New York City
price	used bicycle	new jumbo jet
time	half minute	year
length	soccer field	twice around equator
area	Liechtenstein	Pacific Ocean
height	1 cm	Mt. Everest
foot race	4-cm sprint	marathon
volume	1 drop	1 barrel of oil

discovered per exploratory hole fell rapidly. In the third, relatively long phase, which began in 1951, the volume of petroleum discovered per exploratory hole remained at a low but stable level: 3,454 oil and gas fields were discovered containing 12.9 billion BOE. This is 86% of the total number of fields discovered in the basin through the end of 1974 but only 35% of the recoverable oil equivalent. The volume of petroleum discovered per exploratory hole during this period was low because the vast majority of these fields were small; 73.8% of them contained less than 1 million BOE each.

What the sudden drop-off means

There is a simple reason why the larger fields in the Permian Basin were discovered early: they underlie correspondingly larger surface areas.

The total surface-projection area of the 70 largest fields discovered is 11,074 km^2, which means that their average surface-projection area is about 158 km^2. At the other extreme, the total surface-projection area of the 2,795 fields that each contain less than 1 million BOE is 2,279 km^2; thus, their average surface-projection area is 0.82 km^2. On the basis of pure chance alone, the probability of a single exploratory hole discovering a given one of the larger fields is 193 times that of discovering a given one of these small fields.

Each size category of oil and gas fields, then, has its own characteristic declining discovery-rate curve; and the larger the fields in the size class are, the more rapid is the decline. The overall discovery rate is a weighted average of the discovery rates for the individual field-size classes. The actual decline in the overall discovery

rate in an area is now seen to be very sensitive to the distribution of field sizes in the area.

Enough information related to the rate of discovery of petroleum over the entire conterminous U.S. is available to indicate that the basic reasons for the shape of the rate curve are the same as those for the Permian-Basin curve. Of the more than 20,000 oil fields that have been discovered in the U.S., only 275 are expected to produce more than 100 million barrels of crude oil each (*Oil and Gas Journal* 1976). At the end of 1975, these largest 275 fields had produced 66.5 billion barrels of crude oil—59% of all the crude oil produced in the U.S. to that date—and contained about 90% of all the proved reserves in the U.S. Two hundred and twenty-three of these fields (81%) were discovered before 1950, which is approximately the time when the

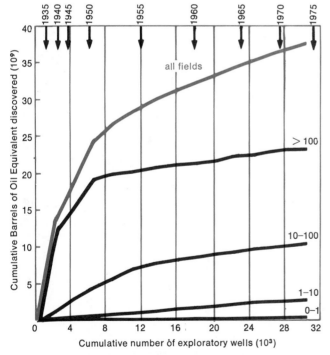

Figure 4. When the amount of petroleum discovered in the Permian Basin is broken down into classes according to the sizes of the individual fields in which the petroleum occurs, and amounts in these classes are plotted against the total cumulative number of exploratory wells, it becomes clear that the largest fields were discovered early. The 4 black lines show the amounts of petroleum discovered in fields of 4 size classes—those containing <1 million BOE, 1–<10 million BOE, 10–<100 million, and ≥100 million—at intervals in the exploration history of the basin. Note that although relatively little additional petroleum has been discovered in very large fields since 1950, petroleum from the largest class of fields still constitutes the major part of the total amount discovered.

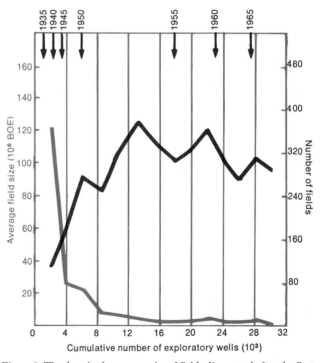

Figure 5. The drop in the average size of fields discovered after the first phases of exploration is vivid here. The period from 1921 to 1974 has been divided into 14 intervals, during each of which approximately 2,000 exploratory wells were drilled. The intervals fall into 3 phases: an initial phase—ending by 1938—when the discovery rate was high, a transitional phase during which the discovery rate fell sharply, and a long phase—beginning in 1951—when the discovery rate has been relatively low but stable. The actual number of fields discovered per drilling interval in the history of exploration of the Permian Basin has fluctuated but has not shown a clear trend up or down. The success rate has not changed despite the fact that wells have become somewhat deeper over time.

high rate of crude-oil discovery ended in the U.S. We therefore conclude that the step-function form of the U.S. discovery-rate curve is a consequence of the highly skewed size distribution of oil fields and the fact that the few largest fields were discovered early.

One cannot help but wonder how long the present low but stable rate of discovery can continue. To answer this question, we again turn to the analysis of the rate of discovery of oil and gas in the Permian Basin. There, the stable low rate of discovery has been maintained for a long period of exploratory drilling by the discovery of a large number of small fields.

The number of barrels of crude oil found per foot of exploratory drilling in the U.S. can increase significantly only if a substantial number of large oil fields are found in the future. Because the low discovery rate has already persisted for 291,000 exploratory holes drilled from 1951 through 1977, it seems unlikely that a substantial number of large fields have been overlooked. On the other hand, the discovery rate may not fall for a long time, because numerous small fields are still being discovered at a fairly constant rate. Of course, the continuation of this relatively constant rate of discovery will not prevent a decline in domestic crude-oil reserves.

The wide range of sizes in oil fields is not peculiar to the Permian Basin or to the U.S.; it is a global phenomenon. Approximately half of all the recoverable crude oil that has been discovered in the world is contained in the 33 largest fields (Nehring 1978). These fields range in size from 83 billion barrels to 5 billion barrels recoverable oil. Of these fields, 25 are in the Middle East, 2 are in the U.S., 2 are in Russia, and there are 1 each in Algeria, China, Libya, and Venezuela. In 1975 these 33 fields accounted for 42% of world oil production.

Thus, we can expect that the world curve will also show a sudden decline within a relatively short period, though it may take some time to recognize the event after it has taken place. And the decline in the discovery rate will be followed in a few years by a decline in the production rate, whose consequences are unlikely to be either agreeable or small.

References

Commoner, B. 1976. *Poverty of Power.* Knopf.

DeGolyer and MacNaughton. 1978. *Twentieth-Century Petroleum Statistics.* Dallas: DeGolyer and MacNaughton.

Harris, DeV. P. 1977. *Mineral Endowment, Resources, and Potential Supply: Theory, Methods for Appraisal, and Case Studies.* Tucson: Minresco.

Hubbert, M. K. 1967. Degree of advancement of petroleum exploration in United States.

Am. Assoc. Petroleum Geologists Bull. 51: 2207–27.

——. 1974. *U.S. Energy Resources: A Review as of 1972,* Part 1. A National Fuels and Energy Policy Study, U.S., 93rd Cong., 2nd sess., Senate Comm. Interior and Insular Affairs, Comm. Print. Serial No. 93-40 (92-75).

Nehring, R. 1978. *Giant Oil Fields and World Oil Resources.* Rand Co. Rept. R-2284-CIA.

Oil and Gas Journal. 1976. Here are the big U.S. reserves. Vol. 74(4): 118–20.

"I have a feeling it's too soon for fossil fuels around here."

Ogden H. Hammond
Robert E. Baron

Synthetic Fuels: Prices, Prospects, and Prior Art

Methods for synthetic fuel manufacture are varied and complex. What will be the impact on society if these technologies replace our present supply technologies?

Synthetic fuels result from the conversion of naturally occurring carbonaceous raw materials to fuels resembling petroleum or natural gas. Since the federal government is actively considering a $2 billion commercialization program to develop synthetic fuels as a substitute for domestic oil and gas, it is appropriate to consider the history of synthetic fuels as well as the raw materials and technologies available for their manufacture. Table 1 shows the raw materials most often considered for synthetic fuels manufacture. In order to place synthetic fuels in the proper perspective, one must examine not only the history and technology of synthetics but also the history of natural petroleum and gas.

Man has been using petroleum, natural gas, and synthetic fuel raw materials since prehistoric times for various purposes. The ancient Chinese were the first to make practical use of flammable natural gas (20). Several thousand years ago, they transported the gas through bamboo pipes to the seashore where it was used to evaporate seawater and produce salt. Early Western peoples are believed to have used petroleum

Ogden Hammond received his S.B., M.S., and Ph.D. from M.I.T., all in chemical engineering. He is currently a research scientist at the M.I.T. Energy Laboratory, a lecturer in the department of chemical engineering, and a teacher in M.I.T.'s Innovation Center. His research and publications have included studies of the economics and equilibria of synthetic fuels processes. Robert Baron received his B.S. in chemical engineering from Worcester Polytechnic Institute and S.M. in chemical engineering from M.I.T. A member of the staff of M.I.T.'s Energy Laboratory, he is currently involved in fluidized bed combustion research and advanced power cycle modeling. Address: Department of Chemical Engineering, M.I.T., Cambridge, MA 02139.

products in warfare. For instance, Greek fire—incendiary materials used in ancient times and the Middle Ages—is said to have contained distilled petroleum or asphaltic fractions. It was projected from tubes or thrown in pots, and some varieties caught fire on contact with water (24, 26).

One of the first records of synthetic fuel production in modern society is in a 1694 patent describing the distillation of "oyle from a kind of stone" (15). At about the same time, John Clayton, an English minister, discovered that coal heated in a closed vessel produced "a spirit which issued out and caught fire at the flame of a candle" (20). The first practical use of this phenomenon came in 1792 when William Murdock used gas to light his house. By 1815, London's streets were lighted with manufactured gas, and by 1820 gaslight had spread to Paris and to Baltimore (20).

In 1834, Selligue, a Frenchman, began to manufacture water gas (a mixture of carbon monoxide and hydrogen) by the reaction of coal and steam. To improve the low heating and illuminating value, he found it necessary to enrich—or "carburet"—this gas with oil, which he distilled from French oil shales. By 1838, commercial gas production was established, and shale oil had found an additional use as lamp fuel (15).

The first coal-derived synthetic fuels had a serious problem: due to the sulfur and nitrogen in the raw material, they contained H_2S and cyanogen, which rendered them both evil-smelling and poisonous. Eventually, it was discovered that scrubbing and contacting with lime and iron oxide

would remove most of these impurities and that proper redistillation techniques would yield more uniform products.

By 1850, a Canadian-born inventor, Abraham Gesner, had discovered that Trinidad pitch, from the island of Trinidad, and Albert coal from New Brunswick could be heated to yield oil or gas. Furthermore, he discovered that both materials were relatively free of sulfur and other pollutants. Thus, he contended, they were superior fuels for the illumination industry because the cost of gas purification would be minimized (24). Gesner spent several years attempting to develop these reserves. However, the Canadian courts prevented him from developing Albert coal, and he could not interest investors in Trinidad pitch. Also in 1850, in Bathgate, Scotland, James Young commercialized a process for recovering paraffins from "bog-head coal" by long low-temperature distillation; he obtained extensive patents on his process (15).

The Downer organization of Boston, the leader of the industry, was producing "coal oil" at the rate of 900,000 gallons per year by 1858. Recovery rates were as high as 110 gallons per ton of "coal" (24). In reality, coal was rarely used as a starting material. Modern geologists classify Albert coal and Trinidad pitch as types of asphalt (i.e. high-grade tar sands) while "bog-head coal" from Scotland was a peculiar substance with a composition somewhere between that of a very high-volatile bituminous coal and a high-grade oil shale.

Natural petroleum was primarily used for medicinal purposes in the 1850s. It was recovered by sponging it from oil springs with blankets, by trenching, and as a relatively undesirable by-product of wells drilled for the purpose of producing salt. In 1854, a group of investors obtained oil rights on 12,000 acres of land and purchased outright another 100 acres in Titusville, Pennsylvania, with the original intention of producing oil for medicinal purposes. One or more investors conceived the idea that oil could be used as an illuminant, and a leading chemist of the time, Benjamin Silliman, Jr., of Yale University, confirmed this belief. Attempts were made to recover the oil by trenching and other conventional practices, but for a variety of reasons these early efforts met with failure.

"Colonel" E. L. Drake was hired in 1857 to make the venture profitable, and after various attempts at trenching, Drake decided to use the drilling technology already developed by the salt well drillers. In 1859, Drake brought in the first oil well, which produced at the rate of about 8–10 gallons per day. He demonstrated that conventional drilling and boring techniques could be used rather inexpensively to recover petroleum resources that were previously unreachable—thus "coal oil" was no longer needed. By 1873—only 14 years after the first successful efforts to retrieve it—American crude oil production exceeded 420 million gallons per year (24), or about 400 times the synthetic output of 1860.

The manufacture of gas for lighting continued, however, and was advanced by the development of the gas mantle in 1885. Shortly thereafter (1890–1900) the development of the safer and cleaner incandescent bulb sharply curtailed U.S. use of gas for lighting purposes, but the demand for gas in heating had grown sufficiently to prevent any collapse of the industry. By 1920, however, with the price of fuel oil lower than the price of coke, gas producers began to use fuel oil as a raw material. By the 1930s, the development of strong piping led to extensive use of natural gas, and manufactured gas was then used only to satisfy peak loads—but even that use died out in the U.S. with the development of gas storage facilities. By 1963, only .5% of all gas sold in the U.S. was manufactured gas (14).

There have been very few efforts to produce synthetic fuels in the years since Drake's well. A notable exception is the Scottish oil shale industry started by Robert Bell in 1862, which functioned for 100 years before its reserves were exhausted. Production in 1952 was 780,000 barrels of oil (15). Exceptionally rich oil shale and a location close to market (i.e. the British Isles) allowed this industry to compete favorably with the petroleum industry. In Estonia, another small-scale oil shale industry was started in 1921, producing gas, oil, and electric power from relatively rich oil shale, but at times government subsidy has been required to sustain this industry (15). A Manchurian oil shale industry, begun in 1929, reached outputs of 1.3 million barrels per year while under Japanese control during World War II. The Communist Chinese expanded the Manchurian operation, and output during the Korean War is believed to have exceeded 14 million barrels per year (15).

The largest synthetic fuel operation was developed by Germany during World War II. With limited access to petroleum reserves and the survival of the German war machine at stake, there was little choice short of an early surrender. Based on extensive reserves of brown coal (lignite), this industry produced most of its liquid fuels by the Bergius process, a high-pressure catalytic hydrogenation of a mixture of coal and product oil. Production peaked at about 45 million barrels per year (29). In the postwar period, when Germany regained access to petroleum reserves, these processes became uneconomic. By 1958, they had all but ceased to exist (4, 14).

Some of the German technology developed during the war years is currently in use at the Sasol plant in Sasolburg, Republic of South Africa. Here coal is gasified in a Lurgi reactor and subsequently undergoes a Fischer-Tropsch reaction to produce

Table 1. Synthetic fuels raw materials

Raw material	*Description*
Lignite	Sometimes known as brown coal, lignite is the youngest geologically of the various coals. Substantial remains of woody material and considerable water are present in this coal. Heating value is generally considered too low to justify long distance transportation.
Sub-bituminous coal	Large deposits of this coal exist in the American West. Most deposits are extremely low in sulfur, making this coal an attractive fuel for electric power generation, despite the high shipping cost due to its low heating value.
Bituminous coal	Also known as "soft" coal, this coal has a high percent of volatile matter. It usually has a relatively high sulfur content and a high heating value. High-grade bituminous coal is the starting point for coke production for the steel industry.
Anthracite	Also known as "hard" coal, this coal was formerly used for home heating. It is almost pure carbon, low in sulfur and nitrogen, and has a high heating value.
High-grade oil shale	A fine-grained sedimentary rock containing sufficient organic matter to yield over 30 gallons of oil per ton of shale retorted under standard conditions (Fischer Assay). The organic matter is known as kerogen, a waxy solid hydrocarbonaceous substance derived from prehistoric plants and animals.
Medium-grade oil shale	Oil shale yielding 10–25 gallons per ton of shale.
Tar sands	Also known as oil sands, or bitumen-impregnated deposits. Tar sands generally consist of high molecular weight organics and mineral matter in the form of sand or sandstone. Water may be an integral part of the deposit.

gaseous and liquid fuels. This plant began operations in the mid-1950s and is the only Western large-scale, coal-based synthetics plant that has been in operation for a reasonable length of time. The synthetic fuels produced are expensive by U.S. standards, but the Republic of South Africa has been subjected to a number of economic embargos because of its political policies, and its political stability is such that it cannot afford to be subject to an oil embargo. Since it has very limited petroleum reserves, it can only be energy-independent by producing synthetic fuels—regardless of the cost. Other exports (such as gold and diamonds) make such economic issues of relatively low importance.

Hydrogenation

Throughout their history, all synthetic fuel processes have had the same purpose—hydrogenation of carbonaceous materials. As can be seen from Table 2, oil shale, tar sands, and coal have a lower ratio of hydrogen to carbon than methane, CH_4, or gasoline, $(-CH_2-)_N$. Therefore, if either methane or gasoline is to be produced from these raw materials, hydrogen must be added or carbon removed or some combination of both. The basic goal of all synthetic fuel processes is to increase the H/C ratio, remove mineral matter, and reduce sulfur and nitrogen contents to meet emission standards—all at a minimum cost.

The ideal raw material for a synthetic fuel process would have an H/C ratio close to that of the product and contain a minimum amount of oxygen. Unfortunately, synthetic fuel raw materials, as a general rule, contain oxygen and have relatively low H/C ratios. Consequently, they require substantial processing, and the costs of processing are correlated with the increase in the *net* H/C ratio (defined in Table 2) required to convert the raw material to useful fuels. This correlation, which can be justified by a detailed consideration beyond the scope of this article, can be simplistically explained by noting that capital costs are functions of the carbon efficiency (the percentage of carbon in the raw material that is converted) and that the carbon efficiency is related to the net H/C ratio of the raw material. The lower the net H/C ratio of the starting material and the higher the net H/C ratio of the desired product, the more capital equipment will be required. However, relatively expensive raw materials with high net H/C ratios may be substituted for this capital equipment.

All synthetic fuels processes involve hydrogenation, but the method and extent vary. There are three very general approaches that can be followed.

Hydrogenation by pyrolysis (commonly used synonyms of pyrolysis are retorting, coking, carbonization, and destructive distillation). Upon heating in an inert atmosphere, carbonaceous materials decompose to yield gases and liquids relatively rich in hydrogen and a solid carbon residue or coke. This is the oldest of all hydrogenation processes, but its basic principles are still used today by coal, oil shale, and tar sands processes.

Hydrogenation by direct reaction with hydrogen. If carbonaceous material is exposed to high-pressure hydrogen at modest temperatures (preferably in the presence of a catalyst) hydrogen-rich gas and liquids are produced. The Bergius process of Germany is an example of this type of processing.

Indirect hydrogenation using water as a source of hydrogen. At elevated temperatures, water (steam) will react with carbonaceous materials to produce a mixture of CO, CO_2, H_2, and various hydrocarbons (the Lurgi gasifier operates on this principle). Oxygen and carbon are removed from the system as CO_2.

We next consider how these basic approaches have been applied to the development of synthetic fuel processes.

Coal gasification

The conversion of coal to pipeline gas requires a greater increase in the net H/C ratio than any other process described here. High-Btu coal gasification produces methane (CH_4) from coal, which ranges from essentially pure carbon to $CH_{.7}$. One source of the necessary hydrogen is water, and the ideal overall reaction is: coal + water $\rightarrow CH_4 + CO_2$. Present processes cannot accomplish this reaction in a single step; rather, a sequence of reactions is employed:

$$1. \begin{cases} \text{coal} + \text{steam} \rightarrow CO + H_2 \\ \qquad\qquad\quad + CH_4 \\ \text{coal} + \text{oxygen} \rightarrow CO_2 \end{cases}$$

$$2. \ CO + H_2O \rightarrow CO_2 + H_2$$

$$3. \ CO + 3H_2 \rightarrow CH_4 + H_2O$$

One of the most widely used methods of coal gasification is the Lurgi process (Fig. 1), whose development began in Germany about 40 years ago (*29*).

Coal preparation. In this unit, coal is crushed and dried. If necessary a pretreatment step to reduce stickiness is employed. The dry, finely divided coal is fed to the gasifier.

Gasifier. In the gasifier, the coal is mixed with steam and pure oxygen. The result is "raw product gas"—a mixture including carbon monoxide, carbon dioxide, methane, and hy-

---

Table 2. Typical compositions of raw material

	Net H/C atomic ratio[a]	S/C atomic ratio	N/C atomic ratio
methane—CH_4	4	0	0
gasoline—$(CH_2)_N$	2	0	0
Colorado oil shale	1.58	.019	.028
Athabasca tar sands	1.50	.019	.004
bituminous coal	.6	.016	.018
sub-bituminous coal	.5	.007	.016
lignite	.25	.005	.015
anthracite	.05	.004	.001

[a] This assumes that all oxygen and sulfur present in the coal are in the form of H_2O and H_2S; all such hydrogen is excluded from the calculation.

SOURCE: *Handbook of Chemistry*, 10th ed. Norbert A. Lange, ed. McGraw-Hill, 1967. *Seventh World Petroleum Congress Proceedings*, Vol. 3, Elsevier Publishing Co. Ltd., 1967.

drogen as well as nitrogen and sulfur impurities, entrained coal, ash fines, and complex hydrocarbons. The reaction of coal with oxygen in the gasifier supplies needed heat, since the reaction of coal with steam is endothermic at temperatures high enough for reasonable reaction rates.

Quench. The raw product gas is passed through quench and heat recovery. This permits separation of heavy oils or tars that might have formed and lowers gas temperature to levels required for the subsequent operations.

Shift conversion. The ratio of hydrogen to carbon monoxide in the product gas is below one. The shift converter changes this ratio to that required by the methanation reaction ($CO + 3H_2 \rightarrow CH_4 + H_2O$) by the water gas shift reaction ($CO + H_2O = $ $CO_2 + H_2$). Since the equilibrium for this reaction is not affected by pressure and is favored at low temperatures, mild operating conditions are desirable. As a result the reaction is generally carried out in two stages: in the first stage temperatures are usually 800–900°F to obtain high initial reaction rates; the second stage operates at 700–750°F to increase yields (*19*). This reaction is relatively rapid, and since it occurs under mild conditions, is relatively inexpensive.

Gas purification. The gas is next fed to a purification unit for removal of H_2S and CO_2. H_2S must be removed prior to methanation because it is a virulent poison for the methanation catalyst. CO_2 is removed to increase the ratio of hydrogen to carbon and to remove oxygen from the system. This is the step in which carbon and oxygen are removed in the indirect hydrogenation approach. Approximately 50% of the carbon fed to the system is removed for high-Btu gas production.

Compression. Most gasifiers operate at relatively modest pressures (10 atm). Since the methanation reaction is favored at high pressure and low temperatures, a compression operation is required. Advanced gasifier designs, which are supposed to operate at elevated pressures, would not require this compression step, but technical difficulties have been encountered to date.

Methanation. The catalytic methanator operates at about 750°F and about 1000 psig (*28*). The reaction produces substantial heat, and if this heat is to be removed by producing high-pressure steam, the reactor must operate above 550°F but not above 900°F, which might cause carbon deposition and catalyst sintering (*13*). A nickel or iron catalyst is commonly used.

Dehydration and purification. After methanation, the gas is passed through a CO_2 scrubber and a dryer to remove the last traces of CO_2 and H_2O. The product gas then has a heating value of approximately 950 Btu per standard cubic foot (Btu/scf) and is suitable for distribution by pipeline.

The coal gasification process outlined still has many problems to which there are no solutions at present— only alternatives which avoid certain expenses and inefficiencies but introduce new ones. Ideally, a gasification process would yield more methane directly from the gasifier. This would be advantageous for three reasons. First, the reaction in catalytic methanation is characterized by the release of a large amount of heat. Since reaction rates increase with temperature, a hot spot could form on the catalyst and get hotter and hotter until the catalyst is destroyed. Methane acts as a diluent, and if there is enough present in the gas, there is less probability of this "thermal runaway" situation. Second, low temperature catalytic methanation is thermodynamically inefficient. The heat that is lost at a relatively low temperature during methanation was supplied by the reaction of coal with oxygen at a much higher temperature in the gasifier. Less catalytic metha-

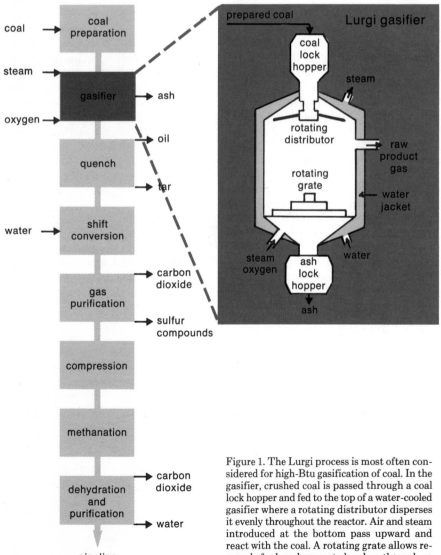

Figure 1. The Lurgi process is most often considered for high-Btu gasification of coal. In the gasifier, crushed coal is passed through a coal lock hopper and fed to the top of a water-cooled gasifier where a rotating distributor disperses it evenly throughout the reactor. Air and steam introduced at the bottom pass upward and react with the coal. A rotating grate allows removal of ash and unreacted carbon through an ash lock hopper.

nation means less heat loss and higher thermal efficiency. Third, methanation is costly; less methanation means a more economical product.

To yield more methane from the gasifier, one must use a feed coal with a high net H/C ratio (12). Bituminous coals have the highest net H/C ratio as well as a high sulfur content. Because most coal is sold for direct combustion processes that cannot easily use high-sulfur coal, the price of coal drops sharply as the sulfur content exceeds 2% (12). Fortunately, the costs of gasification are not greatly affected by the amount of sulfur present, since it is considerably easier to remove sulfur in a coal gasification process than in the direct combustion process used by industrial boilers or electric power plants. This is true because the concentration of sulfur compounds (H_2S) is much higher in the gasifier than in a combustion process, where the SO_2 is diluted by the nitrogen of the air. However, most bituminous coals with high H/C ratios cake or coke—that is, they tend to become sticky when heated. This occurs because hydrocarbon liquids, which may already exist in the coal or may be formed during breakdown of more complex molecules during heating, are driven to the surface. Most present processes (including Lurgi) cannot use a caking coal without some form of pretreatment, such as pyrolyzing the coal at 400–500°F, a procedure that is not only costly but also reduces the H/C ratio of the coal fed to the gasifier.

The same gasifier that is used to produce high-Btu gas can be used to produce low-Btu gas, if air is used as the source of oxygen. The raw product gas is then purified and used directly—without a shift converter or a methanator. The resulting low-Btu gas is a mixture of CO, H_2, N_2, and small amounts of CH_4. Because of its low heating value, storage and transportation are uneconomic, and it is unsuitable for residential or commercial use because of its toxicity. For electric power plants and industrial boilers, low-Btu gas is an alternative to high-sulfur coal with stack gas scrubbing or to low-sulfur coal.

Coal liquefaction

There are three known approaches to coal liquefaction, the conversion of coal to synthetic crude oil.

Coal pyrolysis. Coal may be heated (coked) to produce liquid and gaseous hydrocarbons and coke. U.S. bituminous coal is normally heated to 1800–2500°F with liquid yields of 1–10 gallons per ton of coal. Operations at lower temperatures (800–1400°F) result in an increase in liquid yields (15–20 gallons per ton) (17). Since these liquids usually contain unacceptable levels of nitrogen and sulfur, hydrotreating is required to produce the final synthetic crude. (Hydrotreating is a process in which hydrogen is reacted with the raw liquid fuel. Sulfur and nitrogen in the fuel are removed as H_2S and NH_3. The process was originally developed to upgrade natural petroleum crudes but may be used for shale, tar sands, or coal syncrudes. Unfortunately, the cost of hydrotreatment increases as the H/C ratio of the raw liquid fuel decreases, because the hydrogen tends to raise the H/C ratio of the fuel. In general, raw coal syncrudes tend to have a lower H/C ratio than raw oil shale or tar sands syncrudes and natural petroleum crude.) Pyrolysis produces substantial amounts of by-products, including gas and char, and for pyrolysis to be economic, a reasonable market must be found for the by-products. Processes currently under development include COED Process, at FMC Corp.; TOSCOAL Process, by the Oil Shale Corp.; and Garrett Pyrolysis Process, at Garrett Research and Development.

Fischer-Tropsch synthesis. In 1933, Franz Fischer and Hans Tropsch demonstrated that in the presence of certain catalysts carbon monoxide and hydrogen will react to form a mixture of aliphatic hydrocarbons and some oxyhydrocarbons. (Aliphatic hydrocarbons have the general formula C_NH_{2N+2}, whereas oxyhydrocarbons are compounds of carbon, hydrogen, and oxygen, such as methanol, acetic acid, and ethers.) This finding was followed by the development of processes similar to the Lurgi process, diagrammed in Figure 1, with the exception that the methanation step is replaced by Fischer-Tropsch synthesis. Fischer-Tropsch synthesis—$nCO + (2n + 1)H_2 \rightarrow C_nH_{2n+2} + nH_2O$—generally uses an iron catalyst and operates at about 400°F and high pressures. One of the most successful commercial operations using this technology is the Sasol plant in South Africa.

Direct hydrogenation. Coal is crushed, dried, and slurried in a solvent oil, and the slurry is then fed with hydrogen gas to a low-temperature (850°F) high-pressure (approximately 4000 psig) catalytic hydrogenation reactor. The liquid product is separated from the gas, the gas scrubbed to remove H_2S prior to recycling, and the liquid treated to remove ash. The final product is a synthetic oil. The choice of solvent oil in direct hydrogenation processes is critical to both reaction rates and extent of reaction. Available evidence indicates that the coal liquefaction mechanism involves complex Lewis Acid–Lewis Base interactions between the coal and the organic solvent (2).

Unlike coal gasification, the cost of coal liquefaction by direct hydrogenation is highly dependent on the amount of sulfur in the coal feed. The severe conditions required for sulfur removal cause hydrogenation of the raw liquid fuel (as discussed under hydrotreating), which results in very large hydrogen consumption and correspondingly high costs. Coal liquefaction processes are in a relatively early stage of development; those now under development include H-coal, at Hydrocarbon Research, Inc., and Synthoil, by the U.S. Bureau of Mines. The sketchy economics available suggest that oil shale-derived synthetic crudes will be cheaper (18).

Oil shale and tar sands

When oil shale is heated (pyrolyzed) in an 800–950°F retort, a carbonaceous material in it called kerogen decomposes to yield a raw synthetic crude oil (called shale oil), gas, and a solid residue (known as coked shale) that contains residual carbon and shale rock. The retorting operation is endothermic (requires heat), and three general methods of supplying heat are used. (For a detailed review of engineering approaches to retorting, see ref. 17, from which the following material was gathered.)

1. The heat can be supplied by the combustion of the by-product gases and coked shale in a separate stage of the retort. Examples of this approach now under development include Gas Combustion Retort, U.S. Bureau of Mines; Union Retort, Union Oil Co.; Paraho Kiln, Development Engi-

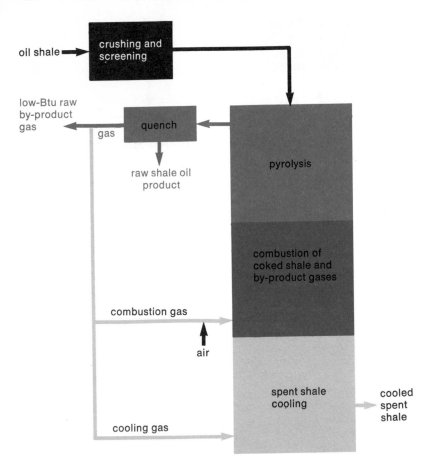

oil shale → crushing and screening

low-Btu raw by-product gas

gas ← quench

raw shale oil product

pyrolysis

combustion gas

air

combustion of coked shale and by-product gases

spent shale cooling → cooled spent shale

cooling gas

Figure 2. Oil shale is crushed and fed to the retort to produce a mixture of shale oil and low-Btu by-product gas. A quench is used to condense the shale oil, and some of the gas is recycled and combusted with the coked shale to provide the heat for pyrolysis. Some of the by-product gases are used to cool the combusted shale, and cooled spent shale is extracted from the lower portion of the retort.

Because water is required in direct oil shale processing as well as to control dust and reduce the alkalinity of the spent shale, water availability becomes an important and perhaps limiting consideration for any oil shale venture. One to three barrels of water are required per barrel of shale oil produced (17) by conventional retorting processes; costs of pipelining the necessary water from the Mississippi River to the Green River formation (the intersection of Colorado, Utah, and Wyoming) have been estimated to be reasonable if no legal or environmental problems are encountered and if a large operation is established (10).

A possible solution to these problems may be found in the recovery of the alkali minerals as a potentially valuable by-product. Superior Oil is pursuing this possibility and has reported an additional benefit—reduction in volume of the processed shale by as much as 50% (17), which may permit the return of the spent shale to the mine.

The technology for conversion of tar sands to synthetic fuels is much simpler than that for converting coal to synthetic fuels. The net ratio of hydrogen to carbon in tar sands is about the same as that of oil shale kerogen. The technologies are also quite similar. Processes currently under development generally employ as a first step a separation unit, using hot or cold water, to reduce the amount of mineral matter processed. The resulting bitumen concentrate may be either heated in a hydrogen atmosphere (hydrocracked) to yield the final product or, more typically, first pyrolyzed and then hydrotreated to remove sulfur and upgrade the oil.

Figure 5 illustrates the basic steps required; however, particular processes may combine two or more of these steps into a single operation. For instance, one process pyrolyzes the raw tar sands in a fluid bed at

neering Co. Figure 2 illustrates this approach.

2. The heat can be supplied by a recycled gas stream heated by an external heater (Fig. 3) using coked shale, by-product gases, shale oil, or any commercially available fuel, depending upon process design and economics. Examples of this approach now under development include modified Paraho Kiln; Steam Gas Recirculation Process, Union Oil Co.; Petrosix Process, Brazilian National Oil Co.

3. The heat can be supplied by externally heated solid materials, such as ceramic balls and spent shale itself. The solid materials may be heated by the combustion of a variety of fuels. Examples of this approach include TOSCO II, the Oil Shale Corp., and Lurgi-Ruhrgas (17), and a variant of it is shown in Figure 4.

Regardless of the retorting method used, the resulting shale oil is then fractionated (distilled), to separate it into various grades, and then hydrotreated to reduce nitrogen and sulfur levels. Typical process yields are about 96% of Fischer Assay (the amount of liquid produced per unit

oil shale processed under standard retorting conditions of 932°F). The Institute of Gas Technology (IGT) hydrotreatment process employs the direct reaction of hydrogen with oil shale, and yields of up to 118% of Fischer Assay are reported (21). IGT obtains these yields by using hydrogen to liquefy some of the carbonaceous residue in the coked shale.

A shale oil industry has for several years been on the borderline of economic feasibility. However, some major technical questions remain unanswered. A large shale oil industry will result in enormous amounts of spent shale: for every barrel of oil produced, over a ton of spent shale must be disposed of. Since the volume of the shale increases by as much as 50% during crushing (9), it is impossible to return all the material to the mine. A further problem is caused by the significant amounts of alkali minerals (principally $NaHCO_3$) that are present in the oil shale (6). Since these minerals are not removed during processing, the improper disposal of spent shale could result in the leaching of such minerals by groundwater, causing a substantial problem in an area that is already quite arid.

Figure 3. An externally heated recycled by-product gas provides the heat to convert crushed oil shale into shale oil and intermediate- to high-Btu raw gas. A recycled gas stream is also used to cool the coked shale. Because the coked shale is not combusted in the retort in this process, carbonaceous matter remains in the shale. Since no air is introduced for combustion, the gases produced—undiluted by nitrogen—have a relatively high heating value.

900°F, and thus in effect combines the separation and pyrolyzing steps. The resulting vapors are then condensed to form a crude synthetic oil that requires hydrotreating to upgrade it to an acceptable syncrude. The remaining coked tar sands are burned in a separate reactor, and a portion of the hot sands are returned to the pyrolyzer to supply the heat of reaction.

The most discussed tar sands deposit is located along the banks of northern Alberta's Athabasca River. A 45,000 bbl/day plant for synthetic fuel manufacture came on stream in 1967. However, further development of this deposit has been seriously threatened by inflating construction costs. It is only with the decision of the Canadian government to invest heavily in the development of this resource that private industry has been able to continue to construct new plants. A major problem with most tar sands processing methods is the removal of mineral fines, which in some deposits are present in relatively large amounts. The Athabasca deposit has, in addition, the obvious problems associated with an arctic climate.

In-situ technologies

In-situ processing is another approach to synthetic fuel production from coal, tar sands, and oil shale, and the procedure is similar for all three materials. First, channels are created in the formation by hydraulic fracturing or explosives (in at least one case the detonation of a nuclear device was considered). A network of injection wells—through which an oxidant, heat transfer agent, or solvent is injected into the formation—

Figure 4. Heat supplied by solid materials heated in an external heater pyrolyzes crushed oil shale to produce shale oil and intermediate to high-Btu gas. The coked shale and solid materials are cooled by recycled gas. As in Figure 3, the gases produced are undiluted by nitrogen.

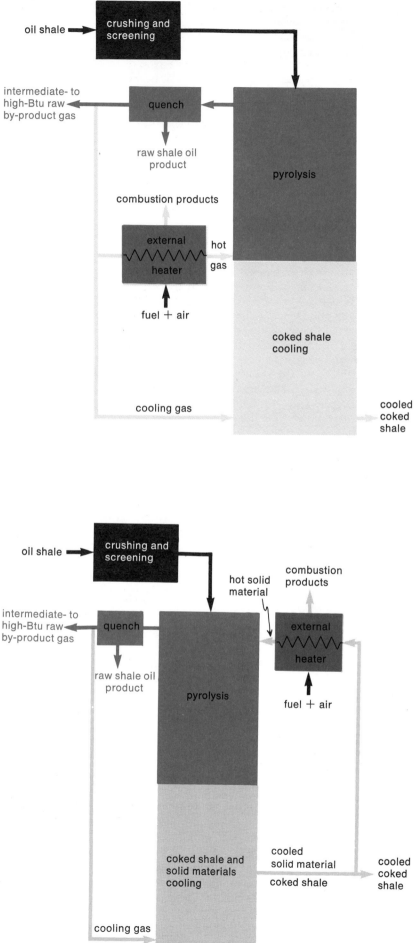

and production wells—through which the product is recovered—is constructed. Next, heat is generated in the formation: the most popular way is to start an underground fire (fire flood) and sustain it by pumping air through the injection wells. Whatever the method, the heat generates either carbonaceous liquids or gases which travel along the channels and are recovered in the production wells.

An alternate approach to using heat is to inject a solvent to dissolve the carbonaceous raw material, which would then travel along a channel and be recovered in a production well. This method has been used with limited success in Athabasca. A combined mining and in-situ process for oil shale is being evaluated by Garrett Research and Development. Their process would mine an underground chamber, blast the ceiling to fill the room with broken shale, and then burn a portion of the shale to supply heat for retorting. The liquids would be recovered from channels drilled in the floor of the chamber.

The major problem with all in-situ processes is communication between injection and production wells. For a production well to be effective, it must contact a channel or crack in the formation. Furthermore, this crack must contact either an injection well

or another crack which in turn contacts an injection well. Only with this type of communication is it possible to inject heat or a solvent into the formation and withdraw the crude synthetic oil. Since the probability of drilling wells that interconnect with cracks or channels is chancy at best, the yields from in-situ processing are usually far less than from above-ground processes.

Synthetic fuel reserves and economics

The relative abundance of synthetic fuel raw materials is indicated in Table 3. For our purposes, *resources* are defined as those known and identified deposits of a given substance regardless of whether they are high or low grade or in an accessible or inaccessible location. Under *total identified remaining resources*, we list the total energy available from these resources, only a fraction of which are sufficiently large, rich, and accessible to warrant commercial consideration. The total energy available from this fraction is listed under *known commercially attractive deposits.* As commercially attractive deposits are developed, the cost of recovering a unit of the given substance begins to increase after a certain point, and therefore only a certain fraction of even a rich com-

mercially attractive deposit is actually recoverable at close to present prices. This fraction is defined as *proved reserves,* and the total energy from them is indicated under *proved reserves recoverable at present prices.* In Table 3, no allowance is made for government subsidies or new discoveries. Since oil shales and tar sands have little or no value unless conversion processes are economically viable and because process economics are not clear, we list as proved reserves recoverable at present prices those portions (as defined in Table 3) of the tar sands and oil shale reserves that would be used first as the raw material for any industry.

The total energy consumption of the United States in 1974 was about .073 $\times 10^{18}$ Btu (*23*). Thus, if coal alone were to supply our energy needs, we would have a 15-year supply at close to current price levels. Similarly, oil shale would yield an additional 16-year supply, for a total available supply of about 31 years at close to present prices.

By contrast, if coal consumption were to continue at its present rate of about 600×10^6 tons per year, we would not see a significant increase in the real price of coal for about 100 years. Although there are tremendous identified deposits of both coal and oil shale

Figure 5. Tar sands may be converted to synthetic fuels by a variety of processes. For instance, in cold water separation, water and a dissolving additive are mixed with tar sands to extract the bitumen, and the resulting mixture is fed to a settling tank where a phase separa-

tion occurs. Since the bitumen-rich phase is less dense than the inorganic phase, it will "float," whereas the inorganic phase, which contains mostly mineral matter and water, will "sink." Because the separation is not 100% efficient, the bitumen-rich phase removed from the top

of the first settling tank is fed to other settling tanks where the process is repeated until a bitumen strain that is relatively free of mineral matter is obtained. This strain is then processed to increase the net H/C ratio.

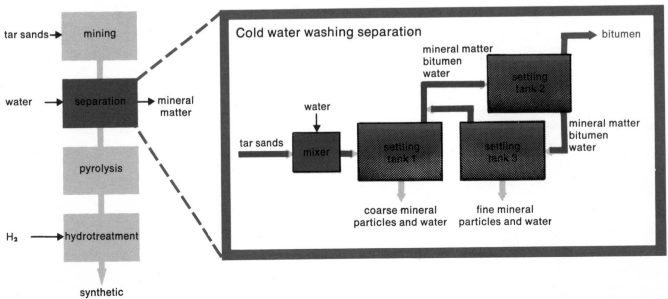

in the U.S., only about 25% (3) of these deposits are both sufficiently rich and sufficiently accessible to consider them commercially attractive at this time. Of that total, only about half the material of a given deposit could be recovered at present prices. Detailed studies of the engineering and economics of underground coal mining by Martin Zimmerman of M.I.T.'s Energy Laboratory suggest that only 10.7 billion tons of coal are recoverable from underground mines without price increases of more than 22% (25). The U.S. Geological Survey reports that of the 118 billion tons of strippable coal reserves, only 45 billion would be economically recoverable (3). It should also be noted that the vast majority of the coal recoverable by underground mining is high in sulfur.

The same problems exist with petroleum. Statistical data show that only 32% of the "oil in place" has been considered to be economically extractable. Thus, as with coal, oil shale, and tar sands, "proved reserves" (reserves judged to be economically recoverable) of petroleum are substantially smaller than commercially attractive petroleum deposits. The oil known to be actually in place in the U.S. is much larger since it includes oil not recovered from old oil fields.

In the preceding discussion we have assumed that synthetic fuel raw materials could be transformed into synthetic substitutes for oil or gas at reasonably competitive prices. Actual economics for the various synthetic fuels processes suggest that this may not be possible—even for that fraction of reserves of raw materials we identified as recoverable at present prices.

To give some historical perspective of the price of energy, manufactured gas sold in 1860 for about $8/MM Btu while coal oil cost approximately $7/MM Btu (in 1860 dollars) (24). Prices of energy in more recent times (after commercialization of petroleum and natural gas) are given in constant 1975 dollars in Table 4.

Table 5 presents recent estimates of synthetic fuel costs in 1975 dollars. Estimates on the costs of synthetic fuel processes may vary. For instance, Table 5 indicates the price of a barrel of syncrude from oil shale will be $15.50/bbl, while *Synthetic Fuels*

Quarterly quotes an estimated price for the Dow West shale oil project of $16.75/bbl if no government subsidies of any kind are available. Syncrudes derived from oil shale and tar sands appear to be the least expensive synthetic fuels primarily because oil shale and tar sands have the highest net H/C ratios of the synthetic fuel raw materials—and, as already noted, capital equipment charges are related to the net H/C ratio.

Early estimates should be treated with a great deal of caution. For example, in early 1973, a National Pe-

Table 3. Fossil fuel reserves

Raw material	Average higher heating value[a] (10^6 Btu/ton)	Total identified remaining resources (Btu × 10^{18})	Known[b] commercially attractive deposits (Btu × 10^{18})	Proved reserves recoverable at present prices (Btu × 10^{18})
lignite	12.5	5.6	1.4 ⎫	
sub-bituminous	17	7.2	1.8 ⎬	
bituminous	22.5	15	3.8	1.1 (all coals)[c]
anthracite	27	.58	.14 ⎭	
high-grade oil shale	6	>4.5	2.3	<1.2[d]
medium-grade oil shale	3	12	0	0
tar sands (U.S.)	~4	.0115	.0038	small
Athabasca tar sands (Alberta, Canada)	~4	4	2	<.4[e]
U.S. natural gas[f]		.5	.49	.39
U.S. petroleum[g]	38	2.5	1.2	.37

[a] Heat obtained when a fuel is combusted at 298°K with stoichiometric oxygen and when all resulting water vapor is condensed.

[b] Lignite and sub-bituminous coal seams more than 10 ft thick, less than 1000 ft overburden; anthracite and bituminous coal seams more than 42 in. thick, less than 1000 ft overburden; oil shale yielding 25–100 gallons/ton, zones more than 100 ft thick; tar sands containing at least 12% bitumen.

[c] 10.7 × 10^9 tons available from underground mining (25); 45 × 10^9 tons available from strip mining (3).

[d] 50% extraction efficiency (7).   [f] Assumes .8 recovery factor (27).

[e] Overburden ratio of less than 1.0 (16).   [g] Assumes .32 recovery factor (27).

Table 4. Cost of energy in 1975$/MM Btu

Year	Electricity *[a]	Natural gas[b]	Oil[c]
1930	30.46	—	.723
1935	30.35	1.355	.702
1940	25.84	1.399	.73
1945	16.11	.960	.62
1950	11.25	.750	.91
1955	9.70	.933	.93
1960	9.05	1.16	.98
1965	8.38	1.11	.88
1970	7.32	1.01	.85
1975 (March)	7.70	1.279	2.00

* As opposed to gas and oil, a Btu of electricity is a unit of *work* and hence worth about three times as much as a unit of heat.

[a] Average price to consumer.

[b] Average price of natural gas to ultimate consumers includes both regulated and unregulated supplies. Intrastate gas in Texas (unregulated) is selling for about $1.90/thousand cubic feet.

[c] Wholesale price of crude oil. MM = 1,000,000.

SOURCE: Calculations based on Statistical Supplement to the Survey of Current Business.

troleum Council estimate of the cost of a 270×10^9 Btu/day gasification plant was \$209 million (22). Before the end of that year the price was estimated to be \$313 million by the Federal Power Commission (11). The same plant was estimated to cost \$740 million by mid-1974, but by March 1975, the plant was estimated at about 1 billion 1975 dollars (8). As another example, the 125,000 bbl/day syncrude tar sands plant in Athabasca was originally estimated at \$900 million in mid-1973, but present estimates are approximately \$2 billion (5).

Only part of the enormous increase in cost of synthetic fuels plants is due to inflation. A significant fraction of the increases are the result of oversights which could not be discovered until actual commercial designs were made. As a general rule of thumb, real costs of new and unproved processes tend to double as soon as an attempt is made to construct them.

Because synthetic fuels plants require enormous capital investments, the costs of synthetic fuels are extremely sensitive to the cost of capital goods and the rate of return on investment. The cost of synthetic fuels is also extremely dependent on the stream factor—defined as the amount of actual annual output divided by the designed annual output, assuming continuous operation. The best stream factors, approximately .91, are attained by oil refineries. Economics for synthetics plants assume a similar stream factor, but it is too early to know whether sufficient excess and spare capacity has been included to make this a realistic estimate. If major operating difficulties were experienced, and stream factors were as low as .5, synthetics costs would increase by almost 100%.

Outlook for the future

This is not the first time that man has believed that the reserves of petroleum were running low. At the turn of this century, methanol was seriously considered as a gasoline substitute. Massive new oil discoveries ended that crisis, but new oil discoveries of the size needed to solve our present problem appear today to be significantly less likely. In the history of the United States roughly 106 billion barrels ($.6 \times 10^{18}$ Btu) of oil and about 480×10^{12} cubic feet ($.480 \times 10^{18}$ Btu) of natural gas have been produced. At the 1974 rate of consumption of oil and gas of roughly $.05 \times 10^{18}$ Btu/year, as much as was consumed during the past 115 years would vanish in about 22 years (27).

The immediate question is what has caused the vast increase in energy consumption. It is worth noting that although total energy consumption increased by more than a factor of three from 1930 to 1970, *per capita* energy consumption increased by only 70% in the same period (13). Thus, population growth is as much or more to blame for the high consumption rate as increased greed for energy. If consumption continues to grow at the rate that it has in the past, our present inexpensive energy supplies will be rapidly exhausted.

It must be remembered that we do not consume energy per se. We consume goods such as a comfortable temperature in our houses, or transportation, or light. Energy is only one of the several inputs that produce these goods—the others could be generally classified as capital and labor. To illustrate this point, consider the situation faced by a homeowner. When the price of home heating fuel is low, the homeowner is able to maintain a comfortable temperature in his home by consuming relatively large quantities of fuel. As the price of home heating fuel rises, the owner considers insulation in order to reduce his home heating bill and hence his fuel consumption. In effect, he is using a capital good (insulation) and the services of someone to install that good (labor) to substitute for the heating fuel (energy). The rational homeowner weighs the long-term costs (depreciation and interests on money expended for purchasing and installing the insulation) versus long-term gains resulting from lower fuel consumption.

The industrial supplier of energy is in somewhat the same position as the homeowner. As long as there is an abundant supply of easily accessible petroleum reserves, he will expend relatively little capital and labor to produce his final energy form (electricity, gasoline, etc.) To produce his energy from alternate energy sources (solar, nuclear, synthetic fuel raw materials, etc.) requires relatively large expenditures of capital and labor and thus is relatively expensive. If the petroleum reserves become sufficiently inaccessible that an equivalent or greater amount of capital and labor is required to obtain

Table 5. Estimated costs for synthetic fuels in 1975\$

Plant[a]	Nominal capacity per calendar day	Capital charges[b]	Operating cost[c]	Coal[d] or natural gas[e]	Total $/bbl	$/MM Btu
tar sands (Syncrude Canada Ltd)	125,000 bbl/day	1.68	.80	.30	15.50	2.78
oil shale (Dow West project)	55,000 bbl/day	1.89	.88	—	15.50	2.77
high-Btu gas from coal (El Paso 4 corners)	244×10^9 Btu/day	2.41	1.15	.35	—	3.91

[a] Plant operating at full capacity for 330 days per year.

[b] Capital charges @ 21.5% (corresponds to 10% discounted cash flow).

[c] Operating costs @ 2.5% of capital invested for taxes and insurance; 4.9% of capital invested for maintenance; labor at \$8/hr.; contingency at 10% of the sum of direct operating cost and cost of feed.

[d] Coal costs @ \$.30/MM Btu, minus by-product credit of \$.75/MM Btu.

[e] Natural gas @ \$2.00/MM Btu.

SOURCE of capital costs: *Synthetic Fuels Quarterly* and Cameron Engineers, Denver, Colorado.

them, then alternative energy sources become viable.

Thus capital and labor may be used to reduce consumption or to increase energy production from alternative sources. Both appear expensive compared to pumping oil from ample reservoirs. It is the job of the engineer and the economist to evaluate these alternatives and to encourage development of the least expensive solutions. There is no shortage of energy, but it is clear that there is a drastic shortage of energy at the prices to which we have been accustomed.

The part synthetic fuels will play in the future cannot yet be determined. It appears certain that there will again be a U.S. synthetic fuels industry of some sort, but how large that industry ought to be and how soon it should develop is unclear. A massive development of a synthetic fuels industry by means of some sort of subsidy appears unwise; if synthetic fuels are subsidized, there will be less incentive to use capital and labor to reduce consumption—and the result will be the rapid exhaustion of our relatively inexpensive synthetic fuel raw materials.

Whatever course of action is followed, the costs to the nation will be high. Enormous amounts of man-made capital goods will be required to take the place of formerly abundant accessible petroleum resources. If the amount of capital goods required are to be produced, consumption of other (consumer) goods must be decreased, and the U.S. standard of living will be significantly affected. If research can develop new technologies that are significantly less expensive than those now conceived of, if there are massive new finds of oil and gas, or if a totally new source of inexpensive energy can be found, we will again be able to pass this problem on to future generations. The probability that we will be that fortunate appears small.

References

1. Abraham, Herbert. 1960. *Asphalts and Allied Substances,* Vol. 1, 6th ed. N.Y.: D. Van Nostrand.

2. Angelovich, J. M., et al. 1970. Solvents used in the conversion of coal. *Indus. and Eng. Chem., Design and Dev.* 9(1).

3. Averitt, Paul. 1973. Coal. United States Mineral Resources, Geological Survey Professional Paper 820, Dept. of the Interior.

4. Bangham, D. H. 1950. *Progress in Coal Sciences.* N.Y.: Interscience Publishers.

5. Berkowitz, Norbert, and James G. Speight. July 1975. The oil sands of Alberta. *Fuel* 54.

6. Culbertson, William C., and Janet K. Pitman. 1973. Oil Shale. United States Mineral Resources, Geological Survey Professional Paper 820, Dept. of the Interior.

7. Duncan, D. C., and V. E. Swanson. 1965. Organic-Rich Shale of the United States and World Land Areas. U.S. Geological Survey Circular 523.

8. El Paso Natural Gas Company. March 13, 1975. Official Stenographer's Report, Federal Power Commission.

9. Evaluation of Coal-Gasification Technology, Part 2: Low and Intermediate Btu Fuel Gases. 1974. National Academy of Engineering.

10. Exxon Report to EPA, Exxon-EPA-1/60/3-74-0096, June 1974.

11. Final Report, The Supply-Technical Advisory Task Force. Synthetic Gas-Coal, National Gas Survey, Federal Power Commission, April 1973.

12. Hammond, Ogden H., and Martin B. Zimmerman. July 1975. The economics of coal-based synthetic gas. *Tech. Rev.* 77(8).

13. Hottel, H. C., and J. B. Howard. 1971. *New Energy Technology.* M.I.T. Press.

14. Kirk-Othmer. 1969. *Encyclopedia of Chemical Technology,* Vol. 10, 2nd ed. NY: Interscience Publishers.

15. Ibid., Vol. 18.

16. Ibid., Vol. 19.

17. Klass, Donald L. August 1975. Synthetic crude oil from shale and coal. *Chem. Tech.*

18. Personal communication with Dr. J. Longwell, Exxon.

19. Shreve, R. Norris. 1967. *Chemical Process Industry.* 3rd ed. NY: McGraw-Hill.

20. McLoughlin, E. V., ed. *The Book of Knowledge,* Vol. 7. 1955. NY: The Grolier Society, Inc.

21. The IGT Oil Shale Process for Above-Ground and In-Situ Hydrotreating, Institute of Gas Technology, September 1975.

22. U.S. Energy Outlook, Coal Availability. National Petroleum Council, Dept. of the Interior, 1973.

23. *Monthly Energy Review,* U.S. National Energy Information Center, Federal Energy Administration, Aug. 1975.

24. Williamson, Harold F., and Arnold R. Daum. 1959. *The American Petroleum Industry: the Age of Illumination 1859–1899.* Northwestern University Press.

25. Zimmerman, Martin B. Long run mineral supply: The case of coal in the United States, Ph.D. diss., 1975, M.I.T.

26. *Encyclopedia Britannica,* Vol. 10. 1969. Encyclopedia Britannica, Inc.

27. Geological Estimates of Undiscovered Recoverable Oil and Gas Resources in the U.S., Geological Survey Circular 725, 1975.

28. Wen, C. Y., P. C. Chen, K. Kato, and A. F. Galli. March 1969. Optimization of Fixed Bed Methanation Processes. Report 7, Engineering Experiment Station, West Virginia University, Morgantown, WV.

29. *Encyclopedia Britannica,* Vol. 15. 1974. Encyclopedia Britannica Inc.

"I've tried it, Ned. Kicking doesn't work. There must be some *other* way to get oil out of shale."

PART 3 *Nuclear Power*

G. L. Kulcinski
G. Kessler
J. Holdren
W. Häfele

Energy for the Long Run: Fission or Fusion?

Factors such as hazards, technological costs, and development time are the significant points of comparison for the two most likely long-term energy sources

The present search for inexhaustible energy sources is characterized by conflicting claims, misinformation about both the potentials and the hazards of the possible sources, and a general frustration about the prospects for the eventual commercialization of any of the proposed schemes. The problems have arisen in part because the proponents and critics of various energy sources have almost always written separate appraisals of what each energy form can accomplish. Therefore, in comparing the two most likely and controversial long-range energy sources—fusion and fast-breeder fission—we have brought together a group of proponents of both kinds of nuclear power, to try to achieve an accurate presentation of the arguments for—and against—each course at this time. We recognize that other forms of energy, such as solar, will make a contribution

G. L. Kulcinski is Professor of Engineering and past director of the Wisconsin Fusion Technology Program at the University of Wisconsin, Madison. In 1978 he was awarded the Curtis W. McGraw award for outstanding early achievement by a young engineering researcher. His research has focused on fusion-reactor design and on the effect of nuclear irradiation on materials. G. Kessler is the director for the research and development programs of the German Fast-Breeder Reactor Program at the Gesellschaft für Kernforschung, Karlsruhe, West Germany. J. Holdren is Professor of Energy and Resources at the University of California, Berkeley, and consultant at the Lawrence Livermore Laboratory in California. W. Häfele is at present the director of the energy program at the International Institute for Applied Systems Analysis, Laxenburg, Austria. This article is derived from a 500-page report of the IIASA. Address: G. L. Kulcinski, Nuclear Engineering Department, University of Wisconsin, 439 Engineering Research Building, 1500 Johnson Drive, Madison, WI 53706.

in the next century, but compared to the potential of nuclear energy, the fraction of the total energy supply that nonnuclear sources can provide will be small, and thus we have not considered those sources here.

We make no claims for having reached a final word on these issues. Indeed, because development of fusion reactors is at a very early stage—not even an experimental device to prove scientific feasibility has yet been produced—our conclusions must be tentative. While many of the general problems associated with the two kinds of reactors are fairly well understood, other more specific problems are much less well understood. (See *ref. 1* for more complete documentation.)

It should also be made clear that we are not comparing fusion reactors to the present light-water moderated fission reactors (LWRs). Uranium resources are not sufficient to fuel the required number of LWRs much after the turn of the century. If the fission process is to provide energy beyond the 21st century, the world must convert from LWRs, which utilize only 1% of the potential energy in uranium, to breeder reactors, which use about 60% to 70%. While the fundamentals of nuclear-energy release are the same for both types of fission reactors, there are significant differences in the technologies required to convert the nuclear energy to electricity.

Both fission breeders and fusion reactors are, in terms of fuel supply, potential providers of very large, almost indefinite amounts of electricity (Table 1). The most likely fuel for fission breeders is uranium 238; that for fusion systems is deuterium and

lithium. Even if the prices of these fuels rise to several times the present values, the total cost of electricity in either system would not be significantly affected. If the substantial amounts of both uranium and lithium that we know are available at about $100/kg are exhausted, extracting either of these fuels at several times more than today's costs will still be economically feasible. Including other possible fuels for both systems in our calculations will only make the outlook more favorable.

The choice between fission and fusion will therefore have to be made on grounds other than fuel resources. Relevant considerations include potential biological and social hazards, costs of research and development, capital costs, technical complexity, and time factors. Before we compare the alternatives in these areas, however, we must look at the present status of fission and fusion reactors.

Where things stand

The nuclear processes involved in fission and fusion breeder reactors are fundamentally different. In fission, the most common reaction used produces plutonium from uranium via the steps

$$^{238}U + n \rightarrow {}^{239}U$$

$$^{239}U \xrightarrow[\beta^-]{23.5 \text{ min}} {}^{239}Np$$

$$^{239}Np \xrightarrow[\beta^-]{2.33 \text{ day}} {}^{239}Pu$$

While ^{238}U is not fissionable by neutrons with energies below 1.4 MeV, ^{239}Pu is fissionable by neutrons of any energy. Since each fission event in ^{239}Pu produces about 3 neutrons, as much—perhaps even more—Pu is produced in the overall reaction than is burned up: thus the name *breeder*. The fundamentals of the production of Pu have been known and utilized for over 30 years.

The most likely reaction to be used in fusion reactors is the one in which deuterium and tritium are combined as follows:

$$D + T \rightarrow n + {}^4He$$

Since there is no significant natural source of tritium on the earth, the tritium supply must be continually replenished by "breeding" it from the element lithium:

$$^6Li + n \rightarrow T + {}^4He$$

$$^7Li + n \rightarrow T + {}^4He + n$$

The capture of the additional neutron from the ^7Li reaction in ^6Li insures that more tritium is produced than is burned up, thus allowing an excess of fuel to be produced. At the present time it is not the breeding of tritium that keeps us from building fusion reactors, but the containment of the D and T for long enough times and at high enough temperatures to produce large amounts of neutrons.

To make some quantitative comparisons between these two means of producing energy, representative designs must be chosen. As a fission system, we have chosen the Liquid-Metal Fast-Breeder Reactor. The LMFBR clearly dominates research and development programs on breeder reactors around the world, making it by far the most likely breeder for commercialization. Historically, fast neutron breeder reactors have been preferred to thermal breeders because of the higher fuel utilization in the fast reactors. Liquid-metal-cooled fast reactors have received much more attention than gas-cooled reactors for reasons partly technical and partly historical, and no prototype gas-cooled fast reactors are under construction at present. Among various existing LMFBR designs we chose the German/Belgian/Dutch prototype fast-breeder reactor SNR 300 because we had full access to all the details of the program (see Fig. 1 for a schematic drawing of the reactor).

Selecting a representative fusion system is difficult, because it is impossible to state with any certainty which configuration will actually lead to a working reactor. At present, most

Table 1. World energy resource picture

World resources	Energy content (Terawatt-years)
Uranium to $100/kg used in light-water reactors	100
Recoverable oil known as of 1976	430
Gas, oil, and coal ultimately recoverable	2,600
Uranium to $100/kg used in fast-breeder reactors	11,000
Lithium to $60/kg used in fusion reactors	11,000

Yearly consumption	Rate (terawatts)
1975 worldwide	8.6
total for 8×10^9 people at 6 kw per capita	48.0

* 1 terawatt-year = 10^{12} watt-years

Figure 1. Schematic diagrams of an SNR-300 fast-breeder reactor and a typical lithium-cooled, deuterium-tritium-fueled tokamak reactor reveal that, while energy is released in different ways in the two kinds of reactors, the rest of the electrical plant is much the same in design and size. Actual nuclear reactions take place in the light blue areas.

Fission reactor: SNR 300 fast breeder

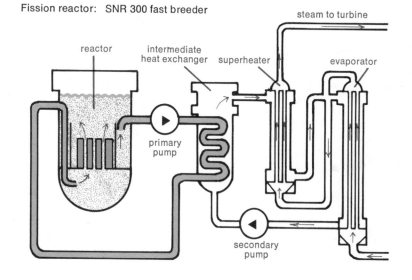

Fusion reactor: typical lithium-cooled DT tokamak

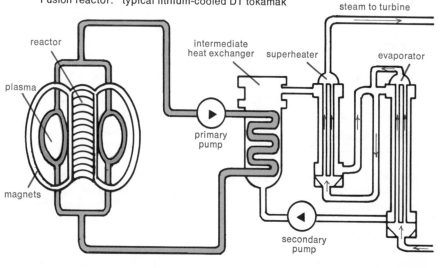

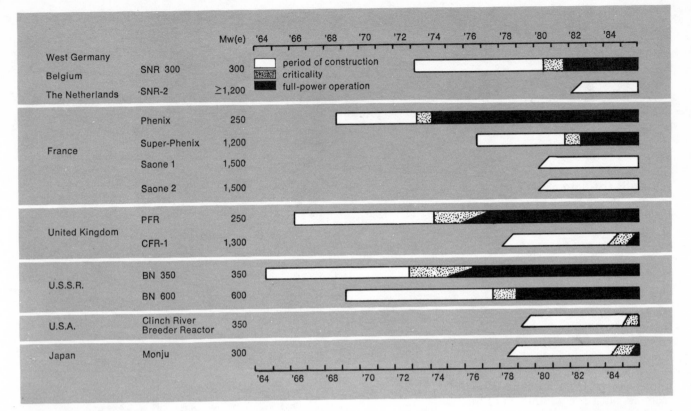

Figure 2. Development of prototype fission fast-breeder reactors is proceeding at different paces in various countries. Note that full-power operation of these reactors does not indicate full-scale commercialization: many problems, such as the closure of the fuel cycle, must be worked out even at this stage. The French Phenix Reactor operated at 85% of full power until August 1976, when engineering difficulties cut this back to about 65%; full-power operation was not restored until the spring of 1978. The U.S. is at present reassessing its commitment to the Clinch River Breeder Reactor.

designs for reactors have been based on what is called the tokamak concept, which seems to provide the greatest promise of success from a scientific standpoint. A tokamak system has a toroidal geometry and complex magnet configuration that make it a very difficult system to design for electricity production. Perhaps some other approach to fusion (e.g., mirrors, laser-driven systems) will eventually lead more easily to a reactor than will the tokamak approach. We have chosen to discuss here the liquid-lithium, stainless-steel–structure tokamak because more extensive and detailed information has been accessible for this design than seems to be available for other approaches. Figure 1 shows a schematic diagram of one of the 12 full-scale tokamaks studied so far.

The status of both fission and fusion systems is best described with regard to three thresholds of feasibility: scientific, engineering, and commercial. The scientific feasibility of fission fast-breeder reactors was demonstrated during the 1950s in the U.S.,

the U.K., and the U.S.S.R. Since that time, the engineering has developed along two lines. The Enrico Fermi Fast Breeder Reactor represents the early line, which used metallic fuels and produced tens to a few hundred megawatts of thermal energy. The second line, represented here by the LMFBR, is that of potential commercial reactors using mixed oxides of plutonium-uranium as fuel. Reactors in this line have a much higher power density, producing up to 1,500 megawatts of power each.

Figure 2 shows the fission fast-breeder projects under development throughout the world thus far. A large set of physics and engineering test facilities are now available, the result of a heavy investment of capital, manpower, and time. The technologies for liquid sodium as a coolant and for mixed oxides as fuels are essentially in hand. Most of the current tests are devoted to proving the safety equipment and demonstrating the fuel-element performance as required in the licensing process for large power reactors of the 1,200 Mw(e)

class. Preparations are now under way for the building of a semicommercial class of 1,200 Mw(e) reactors in France, the U.K., and West Germany. It is possible that these reactor systems will be completed by the late 1980s, and they may be making an input into the electrical generating grids by the 1980s. Engineering feasibility will thus have been demonstrated for fission reactors in France (4,5), the U.K. (6,7), and the U.S.S.R. (8) between 1974 and the mid-80s. The threshold of commercial feasibility is as yet unpredictable.

For fusion power, demonstration of scientific feasibility means creating in an experimental device a combination of fuel density, temperature, and confinement time that would lead to a net output of energy in a reactor. No such feasibility demonstration has yet taken place as of late 1978.

There are fundamentally two ways of confining the deuterium and tritium fuel at high temperatures. The first relies on the use of magnetic fields to keep the hot (about 100,000,000°K)

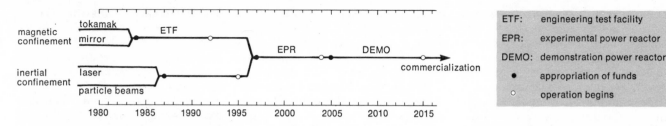

Figure 3. The proposed U.S. Department of Energy program shows that fusion reactors are at a much earlier stage of development than fission fast breeders. The best technological route toward commercialization is not even expected to be decided until the very end of the century. Nevertheless, in the U.S. program, and the comparable Soviet program, it is hoped that engineering feasibility will be demonstrated by the end of this century and that commercialization can be achieved by early in the 21st.

plasma ions away from the reactor components while they react with each other. The tokamak accomplishes this with a toroidal magnetic-field configuration. The second approach to confinement is to use high intensity beams of laser light or energetic charged particles to implode and heat a small pellet of DT to the required reaction conditions. The nuclear reactions take place so fast that it is hoped more energy can be released than was invested in the implosion before the pellet flies apart: thus the term *inertial confinement*. Of the two main approaches to the problem, magnetic confinement has the longer history (research began in the early 1950s) and the greater number of variations, of which tokamaks are one. The idea of inertial confinement for a fusion reactor dates from the early 1960s.

Many proponents of magnetic confinement believe that large DT tokamak devices, now in the late stages of design or early stages of construction, will be the first to achieve scientific breakeven. The rate of progress in magnetic confinement has been truly impressive (see Table 2), and it is hoped that newly constructed devices such as the tokamaks at Princeton and at General Atomic in the U.S. will achieve reactor operation conditions by the early 1980s. Some advocates of inertial confinement believe that laser fusion devices can also achieve scientific breakeven by the early 1980s, but this view is more controversial and is clouded by classification of relevant results.

Once scientific feasibility is achieved with either magnetic or inertial confinement of the fuel, formidable problems of materials and engineering will have to be solved before engineering feasibility can be demon-

strated for a working reactor. This is unlikely to be achieved before the year 2020 even in the United States Department of Energy plan for development of commercial fusion power, which calls for a demonstration reactor to be completed by 2015 (see Fig. 3). Commercial feasibility will not be assured even when such a demonstration reactor exists, and fusion will probably not be proved commercially feasible until early in the 21st century. A contribution of more than 10% to the electricity used in industrial nations still seems unlikely before the years 2040 to 2050.

The materials problem

One of the major technical factors limiting the efficiency and economic viability of both fission and fusion breeder reactors is the degradation of materials due to radiation damage in the reactor environment. The limited lifetime of components in a nuclear plant affects the economies of nuclear power plants in six major areas: (1)

thermal efficiency is limited, because reactors are restricted to lower operating temperatures; (2) operating time is limited, because time is required to change damaged components; (3) capital costs must include outlays for remote handling equipment, and (4) operating costs must cover replacement components and the installation labor; (5) the discarded parts add to the volume of radioactive wastes that must be handled and stored; (6) finally, demand for scarce elements used in the most highly damaged components will be very large.

Fission and fusion structural components and fuels share some of the same intrinsic radiation-damage problems. Irradiation-induced swelling of the fuel cladding and core structural material is probably most important in the LMFBR because of the close tolerances required for coolant flow. High-temperature helium-induced embrittlement will definitely be a greater problem for

Table 2. Plasma parameters achieved in tokamaks*

Year	Confinement time τ_E (sec)	Ion temperature T_i (Kelvin)	Density × confinement time $n\tau_E$ (sec/cm³)	Sustainment time (sec)
1955	10^{-5}	10^5	10^9	10^{-4}
1960	10^{-4}	10^6	10^{10}	3×10^{-3}
1965	2×10^{-3}	10^6	10^{11}	2×10^{-2}
1970	10^{-2}	5×10^6	5×10^{11}	10^{-1}
1976	5×10^{-2}	2×10^7	10^{13}	1
1978	5×10^{-2}	6×10^7	2×10^{13}	1
Needed for a reactor	1	10^8	10^{14}	$\gtrsim 10$

* Data prior to 1978 from B. Pease, Culham Laboratory, U.K.

DT fusion than for fission reactors. This is mainly due to the higher-energy neutron spectrum from the DT reaction, which produces more (n,α) reactions in metals. Irradiation creep, the slow deformation of a structural component under stress at high temperature in combination with displaced atoms, will be a major problem in both types of reactors, because of the high displacement rates in the LMFBR and the high thermal stresses in a fusion reactor. Metal fatigue is also likely to be more severe in fusion reactors, especially in inertial confinement systems.

These problems have been studied for over 20 years for the LMFBR. The result has been the choice of a mixed (Pu and U) oxide fuel, 316 stainless steel as a cladding and core structural material, and B_4C as a control rod material. It is quite probable, however, that even the performance of these materials will not be sufficient for a completely economical breeder economy, and carbide fuels along with alloys containing a high percentage of nickel are being investigated for possible long-term application.

The process of selecting the optimum structural materials for fusion reactors is, by comparison, much more diverse, because the technology is in its infancy (9). All of the alloys proposed for use in fusion reactors suffer from one or more serious deficiencies in that environment. Aluminum alloys are restricted to quite modest temperatures. The refractory metals niobium and vanadium allow higher-temperature operation with less irradiation-induced embrittlement, but they are extremely susceptible to pickup of interstitial impurity atoms, which also causes embrittlement, and at present there is no commercial industry to supply a mature fusion economy. Molybdenum alloys are probably the best suited of the refractory metals, but major technological advances are required before large-scale reactor components made of these alloys can be effectively joined in a vacuum-tight configuration. Titanium alloys allow reasonable operating temperature, and there is very little long-lived radioactivity induced in them, but very little is known about their behavior in an intense neutron environment.

Based on present knowledge, early fusion reactors will almost certainly use special stainless, nonmagnetic (austenitic) steel alloys. These alloys have been the subject of much research, and a great deal is known about their mechanical, physical, and thermal properties in high-temperature liquid-metal environments. Nevertheless, it is now widely accepted that most of the reactor components will not last the lifetime of the power plant. The necessity to replace damaged components quickly in a very high radiation environment will put a severe strain on the design of a fusion power plant.

Without even taking into account the degradation of parts, the low power density (about 1 Mw/m^3) in the blankets of all fusion reactors (not just tokamaks) means that the requirement for materials in a fusion reactor is likely to be in the range of 25 t/Mw(e) of steel, compared with 3 t/Mw(e) in the LMFBR (see Table 3). In addition, the replacement of damaged structural components may require about 0.2 t/Mw(e)-yr of steel for fusion, compared to 0.07 t/Mw(e)-yr for fission breeder reactors. Thus, about four times as much of what are in some cases rather scarce materials would be required to generate the same energy in a fusion economy as in a fission economy. Careful attention will have to be paid to methods of reducing those requirements for fusion, or they could prove to be the limiting factor to the amount of energy that can be produced by fusion, despite essentially unlimited fuel resources.

Materials problems, then, are much more diverse and severe in fusion reactors than in fission reactors. Without intensive long-range development programs in this area, fusion may never transcend the engineering-feasibility phase.

Radioactivity in reactors

The amount of radioactivity in the various nuclear systems has not always been calculated consistently, and thus comparisons have almost always been controversial. We are speaking here of the potentially hazardous radioactivity associated with isotopes *inside* the reactor. While only isotopes *outside* the reactor can cause harm to life, the potential hazard of the material within the system is an important point of comparison because it represents the absolute maximum problem that can occur and because it is very difficult to calculate the fraction of isotopes that could be released for each accident that might occur.

For any comparison to be useful, one must consider, on a per-unit-of-energy basis, total number of radioisotopes produced, their relative toxicity, and their half-life. Once the normalized radioisotope inventories have been calculated, they can be divided by the maximum permissible concentrations (MPCs) allowed in the environment to yield the amount of air or water that must be used to dilute radioisotopes to safe levels. The value of the radioisotope inventory divided by the MPC is called the Biological Hazard Potential (BHP).

Table 3. Critical materials requirements for the nuclear islands of fission and fusion breeder reactors

	Initial requirement in t/Mw(e)	Average makeup in t/Mw(e) · yr *	Total commitment over lifetime of nuclear island in t/Mw(e)
Fission			
Steel	3.1	0.072	4.8
Sodium	2.81	—	2.8
Control Mat.	0.001	0.0004	0.01
Total	6.0	0.72	7.6
Fusion†	25	0.20	30

* For operation at 80% capacity over lifetime of plant.
† Average for UWMAK-I, UWMAK-II, UWMAK-III, PPPL, ORNL, and BNL designs.

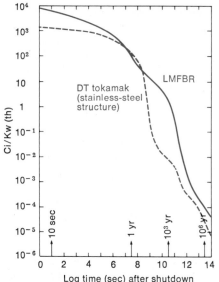

Figure 4. The radioactivity inventory for fission and fusion reactors decreases very slowly after shutdown. The difference between the levels of radioactivity in the two kinds of reactors becomes significant 10 years after shutdown; from then on, the inventory of the fusion system is from 10 to 500 times lower than that of the fission system.

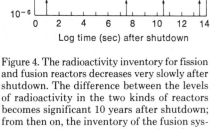

Figure 5. The inhalation hazard from the radioactive inventories of fission and fusion breeder reactor systems is measured in terms of the amount of air required per Kw(th) to dilute the inventory to the maximum permissible concentration (MPC) level set by federal regulatory agencies. This level is well below the threshold at which exposure is known to affect health. Fuel reprocessing lowers radioactive inventory of the fission system but the hazard still remains significantly higher than that of a fusion system.

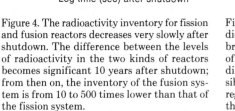

Figure 6. The ingestion hazard from the radioactive inventories of reactors is likewise measured in terms of the amount of water required to dilute the inventory to the MPC level. For water, the MPC refers to the level of radioactivity allowed in drinking water for the public.

Figure 4 compares the radioactivity of the isotopes within fission and fusion breeder reactors as a function of time after the plant has been shut down. Figure 5 compares the amount of air that would be needed to dilute the inventory of radioisotopes to the maximum permissible concentration at various times after shutdown. Such a comparison is mainly applicable to an air release immediately following an accident. Figure 6 does the same for a water dilutent. This could be representative of the amount of ground water that might be required to dilute the radioisotopes from dismantled reactor components stored underground. These figures are of course very large, because they represent absolute maximum release and they are not at all corrected for the *probability* that any of the radioactive inventory will be released into the environment.

As can be seen in these figures, the inventory of radioisotopes for the reference systems is essentially the same (within a factor of 5) for the first 10 years after shutdown. Thereafter, the inventory in the fusion system is a factor of 10 to 500 lower than in the LMFBR for the next several hundred thousand years. The maximum potential hazard of the inventory of radioisotopes as airborne contamination in the fusion reactor is a factor of about 30 to 40 lower than for fission reactors up to the point of reprocessing. The advantage then drops to only a factor of 5 for fusion, but it increases thereafter, becoming a factor of 100,000 after 1,000 years. The potential hazard for water contamination by the fusion reactor inventory starts out a factor of 10 lower than for the LMFBR and remains so up to the point of reprocessing. Thereafter, it increases from a factor of 2 lower to a factor of 200 after 1,000 years of decay. This advantage drops to only a factor of 10 after 10^6 years.

The economic incentive to reprocess fission fuels soon after discharge in large centralized facilities means that large amounts of high-level wastes must be handled, transported, and eventually solidified for long-term storage. In principle, one could integrate the fuel cycle and the reactor, but no major LMFBR program has moved in this direction thus far. At least for the present, then, the separate hazards of the reprocessing and refabrication plants must be included in calculations of the risks that are associated with fission fast-breeder systems.

Fusion systems, on the other hand, have an integrated fuel cycle (tritium separation); and after appropriate compaction, structural steel and other waste material can directly be stored as solids. This tends to reduce the potential for a release of radioisotopes to the environment and could lessen the hazard potential associated with the final transportation of fusion reactor wastes to the ultimate storage facilities.

Finally, the use of vanadium or titanium alloys could increase the above advantages for fusion by even several more orders of magnitude over the LMFBR because their radioisotopes have much shorter half-lives. However, the uncertainty that such alloys can indeed be used in economic fusion power reactors is much larger than for stainless steel. Hopes remain that structural materials for fusion with even more favorable activation properties than steel, such as low radioisotope production rates and short half-lives, will eventually be shown to be feasible.

Normal release of radioisotopes

In the normal course of operation of a nuclear reactor, some quantity of radioisotopes will be released from the plant into the surrounding environment. The main radioisotopes of concern here are tritium, krypton 85, iodine 129, and the alpha emitters for the LMFBR; tritium alone for fusion reactors. Possible accident pathways are different for different systems, and the release of these isotopes can only be estimated once very specific designs have been completed. The scientific community has been in the process of analyzing the detailed LMFBR designs (and their associated fuel cycles) for the last 5 years. Unfortunately, detailed fusion reactor designs are at least 20 years away.

The point of ultimate importance, the fraction of those regularly released isotopes that are actually absorbed by humans, is difficult to calculate for the radioisotopes from both fission and fusion. Such a calculation requires a knowledge of human residence time in the contaminated area, chemical uptake in the various food chains, intake of contaminated air, food, and water, and so on. Some of these assessments have already been made for isotopes such as tritium, iodine, and cesium, but very few have been made for the metallic isotopes or for the actinides.

The release of radioisotopes throughout the entire LMFBR fuel cycle is summarized in Figure 7. In order to meet the recently proposed release limits set by the Environ-

mental Protection Agency for large-scale power plants, annual confinement factors of 10 for krypton and 200 for iodine 129 must be achieved; that is, not more than 1/10 of the krypton isotopes flowing through the plant annually may be released and not more than 1/200 of the iodine 129. For transuranium alpha emitters, the factors are 2×10^9 in the reprocessing plant and 2×10^{10} in the refabrication plant. If a person were to stand at the fence of a facility meeting these standards and obtain all of his air and water from that area, his exposure would be well within EPA limits.

It appears to be no problem to achieve such confinement factors for krypton and iodine. Confinement factors of 10 for krypton are considered to be technically feasible today, whereas

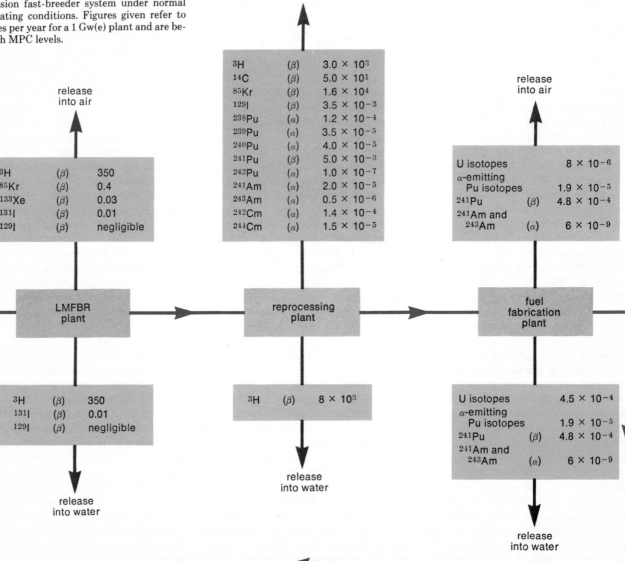

Figure 7. Some radioisotopes are regularly released into the environment from each part of a fission fast-breeder system under normal operating conditions. Figures given refer to curies per year for a 1 Gw(e) plant and are beneath MPC levels.

release into air

^{3}H	(β)	3.0×10^3
^{14}C	(β)	5.0×10^1
^{85}Kr	(β)	1.6×10^4
^{129}I	(β)	3.5×10^{-3}
^{238}Pu	(α)	1.2×10^{-4}
^{239}Pu	(α)	3.5×10^{-5}
^{240}Pu	(α)	4.0×10^{-5}
^{241}Pu	(β)	5.0×10^{-3}
^{242}Pu	(α)	1.0×10^{-7}
^{241}Am	(α)	2.0×10^{-5}
^{243}Am	(α)	0.5×10^{-6}
^{242}Cm	(α)	1.4×10^{-4}
^{244}Cm	(α)	1.5×10^{-5}

release into air

U isotopes		8×10^{-6}
α-emitting Pu isotopes		1.9×10^{-5}
^{241}Pu	(β)	4.8×10^{-4}
^{241}Am and ^{243}Am	(α)	6×10^{-9}

release into air

^{3}H	(β)	350
^{85}Kr	(β)	0.4
^{133}Xe	(β)	0.03
^{131}I	(β)	0.01
^{129}I	(β)	negligible

| **LMFBR plant** | **reprocessing plant** | **fuel fabrication plant** |

^{3}H	(β)	350
^{131}I	(β)	0.01
^{129}I	(β)	negligible

release into water

| ^{3}H | (β) | 8×10^3 |

release into water

U isotopes		4.5×10^{-4}
α-emitting Pu isotopes		1.9×10^{-5}
^{241}Pu	(β)	4.8×10^{-4}
^{241}Am and ^{243}Am	(α)	6×10^{-9}

release into water

confinement factors of 10^3 have already been achieved by cryogenic methods in large-scale prototype test facilities (10,11). Iodine confinement factors of 200 have already been demonstrated, and higher iodine confinement factors are within reach (12). For the alpha emitters, the confinement factors mentioned will be more difficult to obtain. Even though these values have already been achieved in small pilot plants, they still have to be demonstrated in large commercial facilities (10,13).

For fusion reactors, concern lies almost entirely with tritium, and since the fuel fabrication and reprocessing units are confined to the reactor building we have only one location to consider. The throughput of tritium (both in the dynamic fuel handling system and the tritium breeding blanket) is estimated to be about 250 MCi/Gw(e)-day. If we assume that the release rate is limited by the requirement that the tritiated water concentration cannot produce a dose greater than 5 mrem/yr in an individual who takes all of his drinking water from the plant discharge, no more than 300 Ci/day should be released. This means that the control factor has to be roughly 1 part per million per day, which is 3 orders of magnitude higher than the control achieved in present-day LWRs and a factor of 100 higher than assumed for present LMFBRs. But such improvements, while not easy, should be attainable in the next 20 years. It must be noted that while the confinement factors for both LMFBRs and fusion systems seem attainable, it remains to be seen what the cost burden will be for these controls.

Accidental release of radioisotopes

Accidents in fission and fusion breeder reactors can be divided into two kinds. The first kind we call design-basis accidents, since they are a consideration in the design of the system. Included are both *realistic* accidents—those that are known to be conceivable, though some chains of accidental events are included that are of low probability—and *hypothetical* accidents—chains of accidental events that have not been proved to be inconceivable but have a probability of occurrence that is below a given level. Realistic accidents are taken into account in the

choice of core safety parameters and the design of the shutdown system. Hypothetical accidents are taken into account in the design of the cooling system and of the surrounding containment system. The second kind of accident is that caused by sabotage or acts of war. The precautions required to prevent these are more procedural than technological.

Early concerns about the safety of LMFBRs focused on control characteristics and the possibility of core recompaction (that is, the reassembling of the fuel elements in critical configuration after an accident). Accidents in this class begin with boiling of sodium in the cooling system and local fuel melting. The recompaction could be due to pressure pulses resulting from sodium-vapor explosions in the core or other unique coolant flow schemes resulting from an accident. These concerns were accentuated by the emphasis on compact cores and metallic fuel elements in breeder designs of the 1950s and early 1960s. Large cores and mixed-oxide fuels are typical of all prototype and commercial LMFBR designs in the 1970s, and it is now known that the control characteristics are substantially similar to those of the LWR. Moreover, a large and growing body of theoretical and experimental evidence supports the view that the propagation of local fuel failures in a way that leads to recompaction in the large-core, mixed-oxide-fueled LMFBR would require combinations of events and degrees of spatial and temporal coherence that are not physically realistic (1) (e.g. the probability of such a series of events occurring is less than once every million years).

The large LMFBR prototypes that are in operation in France and in advanced stages of construction in West Germany have undergone licensing reviews as stringent with respect to safety as the ones that are applied to LWRs. The designs of these large LMFBRs take into account the possibility of failure of two independent shutdown systems following a hypothetical large insertion of reactivity or coast-down of the main sodium pumps. The calculated consequences of melting and core disassembly in these maximum hypothetical accidents define the design characteristics of the containment systems required for licensing (strength of re-

actor vessel and primary piping; strength and leak rates of surrounding double steel and concrete containment structures). In addition to the pressure loads during such an accident, the long-term cooling of large masses of molten and dispersed fuel after the accident must also be taken into account.

While this capability appears to be at hand for the 300 Mw(e) class LMFBR, additional development work is needed for large LMFBR power stations. These design requirements can be met with reasonable technical effort, and as the French and German experience indicates, radiation doses can be restricted to 1 rem or less at the plant boundary in the event that a design-basis accident occurs. The overall conclusion is that the LMFBR can meet the same predetermined safety standards that are applied to other fission reactors. This may or may not also hold for the fabrication and reprocessing plants associated with the LMFBR.

Fusion-reactor safety analysis is necessarily in a much more preliminary phase, because the technology cannot be described in detail. Examination of stored energies and potential pathways for energy release in tokamak designs indicates that sudden failures of the magnet support and vacuum systems could produce enough mechanical energy to severely damage the reactor. Loss of coolant or coolant flow coupled with failure to shutdown the fusion reactor could also cause local interior structural damage. However, the characteristics of fusion plasmas and the very small amount of fuel present in the reaction chamber at any time mean that reactivity accidents will not be an important concern.

Decay heat due to neutron activation of structural materials is small enough in most designs to be much easier to handle than in fission reactors. For fusion-reactor designs where liquid lithium serves as breeding medium and coolant, the very large chemical energy stored in this coolant and the high flame temperature of the lithium-air and lithium-water reactions probably represent fusion's most important vulnerability to accidents capable of releasing sizable quantities of radioactivity. Both LMFBRs and liquid-lithium-cooled fusion reactors require careful design of steam generators to ensure safety in the event of possible leaks that would bring water into contact with liquid metal.

The risks for many of the possible accident pathways in fusion reactors can be minimized by intelligent design, which includes the possibility of breeding tritium in ceramic lithium compounds (which would lower the inventory of tritium by one or two orders of magnitude in the region surrounding the plasma) and cooling with pressurized helium instead of liquid lithium to reduce the fire potential. Such an approach could reduce the inventory of tritium that could be released from the breeding area in an accident, though other sources of tritium in proximity to the blanket (e.g. vacuum pumps) will not be affected by the change to solid breeders. Enthusiasm about the potential flexibility in design of fusion reactors must be tempered with the recognition that there may be important trade-offs—for example, the possible need to use toxic and relatively scarce beryllium for neutron multiplication if some solid breeding compounds are employed in realistic blanket designs.

Designers of fusion systems can anticipate that the approaches they devise to control energy release from magnets, vacuum systems, coolant, and so on will be subjected to much the same critical scrutiny and demand for high reliability experienced now in fission-reactor licensing proceedings. It is too early to say how difficult it will be to survive such scrutiny, but the authors feel that it will be possible if the full attention of the engineering community is given to the problem.

Comparison of fission breeders and fusion reactors can also be made with respect to the consequences of events worse than the design-basis accidents—accidents resulting from acts of war or sabotage. Analyzing the Rasmussen Report of the U.S. Nuclear Regulatory Commission reveals that hypothetical release of a substantial fraction of the fission products and 0.5% of the actinides in an LWR—the same isotopes that are of concern in an LMFBR—would produce roughly 100 times more early deaths under adverse meteorological conditions than the release of 10 kg of

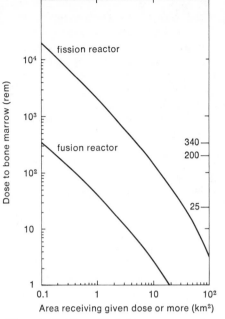

Figure 8. Plotting critical dose versus area for severe releases from fission and fusion breeder systems shows that fusion has a significant advantage over fission. To suggest the meaning of these figures: 1% of a population receiving a dose of 200 rem will die within 60 days; 50% of a population receiving a dose of 340 rem will die within 60 days. Twenty-five rem is designated by the International Commission on Radiation Protection as the emergency dose limit: that is, one-time exposure to this amount of radiation is not considered harmful.

tritium from a fusion plant. This much tritium could represent 10–20% of the total inventory (Fig. 8). Much in need of further investigation is what fraction of the activation products in a fusion reactor and of the actinides in an LMFBR could be released in such hypothetical events, as these fractions significantly affect the outcome. Comparative examination of delayed as well as early casualties is also needed.

A further caveat is in order. In a fusion reactor, most of the fuel cycle is within the reactor-containment structure. For the LMFBR there must exist, in addition, fuel-reprocessing and fuel-fabrication plants, which means that large amounts of highly radioactive wastes must be handled and transported. Analysis of the potential hazards in these parts of the fuel cycle is not yet nearly as refined as that for the LMFBR itself.

Furthermore, both LMFBR and fusion reactors will require facilities for

long-term radioactive-waste management, for which accident analysis will also have to be done. The fact is that, after appropriate compacting, radioactive structural steel and other waste materials from fusion reactors can be directly stored as solids instead of being chemically reprocessed. This gives fusion systems a large advantage in terms of potential for release of radioisotopes to the environment during both transportation and storage.

Safeguarding nuclear reactors

Both fission and fusion breeder reactors hold dangers beyond their own operation and malfunctioning—namely, the possibility that materials and technology will be diverted from their peaceful use in the production of energy to the production of weapons. The hazards are not, however, the same for both kinds of system.

As fission power spreads, the associated spread of bomb-related material is more of a threat than the spread of bomb-related knowledge. (This is a problem of all forms of fission, not just breeders.) If fusion power spreads, and if the inertial-confinement approach predominates, the associated spread of bomb-related knowledge will be more important than the direct spread of bomb-related material. Neutrons from any DT fusion reactor could be used by the operators to produce fissile material, however, and thus both types of nuclear reactors pose an indirect threat of the spread of bomb-related material.

Misuse of nuclear material as a radiological rather than an explosive weapon is a threat associated with both fission and fusion. Here fusion's hazard is from the tritium, which is used in all approaches, not just the inertial-confinement approach. With respect to airborne dispersal of plutonium or tritium reactor inventories, fusion appears to have a quantitative advantage over fission of 2 to 5 orders of magnitude (see Table 4), depending on the chemical form of the materials. With respect to waterborne dispersal, the fusion reactor has only a slight advantage.

For fission power, safeguarding of fuel transport is a greater problem than safeguarding power stations or reprocessing plants. As we have mentioned, the fission fuel cycle is spread out, and it will remain so unless collocation of some parts of the fuel cycle becomes standard procedure. For fusion, the tritium for the most part remains in the power stations; transportation is necessary only when new power stations are being started up. The centralization of the problem for fusion makes the protection against theft easier than for a fission economy.

The detection of diversion of nuclear material is addressed by the Non-Proliferation Treaty, which calls for more international than national controls. For a state, the most direct route to the fabrication of a few crude nuclear explosive devices is probably the construction of centrifuges which could separate out the fissile isotope ^{235}U from the natural uranium; it is not the deployment of economically significant civilian nuclear fission power.

Establishing an equilibrium of any kind between risks, benefits, and costs is, in the first analysis, a step that entails social and political considerations as well as technical insights. At the present time the arguments seem to be mainly in the social and political arenas, as the technical issues are mostly solved.

Commercialization

If commercial feasibility of fission fast-breeder power stations is indeed attained between 1990 and the year 2000, this will represent roughly 50 years in the scientific and engineering feasibility stages. Approximately the same amount of time is anticipated for fusion reactors (although the uncertainty is greater), but since serious research into the scientific feasibility of fusion reactors started later, commercial feasibility of large-scale power stations will probably not be achieved until after 2020.

The economic viability of fission fast breeders requires the services of a separate fuel cycle. But such a fuel cycle can be developed on a technically and economically significant scale only when irradiated fuel is available in significant quantities from operating reactors. Thus 10 to 15 years must be added to the time required to achieve operation of commercial reactors for the comple-

tion of the full fission-breeder system. On the other hand, no separate fuel cycle will be required after a large number of fusion reactors are constructed, and therefore the point of commercial feasibility should be easier to define. Both fission and fusion breeder systems will require facilities for final waste disposal, however, and this will extend the complete time frame in both cases.

Three generations of reactors seem to be necessary before commercial feasibility for both kinds of energy sources can be demonstrated: experimental power reactors, producing 10 to a few 100 Mw(th); prototype demonstration reactors, 250 to 500 Mw(e); and semicommercial reactors, 1,000 to 1,500 Mw(e). Along with these

major facilities, a large number of smaller but equally important facilities need to be developed to test various aspects of the physics, engineering, materials, and safety for both kinds of systems. Materials-testing facilities can be particularly costly and time-consuming to the overall program development. There is hardly any way to circumvent these procedures, as each generation of reactors requires higher performance characteristics, which are difficult to test in facilities existing up to that point.

In order to be useful, fast breeders and fusion reactors must fit into existing schemes and rules of electricity production. Demonstration of availability, maintainability, and repair-

Table 4. Summary of the radiological hazards of plutonium and tritium (normalized to a 1 Gw(e) power plant when necessary)

	Pu (LMFBR)	Tritium (DT Tokamak)
Inventory of reactor (kg)	3,400	25[a]
Annual flow outside reactor (kg/y)	1,600	27[b]
Hazards in Air		
Maximum permissible concentration (MPC) in air (Ci/km^3)		
Insoluble ^{239}Pu or T_2 gas	0.001	40,000
Soluble ^{239}Pu or tritiated water vapor	0.00006	200
Biological hazard potential (BHP) per gram of element (km^3 of air required to dilute 1 gram of radioisotopes to maximum permissible concentration)		
Pure ^{239}Pu or elemental T	63–1000	0.25
Reactor[c] grade Pu or T in tritiated water vapor	300–5000	50
Total amount of air required to dilute reactor inventory to the MPC level (10^6 km^3)		
Least harmful form of radioisotopes[d]	1,020	0.006
Most harmful form of radioisotopes[e]	17,000	1.25
Hazards in Water		
MPC (Ci/km^3 of water for soluble forms)	5,000	3,000,000
BHP (m^3 of water required to dilute 1 gram of radioisotopes to the MPC level)		
Pure ^{239}Pu insoluble compound	12,500	
Reactor grade Pu in soluble compound	62,500	
Tritiated water		3,300,000
BHP of reactor inventory in km^3 of water required to dilute all radioisotopes to MPC levels	210	83

[a] Roughly one half in the cold storage for backup and the other half actively circulating in reactor.
[b] At a tritium breeding ratio of 1.25.
[c] Contains ^{238}Pu, ^{239}Pu, ^{240}Pu, ^{241}Pu, ^{242}Pu.
[d] Reactor grade Pu dispersed in insoluble form, tritium dispersed as T gas.
[e] Reactor grade Pu dispersed in soluble form, tritium dispersed as vapor.

ability is in itself a complex procedure that requires time. In fact, the rules and fundamental data underlying the licensing process must be developed almost in parallel with the reactors and facilities that are to be licensed. Public acceptance has been shown to be a problem distinct from—and perhaps harder to resolve than—licensing.

In the U.S., because its development program is so broad and stretched out, more than $10 billion are expected to be necessary for reaching commercial maturity of the LMFBR (see Fig. 9)—about 6 times the amount already spent. By contrast, in European countries the development programs seem to be narrower and more compact in time and thus are likely to be considerably cheaper. The difference in funding points to the degree of flexibility that such programs seem to have. In any event, the parallel development programs throughout the world contribute significantly to each other—which, in fact, may explain the seemingly lower cost of the European programs.

The situation for the fusion program is much less well defined, but recent projections in the U.S. program reveal that it may require $20 to $25 billion to bring fusion through the demonstration power-reactor phase, and it is not unreasonable to expect that another $5 to $10 billion will be required to progress through the commercialization stage. In contrast to the fission-breeder program, the European fusion program is much smaller and of longer duration. The Soviet program is approximately the same as the U.S. program in level of effort now, and it is expected to keep pace with the U.S. program. Therefore, it is reasonable to expect that, worldwide, as much as $50 billion may be required to reach commercial feasibility of fusion. The same benefits of international cooperation in fusion research as in fission are expected to allow for considerable flexibility in reactor design and should increase the probability of long-term success.

Energy for the long run

From this brief analysis we now come to several conclusions. First, it is of central importance that both fission breeders and DT fusion have the potential, in terms of fuel supply, of

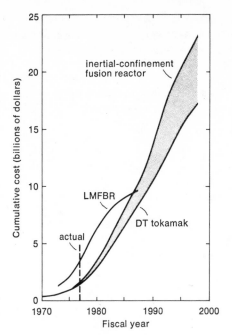

Figure 9. The projected cumulative cost of bringing fusion power to the commercial stage in the U.S. is higher than that of the LMFBR (excluding the investment in the present light-water reactor industry, with which the fast breeder shares some technology and test requirements). The curve for fusion reactors using inertial confinement is based on a rate of fiscal support for research at 33% of the level of support for the magnetic confinement approach. Costs calculated represent government funding to the end of the century, by which time private industry should be able to take over.

providing very large, almost indefinite, amounts of electricity. In this respect, there is no difference between them that is of any practical significance.

To solve the fuel supply problem for the indefinite future would be an enormous benefit, but we can see from this study that the benefit has its price. Both for fission breeders and for fusion, the price includes a heavy investment in research and development before the commercial stage is reached, continuing high capital costs for the commercial reactors and supporting plants, and a commitment to maintain a high degree of meticulousness and vigilance in the construction and operation of these facilities and in the sequestering of their wastes.

While there are significant differences in the basic physical processes of fission and fusion, the presently envisaged technologies for using these processes in electricity production

have much in common: lack of any significant air or water pollutants; complex large-scale engineering based on large, central-station power plants; material damage and activation by neutrons; the need to contain inventories of radioactivity within the plant and to manage radioactive wastes beyond the lifetime of the reactors; and for many present designs, use of liquid-metal cooling and heat-transfer technologies.

In principle, however, the nature of the fusion process allows for a much greater degree of flexibility in the technologies used to harness it. This flexibility, if explored and utilized, offers the possibility for fusion based on the deuterium-tritium cycle to be quantitatively superior to fission in important environmental and social respects, despite the qualitative similarities already mentioned. Specifically, fusion has the potential for quantitative advantages in the form of lower hazard potential in its radioactive inventory (and, accordingly, smaller predicted consequences of hypothetical large releases); lower radioactive decay heat; smaller hazard potential and shorter hazard lifetime associated with radioactive wastes; less shipment of dangerous material outside the reactors; and small hazard potential for use of tritium as a radiological weapon (compared with plutonium in fission reactors).

There are, of course, qualitative differences in accident pathways in fusion and fission and these need to be better understood before more definitive statements can be made. With respect to the spread of the capability of making nuclear bombs, we conclude that fission spreads relevant material more than knowledge and fusion spreads knowledge (related to the inertial confinement approach) more than material.

It must be emphasized that achieving the potential environmental advantages of fusion in a practical system will require giving high priority and prolonged attention to environmental characteristics from the earliest stages of designing fusion systems. The advantages will not materialize automatically simply because fusion is fusion. Fusion systems can be envisioned that will not have the most important environmental advantages over a fission-breeder system.

The environmental advantages of fusion that are achieved will have to be weighed against the cost of achieving them. No such weighing can be done today, because the technology has so far to go and because the value that society will place on such advantages has yet to be determined.

The LMFBR, by contrast, has passed the thresholds of scientific and engineering feasibility, with commercial feasibility still to be demonstrated. Herein lies a dilemma of timing. The LMFBR meets the fundamental requirement for a long-term energy source—namely, a nearly inexhaustible fuel supply—but the timing of its development has been such that the LMFBR's commercial feasibility, as well as its environmental and social characteristics, are being judged against the yardstick of existing short-term or transitional energy technologies such as oil, natural gas, light-water reactors, and coal. Such comparisons are relevant for helping to determine the appropriate timing for commercial introduction of a technology such as the breeder, but since oil, gas, LWRs, and coal do not meet the basic fuel-supply criterion for long-term sources, they are not suitable yardsticks for judging the LMFBR's viability and desirability as a way to meet the long-term needs.

Just as part of the present predicament of the LMFBR arises from evaluating a long-term source against short-term competitors, there is a related pitfall that could damage the future of fusion. The pitfall is that the desire to bring fusion to commercial fruition in time to compete in the transition time frame (say, in the period 2000 to 2030) may lead fusion programs around the world to place a disproportionate emphasis on early engineering feasibility at the expense of potential environmental advantages. If fusion technology is steered too early in the direction of doing whatever seems necessary to produce commercial power as quickly as possible, the field may be shaped for a long time to come by approaches that exclude the environmental benefits which represent fusion's greatest asset as a long-term energy source.

It is essential, therefore, to keep separate in technology assessments the differing requirements of the short-term, transitional, and long-term

phases of the energy problem. The most significant comparison of long-term sources is with each other, and that is the pertinence of our comparison of LMFBR and DT fusion here. As the needed information becomes available, such comparisons should be extended to include large-scale use of solar energy and perhaps fusion and fission fuel cycles other than the DT tokamak and the plutonium-burning LMFBR.

References

1. W. Häfele, J. P. Holdren, G. Kessler, and G. L. Kulcinski. 1976. *Fusion and Fast Breeder Reactors,* RR-77-8. Laxenburg, Austria: Int. Inst. Appl. Sys. Anal.

2. Uranium Resources, Production and Demand. A Joint Report by the OECD Nuclear Energy Agency and the IAEA, December 1977. Paris: Organization for Economic Cooperation and Development.

3. H. E. Goeller et al., eds. 1974. *Survey of Energy Resources, 1974.* London: World Energy Conference.

4. G. Vendryes et al. 1977. *Status and Prospects of the French Fast Reactor Programme,* IAEA-CN-36/219. Salzburg: International Conference on Nuclear Power and Its Fuel Cycle.

5. J. Sauteron et al. 1977. *The Technology of Fast Reactor Fuel Reprocessing,* IAEA-CN-36/567. Salzburg: International Conference on Nuclear Power and Its Fuel Cycle.

6. N. L. Franklin et al. 1977. *Status and Programme for the Fast Breeder Reactor in the UK,* IAEA-CN-36/54, Salzburg: International Conference on Nuclear Power and Its Fuel Cycle.

7. R. H. Allardice et al. 1977. *Fast Reactor Fuel Reprocessing in the UK,* IAEA-CN-36/66. Salzburg: International Conference on Nuclear Power and Its Fuel Cycle.

8. O. D. Kazachouski et al. 1977. *The USSR Reactor Development Programme and Its Present State,* IAEA-CN-36/356. Salzburg: International Conference on Nuclear Power and Its Fuel Cycle.

9. G. L. Kulcinski. 1976. Radiation Damage, the Second Most Serious Obstacle to Commercialization of Fusion Power. In *Proc. Int. Conf. Radiation Effects and Tritium Technology for Fusion Reactors,* Conf.-750989, vol. 1.

10. R. Papp. 1977. *Brennstoffmengen, Aktivitätsinventare und radioaktive Freisetzungen bei der Energieerzeugung auf der Basis von Leichtwasserreaktoren und schnellen Natriumbrütern,* KFK-2453. Karlsruhe, West Germany: Karlsruhe Laboratory.

11. R. Ammon et al. In press. Entwicklung der Kr-85-Abtrennung aus dem Abgas der grossen Wideraufarbeitungsanlage. In *Jahreskolloquium 1977 des Projektes Nukleare Sicherheit,* KFK-2570, Karlsruhe, West Germany: Karlsruhe Laboratory.

12. J. G. Wilhelm et al. 1976. *Head-End Iodine Removal from a Reprocessing Plant with a Solid Sorbent.* In *Proc. of the 14th Air-Cleaning Conf.*

13. G. Kessler et al. 1977. *Safety of the Liquid Metal Cooled Fast Breeder Reactor and Aspects of Its Fuel Cycle,* IAEA-CN-36.572. Salzburg: Int. Conf. on Nuclear Power and Its Fuel Cycle.

Alvin M. Weinberg

The Maturity and Future of Nuclear Energy

Views

The most serious question now facing nuclear energy is its acceptance by the public

It was in July 1942 that Arthur H. Compton, Director of the wartime Metallurgical Laboratory at The University of Chicago, announced to the staff that the multiplication constant, as measured in a graphite-uranium lattice exponential experiment, had been shown to exceed unity. What had been presumption and hope suddenly became reality: no one of the some seventy-five people assembled in Eckhart Hall any longer doubted that the chain reaction should be seriously pursued. The actual demonstration, on 2 December 1942, exciting though it was, in reality proved to be something of an anticlimax. One distinguished theorist at the Metallurgical Laboratory was so certain of the success of the experiment that he offered not to attend it, although he later relented.

The Hanford plutonium-producing reactors were being designed even before the first chain reaction had been demonstrated: a report by Wigner and his associates (*1*), outlining the Hanford design, was issued in January 1943. Construction of the reactors began late that year, and the first one—rated at 250,000 kilowatts—went critical in September 1944.

Alvin M. Weinberg, former Director of the Oak Ridge National Laboratory, and now Director of the Institute for Energy Analysis, has devoted most of his life to the development of nuclear energy. This paper was given when Dr. Weinberg received the Heinrich Hertz Energy Prize in Karlsruhe, Germany, 12 June 1975.
The author gratefully acknowledges the help of D. F. Cope, P. R. Kasten, A. M. Perry, and D. L. Phung in the preparation of this article. Address: Institute for Energy Analysis, Oak Ridge Associated Universities, Oak Ridge, TN 37830.

Table 1. Kinds of reactors

Advanced Gas-cooled Reactor	AGR
Canadian Deuterium–Natural Uranium Reactor	CANDU
Gas-Cooled Fast Breeder Reactor	GCFBR
High Temperature Gas-cooled Reactor	HTGR
Liquid Metal Fast Breeder Reactor	LMFBR
Light Water Breeder Reactor	LWBR
Light Water Reactor	LWR
Molten Salt Breeder Reactor	MSBR
Steam Generating Heavy Water Reactor	SGHWR
Submarine Intermediate Reactor	SIR

Following completion of design of the Hanford reactors, the Metallurgical Laboratory staff turned its attention to the possibility of using fission energy for generation of electric power. At the so-called New Piles Seminars, most of the reactor types that later emerged as power plants were described, at least conceptually: the pressurized water reactor, the pebble-bed reactor of Farrington Daniels (which was the forerunner of the present-day high-temperature gas-cooled reactor), the fast breeder, the aqueous homogeneous breeder, heavy water moderator reactors. These various proposals were presented by proponents and were criticized by advocates of other avenues of development, as lines already had been drawn among those who preferred different reactor types.

To what extent has nuclear energy followed the paths laid out in those wartime discussions at the Metallurgical Laboratory? What elements in today's nuclear scene were overlooked or underestimated? What can we foresee for the future, perhaps the next thirty years? This reexamination of nuclear technology comes at a time when energy is perhaps the world's most important public issue. With oil at $11 per barrel, nuclear energy is a major means of alleviating the global energy shortage, yet many voices are questioning the commitment to nuclear energy.

Original aims of nuclear energy development

Initially, three aims dominated our thinking about nuclear power reactors:

1. *Compactness.* Since the first power reactors were intended to propel a submarine, they had to be compact, which required the use of enriched uranium and either light water moderation or no moderation. The response to the requirement for a compact energy source was the pressurized water reactor and the beryllium-moderated, oxide-fueled, sodium-cooled Submarine Intermediate Reactor (SIR) (see Table 1 for reactor abbreviations).

2. *Conservation.* Because little was known about uranium reserves, much attention was focused, even in the earliest days, on breeder reactors as a means of conserving uranium. Indeed, the first reactor to produce electricity, the Experimental Breeder Reactor I, was a fast breeder prototype, and the second, the Homogeneous Reactor Experiment, was a thermal breeder prototype.

3. *Cheapness.* Since conventional electric energy was generated for

less than 5 mills per kilowatt-hour, central nuclear power plants were supposed to produce electricity at this price or less. Cost consciousness led the early designers to focus on systems with intrinsically low fuel-cycle charges—i.e. reactors that used natural uranium and were therefore moderated by either heavy water or graphite.

In the early days at the Chicago Metallurgical Laboratory, nuclear power was conceived primarily in the first two contexts: as a means of propelling a submarine and as a long-term energy source—the breeder. Nuclear energy merely as a short-term economic source of power was not viewed at the time as being of central importance, perhaps because energy from fossil fuel was then so inexpensive in the United States. In the ensuing years, these original logical threads—compactness, conservation, cheapness—that in the early days seemed distinct have become interwoven and indistinct. Perhaps the most surprising development has been the emergence of the light water reactor (LWR) as a dominant type for central power, rather than being used simply as a compact source of power. There are throughout the world, operating or being built, more than 200 light water reactors.

That the LWRs have achieved such preeminence as competitive sources of central electricity, despite their use of enriched uranium—and their original invention for a different purpose—is attributable to three factors, aside from the increased cost of fossil-based energy: (1) Naval pressurized water reactors were the only U.S. power reactors available when it was decided to construct a commercial power plant at Shippingport, Pennsylvania. (2) LWRs had proved to be relatively cheap to build. (3) Perhaps most important, the cost of a unit of separative work had fallen almost threefold between 1945 and 1960. At the time Daniels proposed the pebble-bed pile in 1945, it was absurd to consider burning ^{235}U (uranium-235) costing $35 per gram, but it was not absurd to burn enriched uranium in a reactor with low conversion ratio when ^{235}U cost about $12 per gram. Thus, commercial success of the light water reactor reflected the success of the gas-

eous diffusion process, as well as the success of oxide fuel that made high burnup possible. Indeed, in the original metal-fueled reactors, 3,000 megawatt-days per ton (Mwd/t) was a great achievement; today, LWRs with oxide-fuel elements reach 30,000 Mwd/t.

The original belief that only reactors with good fuel economy and natural uranium could be expected to compete commercially was invalidated by the success of the diffusion plants and the relatively lower capital costs of the light water systems as compared with the graphite or the heavy water systems. Yet, the final results are not entirely known. We speak of the maturity of nuclear energy, but commercial fission reactors are not yet old enough for us to determine accurately the relative costs of the different reactor types. The intrinsically lower fuel-cycle cost of the heavy water systems at one time was easily overbalanced by its higher capital cost, except when fixed charges were low. But with all prices escalating so rapidly, it is no longer as clear that LWRs can claim an overall advantage.

The race to commercial power which started in 1945—with high neutron economy (heavy water moderated) reactors leading, at least in concept—shifted in the 1960s to LWRs, and possibly HTGRs, and is still unfinished: Pickering, a heavy water station in Canada, is enjoying impressive success. It is probably a race without a clear-cut winner; the reactor chosen in a given circumstance will depend, in addition to price, on many other factors.

I mention this history to draw a moral: in big reactor-engineering developments, fundamental advantages—at least as perceived a priori—do not necessarily determine the ultimate course of events. Developments acquire a force and momentum of their own which can preempt the direction of a technology. Newer technologies based on better fundamental principles can be bypassed by the momentum of the older system. The light water reactors were *not* invented because we believed they would make good central power plants. The momentum of the LWR development and the improvements in gaseous diffu-

sion technology (visualized particularly by Karl Cohen) overcame the basically poor neutron economy of the light water reactors and made them much more attractive as central power stations than most would have expected when the *Nautilus* prototype was built in 1953.

Breeders, fast and thermal

The distinction between breeder and burner reactors also is being eroded. I recall listening at the Metallurgical Laboratory in 1944 to Enrico Fermi's first description of the fast breeder, when he emphasized the high neutron yield per fission in ^{239}Pu (plutonium-239), and for Fermi the breeder meant the fast neutron breeder. Eugene Wigner viewed the matter more broadly; though he and Harry Soodak were the first to design a sodium-cooled fast breeder, Wigner also espoused the Th-^{233}U (thorium–uranium-233) thermal neutron breeder. Since poisoning, i.e. neutron absorption, by fission products was fatal to the thermal breeder, Wigner proposed the idea (also suggested by Harold Urey) of fluid fuel from which the fission products, and notably ^{135}Xe (xenon-135), could be milked continuously. Some thought even then was given to the possibility of breeding, or near breeding, in heterogeneous, heavy water moderated reactors. We spoke of artificial natural uranium, meaning a one percent mixture of ^{233}U in ^{232}Th, which would barely breed.

As for the light water, closely packed reactors, we knew toward the end of the war that the multiplication constant would approach unity in a natural uranium lattice (because of the high-interaction fast effect), but it seemed unlikely that the conversion ratio in a Th-^{233}U version of such a system could exceed unity. However, James B. Conant, in a letter to Wigner, raised exactly this possibility around 1943: he asked whether Germany could build a stockpile of ^{233}U bombs by breeding in the ^{233}U cycle. Our answer was no, but largely because we did not then suspect the neutron yield in the fission of ^{233}U to be approximately 2.3 rather than 2.1.

Thirty years later, the fast breeder remains the primary goal of much

Table 2. Approximate thirty-year commitment of uranium oxide to 1,000-Mwe reactors

	LWR	CANDU	HTGR
Capacity factor (percent)	70	80	70
No recycle (tonne)	5,000	4,100	4,500
With recycle (tonne)	3,200	2,000	2,500

Table 3. Conservation coefficients (*C*) for various breeders

	Breeding ratio	Doubling time (years)	Specific inventory (kg/Mwe)	C
LMFBR (PuO$_2$)	1.2	15	3	0.022
LMFBR (PuC)	1.3	10	3	0.033
GCFBR (PuO$_2$)	1.4	10	4	0.025
MSBR (Th)	1.07	20	1.5	0.033

$400/kg—for the entire recycle were quoted. If these prices indeed prevail, then the original incentive to use fluid fuel remains valid. The Molten Salt Reactor Experiment demonstrated that ^{233}U could be recovered from the salt with remarkably little difficulty and at trivial cost.

I can add little to what has been said many times about the relative merits of thermal and fast reactors of our nuclear development but in an embodiment that none in the early days could have conceived: oxide fuel, loose—not close—coupling between fuel and chemical recycle, 3 to 5 tonnes of plutonium per 1,000 megawatts electric (Mwe) instead of the 100 or so kilograms specified in the original design.

The ultimate course of the thermal breeder remains uncertain. The Light Water Breeder Reactor (LWBR), which had been dismissed by many because its fuel cycle was too expensive, in today's context of escalating capital costs must be reexamined. And the artificial natural uranium, heavy water reactors, now emerging as a variant of the CANDU with organic cooling, again must be placed in contention. In the meantime, the fluid-fueled breeders are at the moment considered less seriously than the LWBR. Thus, the thermal breeders, or near breeders, may emerge more as outgrowths of the LWR and CANDU burner reactors than as entirely new engineering concepts such as the fluid-fuel systems. Yet I cannot concede that this trend is necessarily correct. The fuel recycle for heterogeneous reactors continues to look more and more complex. At the May 1975 Johnson Foundation-Cornell University Wingspread Conference on Advanced Nuclear Converters and Near Breeders, costs as high as $200/kg—even

as breeders. Those who look upon the breeder primarily as a means of easing the transition to some steady state, no-growth society place much stress on breeding ratio and doubling time (i.e. time to double the original fissile inventory); those who look upon the breeder simply as the energy source needed to maintain our society at a more or less steady state focus more on the economics and reliability of the reactor; its doubling time is not of primary importance as long as the reactor is self-sustaining.

We do not know how long the transition to a steady state will require, and therefore it is impossible to specify what doubling time, or what breeding ratio, will be needed to match the growth of the energy economy. We do know that our cheap uranium resources are limited; for example, the U.S. resource below $65 per kg is estimated at not more than 3.5×10^6 tonnes. This could support about 1,100 LWRs (with full recycle), 1,700 CANDUs, or 1,400 HTGRs for thirty years at the capacity factors indicated in Table 2 and 1,000-Mwe rating.

The total amount of uranium that must be extracted from the ground to fuel a system based entirely on breeders, if use of nuclear energy grows linearly, is inversely proportional to the conservation coefficient *C*: $C \sim$ (doubling time $\times$ spe-

cific inventory)$^{-1} \sim$ breeding gain $\times$ (specific power)^2. The conservation coefficient for various breeders is given in Table 3. The conservation coefficient of the proposed molten salt thermal breeders can match those of the fast breeders, if the specific inventory can be kept low. It is on this account, among others, that a case can be made for keeping the light of thermal breeding flickering, even though most of the world is committed to the fast breeder.

For a break-even breeder, such as the LWBR, or for advanced converters, like the CANDU-Th reactor, the uranium requirements (as just defined for breeders) cannot be expressed in terms of a conservation coefficient, because these reactors cannot create surplus fuel to start up additional reactors. Uranium is required—entirely for the LWBR, or primarily for the CANDU-Th near breeder—as feed to enrichment plants to provide initial inventories of fissile fuel to expand the nuclear capacity. In both cases, the thirty-year lifetime uranium commitment is approximately 1,500 tonnes of uranium oxide (U$_3$O$_8$) for 1,000 Mwe; and all or most of that commitment must be provided upon start-up of the reactor. Therefore, if the economy expands fast enough, using these reactors rather than burners does not conserve uranium.

Studies, based on rather high energy-growth scenarios, suggest that cumulative U.S. uranium requirements over the next half century might range from 4,500,000 to 8,000,000 tonnes of U$_3$O$_8$, if we should continue to rely on contemporary LWRs and low-gain HTGRs. At the end of that period, cumulative consumption might increase by 150,000 to 400,000 tonnes of U$_3$O$_8$ each year. In contrast, introduction of the Liquid Metal Fast Breeder Reactor (LMFBR), or its equivalent, during the late 1980s or early 1990s, will cause the cumulative uranium consumption to level off, perhaps between 2,000,000 and 4,000,000 tonnes by the year 2020.

Extensive use of advanced converters would reduce the rate of depletion of uranium resources. However, if nuclear fission is viewed as a permanent means of providing energy, and not one that will ever be displaced by, say, solar or fusion energy, then the advanced converters can at best serve to lengthen the time before we deplete our high-grade uranium deposits. Converters cannot be looked upon as a means of using the residual uranium in the rocks (perhaps 4 ppm). The net self-energy ratio of a LWR with plutonium recycle—i.e. the energy produced divided by all the energy inputs to build, operate, and fuel the reactor—is about unity if the original uranium ore concentration is 10 ppm. Even the advanced converters, which might use uranium five times more efficiently (2), would barely show a positive net energy gain if fueled with 4 ppm uranium.

New issues in nuclear energy

On the whole, the early workers in nuclear energy perceived with surprising clarity most of the major issues of nuclear development. We recognized the importance of breeding, of low-cost power, of waste disposal, of the environmental impact of nuclear reactors. Yet, though reactor safety was never far from our minds, none of us, with the possible exception of Edward Teller, realized how strongly reactor development would depend on the issue of reactor safety. One of my earliest jobs at the Metallurgical Laboratory, in collaboration with Teller, was to estimate the amount of ^{14}C (carbon-14) that would be emitted from an air-cooled graphite reactor. Then in 1948, it was Teller who insisted that a special committee on reactor safety be established.

Thirty years later, with the maturity of nuclear energy, we recognize that these questions of reactor safety and environmental impact, which, in a sense, were originally perceived as secondary to the question of whether the reactor performed properly and whether it would be economical as compared to fossil fuel, have become dominant. Safety of LWRs has received serious attention in the Rasmussen report (3), a landmark in analysis of the possible hazards of light water reactors. It is probable that this mode of analysis which seeks to assess the probability, as well as the severity, of a reactor accident will

become a standard approach for judging the safety of reactors other than the LWR.

The Rasmussen study has received both criticism and praise: often the criticism stems from those who dislike nuclear energy and the praise comes from the nuclear community. Its critics notwithstanding, the study has shown the probability of a reactor accident that would cause immense damage to be extremely small. Whether or not one believes that the probability of fatality from 100 LWRs distributed in the United States is precisely one in 5×10^9 per year (as estimated in the Rasmussen study), it is hard to imagine the estimate's being 100-fold wrong, and even in that case the risk of harm is much less than the danger imposed by other man-made devices.

Reactor safety, ultimately, is characterized by the combination of a very small probability of a serious accident and a large degree of damage. The closest analogy to a reactor, in this respect, is a dam: both produce electricity, and although the probability of accident in either is relatively small, the damage can be very large. More than that, water from a broken dam would inundate land and make it uninhabitable, much as radioactivity from a catastrophic reactor accident would interdict land. The two worst dam-connected disasters in modern history led to the Johnstown, Pennsylvania, flood in 1889 and the massive slide of Mount Toc into the reservoir of the Vajont Dam, near Longarone, Italy, in 1963, forcing a 100-meter high wall of water over the dam (4). About 2,000 persons were killed in each of these dam disasters.

The probability of failure of a dam in California, largely from earthquakes, is perhaps one in 100 to one in 1,000 per year (5); yet residents of California show little reluctance to live downstream from a dam, where the probability of harm is orders of magnitude greater than the probability of harm in a location near a nuclear reactor. Moreover, the consequences of a dam failure can be much more severe; it is estimated that failure of the Folsom Dam above Sacramento, California, would cause 260,000 deaths.

As I compare the issues we perceived during the infancy of nuclear energy with those that have emerged during its maturity, the public perception and acceptance of nuclear energy appears to be the question that we missed rather badly in the very early days. This issue has emerged as the most critical question concerning the future of nuclear energy.

What is safe enough? The public perceives a large dam to be sufficiently safe and hardly gives a second thought to questions of dam safety. A sizable fraction of the American public perceives nuclear reactors to be unsafe, even though reactors pose a lower probability of causing harm than do dams.

We must concede that we can never, for all time, say what is "safe enough." Such questions are judged in particular social, cultural, and economic contexts. In the present situation, with oil in crisis, it seems logical to expect the definition of safe enough to be more liberal than it would be in the absence of such crisis. Countries with vast coal resources can afford the luxury of imposing stricter definitions of safe enough than can countries that have no coal. Still the argument about what is safe enough is not so sharply rational. What was safe enough in 1950 is judged by some not to be safe enough in 1975. As deficiencies are discovered, the nuclear community is committed to correcting them, to improving the safety of reactors and their subsystems. But this surely does not mean that every new improvement must be backfitted onto every reactor. For example, I continue to advocate nuclear parks as a way of improving the safety of reactors and their subsystems (6). This of course does not mean that all reactors that have already been built must be shut down and rebuilt in nuclear parks, nor does advocacy of nuclear parks imply that the risk, as represented by a few hundred reactors that will operate outside parks for perhaps thirty years, is unacceptable.

The next thirty years

It is dangerous to predict the future. (My own predictions of very cheap nuclear power have turned out to be very wrong.) One should take the precaution of not committing one's prediction to writing, or of being elderly and in poor health at the time of the prediction. I cannot claim the latter excuse, and I obviously am committing my views to writing; nevertheless, I make the following predictions with respect to energy and nuclear power over the next thirty years.

1. The rate of growth of energy, at least in the United States, and possibly in the rest of the world, will diminish. The Project Independence Report (7) suggests that if oil remains at $11 per barrel, then the rate of growth of energy use in the United States will diminish by 1985 from the present 4.2 percent per year to about 2.9 percent annually. This rate would be further decreased if active measures of conservation, in addition to the working of the marketplace, were taken. Nevertheless, I would warn most stringently against those who claim that conservation alone can solve the energy problem; it cannot.

2. Competition between the different nonbreeder reactors will have been settled, and each will find its place in an appropriate setting. As for the breeder, I would hope that the competition that gives the world at least seven choices of nonbreeders—two LWRs (pressurized water and boiling water), two heavy water reactors (CANDU and Steam Generating Heavy Water Reactor), and three graphite-moderated reactors (pressure tube [Hanford N-type], AGR, and HTGR)—will be extended to breeders and near breeders. Being an optimist, I hope that by the year 2000 the world will have a choice of six breeders and near breeders—LMFBR, GCFBR, MSBR, LWBR, CANDU-Th, and HTGR.

3. I confess to being less optimistic about fusion (not to speak of geothermal and solar energy), but this is undoubtedly a prejudice that can be expected, if not excused, in one who has devoted his career to the development of fission.

4. Disposal of radioactive wastes in geologic formations, particularly in bedded salt, will be judged acceptable. One important discovery that will help convince those who are ready to listen is the remarkable finding of the spontaneous chain reaction that occurred in Oklo, Gabon, some 2×10^9 years ago. The total energy released during the million years the reactor operated was about 100×10^9 kilowatt-hours. Some 4 tonnes of fission products, and perhaps 1 tonne of ^{239}Pu, were produced in a natural environment under circumstances very probably less favorable to sequestering the fission products from the biosphere than would prevail with disposal in salt. Yet evidence suggests that most of the rare-earth fission products, and the Pu, seem to have remained close to where they were created, though the fission-product gases seem to have escaped (8).

5. I expect the trend toward cluster siting, and especially colocation of fuel fabrication and chemical reprocessing plants, to move vigorously. In the United States this trend is reinforced by the study on nuclear centers now being conducted as required by Title II, Section 207, of the Energy Reorganization Act (9).

Can the people be unscared?

As I look into the future, I expect the nuclear controversy will grow rather than diminish, at least during the next few years. I hope that continued successful operation of nuclear power plants, together with the growing shortage of oil, will blunt the attack on nuclear power and that the public soon will recognize nuclear energy as an all-but-inexhaustible energy source that, when handled properly, is acceptably safe. Yet, as U.S. Senator John O. Pastore has said, "It is easier to scare people than to unscare them." Despite the reassurances that I and other nuclear proponents give, portions of the public remain unconvinced.

That we estimate the actual, as contrasted to the potential, hazards of nuclear energy to be no greater than the hazards of other energy-producing systems—particularly dams—perhaps is the best way to reassure the public. But there are certain essential elements of nuclear energy that are different and that seem to me to be responsible for the peculiar concern over nuclear energy.

First is the newness of the radiation hazard. When electricity was first introduced, many people were concerned about the possible danger inherent in that new form of energy: several thousand people die of electrocution each year, yet no one would consider banning electricity. So the newness of the radiation hazard, at a time when the public in general is much more sensitive to the deleterious side effects of technology, is certainly a factor in the public's concern.

There is perhaps a deeply psychological aspect to the radiation hazard that magnifies the concern over it. In the long course of human history, we have come to terms rather well with those natural hazards that threaten human life—floods, pestilence, famine. These are things we are prepared to deal with. In the past hundred years, electricity has become part of our human experience and an accepted hazard, but radiation, especially with its genetic threat, has not yet gained such acceptance. I would expect our collective psychology to come to terms with radiation, just as it has come to terms with electricity: we accept a great benefit along with a certain risk.

Second is the persistence of the radioactive hazard. Land interdicted by radioactive contamination could be hard to clean up. This aspect of the nuclear hazard has prompted me to urge central siting, insofar as it is possible, so that relatively few areas are placed at risk. Such an option is hardly available in Europe, however. One possibility in Europe is underground siting, which, preliminary estimates suggest, could reduce the absolute damage in the 10^{-9} per year very large hypothetical accident; whether this is worthwhile, considering the extremely small probability of a serious release of radioactivity, is hardly a matter that can be settled scientifically. Again, in making such judgments, one must consider alternatives and alternative risks. I repeat that dam failures can interdict land for a long time, and a dam failure appears to be much more probable than a really serious reactor accident.

Third is the question of diversion of fissile material, a possibility that some opponents of nuclear technology consider to be the strongest objection to nuclear energy. But this is a concern we have lived with since the bomb; it is not a characteristic peculiar to nuclear power plants. We cannot avoid this unpleasant fact: when man moved into a new era with the discovery of fission, he assumed certain risks—among them, the possibility of clandestine manufacture of nuclear bombs—in exchange for a weapon that makes war between the superpowers all but unthinkable.

The fourth general concern people have about nuclear energy is the meticulous attention to detail, not to say rigid security measures, and social stability that it demands. In this sense, nuclear technology is the most demanding technology. Can our societies, our human institutions, cope with the demands of this technology? Obviously one cannot answer such questions with finality.

I have referred elsewhere to nuclear energy as a "Faustian bargain," whereby in exchange for an inexhaustible source of energy, society promises to pay meticulous attention to detail and to maintain a certain degree of social stability (10). Some have interpreted this to mean that nuclear energy is devilish, and therefore we should have nothing to do with it. But those who draw such a conclusion forget that Goethe's Faust was redeemed:

Angels (*soaring in the higher atmosphere, bearing Faust's immortal part*)

'Who e'er aspiring, struggles on,
For him there is salvation.'
And if to him Celestial Love
Its favouring grace has given,
The Blessed Host comes from Above
And welcomes him to Heaven [*11*].

In embarking on the nuclear enterprise we certainly accept on faith that our human intellect is capable of dealing with this new source of energy. But we have much more than faith. We have acquired wisdom about the real technical problems of reactors, and patiently and seriously have dealt with each of the problems as it has arisen. To reject nuclear energy while invoking such magical talismans as fusion, or solar or geothermal energy, seems

to belie an unwarranted technological optimism. Of course we must push on with development of these energy sources. But we must also pursue fission technology responsibly; we must learn from our mistakes and, as we learn, redeem ourselves by wisdom even as Faust redeemed his pledge through struggle, wisdom, and faith.

References

1. M. G. J. Boissevain, M. C. Leverett, L. A. Ohlinger, A. M. Weinberg, E. P. Wigner, G. J. Young. 1943. Preliminary process design of liquid cooled power plant producing 5×10^5 KW. CE-407, Metallurgical Laboratory, Univ. of Chicago.

2. W. B. Lewis. Abundant harnessed energy at low cost and low risk from nuclear fission, Table 3. Elizabeth Laird Memorial Lecture, Univ. of Western Ontario, London, Ontario, 3 Apr. 1974.

3. N. C. Rasmussen. 1974. Reactor safety study: An assessment of accident risks in U.S. commercial nuclear power plants. WASH-1400 (draft). Washington, DC: U.S. Govt. Printing Office.

4. *The World Book Year Book*. 1964. Chicago: Field Enterprises Educational Corp., pp. 294–95.

5. P. Ayyaswamy, B. Hauss, T. Hseih, A. Moscati, T. E. Hicks, D. Okrent. 1974. Estimates of the risks associated with dam failure. UCLA-ENG-7423. Univ. of Calif. at Los Angeles.

6. A. M. Weinberg. 1971. Moral imperatives of nuclear energy. *Nuclear News* 14:33–37.

7. Federal Energy Administration, J. C. Sawhill, Administrator. 1974. *Project Independence*. Washington, DC: U.S. Govt. Printing Office.

8. R. Hagemann, G. Nief, and E. Roth. 1974. Études chimiques et isotopiques du réacteur naturel d'Oklo, Le Phénomène d'Oklo. *Bulletin d'Informations Scientifiques et Techniques*, ISSN-0007-4543: 193:76. Paris: Commissariat à l'Énergie Atomique.

9. Energy Reorganization Act of 1974. 11 Oct. 1974. Public Law 93-438, 93rd Congress, H. R. 11510, Title II, Section 207. Washington, DC: U.S. Govt. Printing Office.

10. A. M. Weinberg. 1972. Social institutions and nuclear energy. *Science* 177:27–34.

11. Johann Wolfgang von Goethe. *Faust*. Translated by G. M. Priest, *Great Books of the Western World*, R. M. Hutchins, Editor in Chief. Encyclopaedia Britannica, Inc., 1952, p. 290.

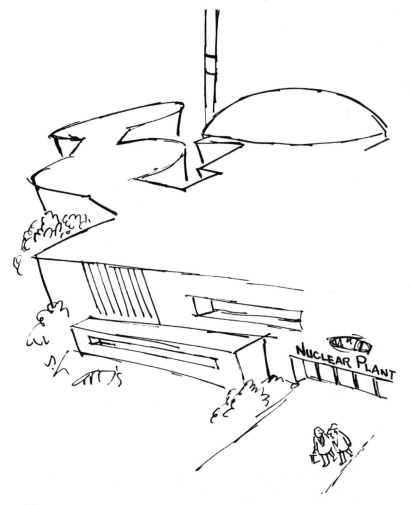

"Of course it's perfectly safe. Any accident would be in complete violation of the guidelines established by the Nuclear Regulatory Commission."

David J. Rose
Patrick W. Walsh
Larry L. Leskovjan

Views

Nuclear Power—Compared to What?

From environmental, economic, and societal points of view, nuclear power appears at present more acceptable for the generation of electricity than coal-burning and other technologies

Few issues have generated so much feeling and dispute as the acceptability of nuclear fission power and the choices to be made between nuclear power and its alternatives. Before the year 2000, one trillion dollars or more will be invested in new electric power plants and accompanying equipment, and the additional total costs—environmental, economic, and societal—among the options will differ widely and may be as large. The main obstacles to the resolution of the debate have been the inadequate data on the costs of alternate energy sources and the lack of effective communication. Several partial assessments of these costs now exist which support our conclusion that nuclear power appears to be not only economically cheaper than its main alternative, coal, but also safer by rational criteria.

We are primarily concerned here with decisions which will principally affect electric power generation at about the end of this century. A nuclear power plant decided upon now could be in

David J. Rose is Professor of Nuclear Engineering at the Massachusetts Institute of Technology. Patrick W. Walsh is a nuclear and environmental engineer currently studying law at the University of Wisconsin. Larry L. Leskovjan is a licensing engineer for the Florida Power and Light Company, Miami, Florida. The authors gratefully acknowledge the assistance of John Finklea, both in relation to the data and to their interpretation. They also thank Dade W. Moeller and James L. Whittenberger for the opportunity to discuss many of the issues contained in this paper. The work discussed here was supported in part by National Science Foundation Contract NSF–ST40016–000, "Priorities for Health Research Relating to Energy System Development." Address: Dr. Rose: Department of Nuclear Engineering, 24–210, Massachusetts Institute of Technology, Cambridge, MA 02139.

full operation by 1986 and in its mid-life in the year 2000. At that time petroleum and natural gas will be so scarce that they will hardly be discussed. The real choices will be among nuclear fission (with breeder reactors in prospect), controlled fusion, bulk solar power, and coal. None of these options is without substantial difficulty.

Several other prospects seem either too limited, too difficult to develop economically, or even too disadvantaged from a technical point of view for inclusion in the above list. Handy, clean geothermal sources are rare, and in the United States they all lie in the west, which may be beneficial regionally but which precludes national development. The economic and/or environmental costs of wind power will be too high, except in restricted localities. The use of ocean thermal gradients poses initial problems comparable to using waste heat from many present power plants, as well as difficulties of ocean siting. Tides yield negligible power.

It is generally agreed that the rate at which all resources are used could be substantially affected by energy conservation practices. These savings could reach a limit of 30–40% of classic (1972) energy projections. Energy conservation through more efficient use would also bring a net beneficial environmental impact, from a reduction of pollution because less fuel is mined, transported, transformed, and consumed and because less partly reacted fuel enters the biosphere in a highly reactive state. Beyond the 30–40% savings, the payoff is unclear because the total impact of conservation on different life styles is not well known. The main obstacles to substantial conservation as a viable

option are the attitudes and personal desires of people and the inadequacy of enforcement by institutions. In the Energy Research and Development Administration's National Energy Plan (1), conservation is accorded a substantial effect before 1985–90. Except for degree and timing, however, conservation offers no escape from the issue of nuclear power vs. its alternatives.

Available resources

It is clear from a study of Table 1, which shows estimates of total remaining and recoverable petroleum and natural gas in both onshore and offshore regions belonging to the United States, that these fuels do not enter the main line of options for the twenty-first century. These estimates, originating mainly in the National Academy of Sciences COMRATE study (2, Sec. 1), are much more pessimistic than most speculations in vogue even a few years ago. The ERDA 28 June 1975 National Plan (1) generally agrees with the COMRATE study, and the figures correspond with predictions made by M. King Hubbert (3).

If we project energy use based on the U.S. national 1974 figure of 73 Q, petroleum and gas will last 25 years, which is beyond the time horizon of conventional economic concern. Conventional economics would advocate expanding development and use of these resources because they are presently still fairly cheap, but this route leads to an economic and political precipice not noticeably advertised by its advocates. Improbable arrangements to import vast quantities of foreign petroleum at an acceptable price will ultimately fail because Middle Eastern petroleum will run out early in the twenty-first century.

Table 1 also shows proven U.S. uranium resources as well as speculated resources at concentrations greater than 200 parts per million U_3O_8 by weight, which could be recovered at reasonable cost. Natural uranium consists of 0.7% ^{235}U, which fissions easily, and 99.3% ^{238}U, which does not. Present-day reactors utilize most of the ^{235}U, augmented by burnup of about 0.2% of the ^{238}U, which has been bred into plutonium in the reactor. Thus they use only about 1% of the mined resource. Breeder reactors could use the major uranium constituent, ^{238}U. The table actually presents four estimates for uranium, the smallest of which, 610 Q, approaches estimated petroleum reserves.

As a consequence, the operation of nuclear fission power plants well into the twenty-first century requires that a breeder reactor be perfected somewhere between the 1990s and 2010, depending on confidence about speculated reserves. Once breeder reactors are functional, uranium resources become relatively unlimited, because the small amounts of fuel needed can even be obtained from seawater (although at a high cost: $1,000–$2,000/kg).

Uranium reserves do not affect the broad nuclear vs. alternatives question, but they do have a bearing on internal strategies within the nuclear sector, because of our optimism that

satisfactory breeder reactors can apparently be developed—e.g. the French "Phénix" demonstration experiment. Table 1 does not show thorium resources, but they are thought to exceed those of uranium and can be used in alternate fission reactor and breeder cycles.

The fuels used in fusion technology are deuterium, which is in virtually unlimited supply in water, and lithium, from which the tritium fuel is bred via neutron reactions in a blanket surrounding the reactor. The lithium resources must also provide for the large inventory needed in the reactor itself, and thus the total actual fuel available may be substantially less than shown in Table 1, but it is still relatively large.

The coal resources listed in Table 1 are thought to be recoverable using existing technology. All coal in place in the United States is about five times as large and constitutes one-quarter of the world's supply. In principle, shale oil resources might approach those of coal, but they are not listed on the table because their recovery appears difficult at present without excessive environmental and other societal costs.

Thus our near-term alternatives seem to be nuclear fission and coal power, and any choice between them must take into account the comparative environmental, societal, and economic costs. Fusion and solar technologies, which are further in the future, can only be discussed briefly.

Hazards of nuclear fission power

Just what is the extent of the health hazards from mining, radioactivity, and reactor accidents that are talked about so extensively? Table 2 presents average data for the United States on a per-plant basis for fatalities and injuries associated with light-water reactor technology. (Note that only normal occupational risks are included; sabotage and related events will be discussed later.) The largest number of fatalities comes from the mining and milling of uranium ore. As with almost all mining operations, these accidents need to be reduced. The main radiation-related deaths occur in the reactor-operation category—about one fatality every nine years is statistically attributable to radiation from the plant. Such fatalities would be caused partly by off-gas emitted from presently operating boiling-water reactors and partly from radiation emitted in refueling and other operations. To correct the off-gas problem, new regulations require the installation of charcoal traps on all boiling-water reactors which will hold the gas until its radioactivity has decreased by a factor of about 100.

The total deaths attributable to the existence and operation of nuclear plants, according to Table 2, are about one every two years. Injuries account for a little over 500 days of absence annually. Studies by other groups, both in and outside the Atomic Energy Commission (now the Nuclear Regulatory Commission), offer similar figures, within a factor of about two. Lave and Freeburg (4) cite 0.12 fatalities per 1,000 MWe plant-year from mining and milling uranium; Sagan (5) estimates 0.93 fatalities and 1,022 days off for the whole system per plant-year; and Hamilton and his colleagues (6) and Starr, Greenfield, and Hausknecht (7) come to much the same conclusion.

Table 2 does not appear to contain the most hotly debated item—the probability of large nuclear accidents, but it is there, under reactor operation. Rasmussen and his colleagues have made the most comprehensive study to date on this topic (8), and in the 1974 draft version they predict a "prompt" death rate of about .0004 persons per reactor-year of operation, a startlingly small number in view of

Table 1. U.S. energy resources, in units of $Q = 10^{15}$ Btu $\approx 10^{18}$ joule (see text for references and qualifications)

Resource	Common units	Energy in Q units
Petroleum	160×10^9 bbl	928
Natural gas	750×10^{12} ft^3	787
Uranium proven	700,000 tons	610* 61,000†
Uranium speculated (total in ore $\geqslant 200$ ppm)	3,000,000 tons?	2,600* 260,000†
Uranium available for breeders	—	almost unlimited
Deuterium for fusion	—	almost unlimited
Lithium-6 for fusion	10^6 tons?	3×10^8
Coal	600×10^9 tons	14,000

* If used in present light-water reactors only.
† If used in breeder reactors.

the loud debate on the topic. Delayed deaths—from cancers and genetic faults, for example—could multiply the total severalfold, perhaps by a factor of 10. A study and critique by the American Physical Society (9) would increase Rasmussen's results by a factor of about 25 for many accident categories, because of the effects on populations downwind from the accident and also from long-term effects. Applying this (possibly pessimistic) increase throughout, we arrive at an expectation value of 0.01 deaths per reactor-year, which forms a small part of the entry 0.117 in Table 2, under reactor operation. The most pessimistic estimate has been made by Kendall (10), who would increase the accident probabilities severalfold over the value 0.01.

As attention shifts toward plutonium-fueled breeder reactors, the toxicity of plutonium becomes central to accidents. Rose (11), in his 1974 summary of nuclear power prospects, and Bair and Thompson (12) discuss the matter. Briefly, plutonium is very bad, but that statement must be made more precise. It goes to the bones and liver, and, like iron, it tends not to be excreted. Studies continue, and most reviews to date indicate that Tamplin and Cochran's idea that single small inhaled particles are vastly more carcinogenic (the "hot particle" theory, 13) is not justified (14). This and other plutonium hazards, as judged in the BEIR report (15), constitute part of the 0.117 entry in Table 2.

Uneasiness about plutonium persists, especially because all its modes of activity are not yet fully revealed. For example, Gofman raises a new issue (16) about inhaled insoluble plutonium oxide. Most lung cancers are bronchiogenic, and if the bronchial cilia are impaired—usually as a result of smoking—not only will plutonium tend to concentrate in the bronchial epithelium, but also its removal will be slow. In addition, "reactor-grade" plutonium contains a mixture of isotopes and is about five times more active than the standard ^{239}Pu. By this reasoning Gofman concludes that the limit for occupational exposure (0.26 μg ^{239}Pu equivalent) represents several lethal doses for heavy smokers and about 0.04 lethal doses for non-smokers, independent of any "hot particle" model. Additionally, according to Gofman the public expo-

sure limit (0.0082 μg ^{239}Pu equivalent) would, if universally reached, increase the present 63,000 annual lung cancer deaths in the United States by a factor of four.

The evidence so far does not seem to favor Gofman's severe view, but it is inadequate. If plutonium does turn out to be so much more hazardous—contrary to most prevailing views—then a serious reassessment will be in order. First, the Th-^{233}U breeder cycle, which requires no plutonium after an initial period, will look more attractive. Second, the viability of the entire nuclear option will depend on how well plutonium can, in fact, be contained. While awaiting further evidence, we believe that the plutonium issue will be tractable.

The problem of nuclear waste disposal was somewhat artificially generated by the U.S. Atomic Energy Commission, because inadequate (even niggardly) funding did not permit the preparation of attractive options and assessments (see Refs. 17 and 18 for more details). The problem has now started to receive its proper share of attention: for example, work is underway at the Battelle-Northwest and Oak Ridge National Laboratories. According to present plans the wastes would be turned into a glassy form with low leachability. Volume would be about 2 m^3/reactor year operation. After about 700 years in undisturbed storage, the activity will have declined to a relatively low but still very significant level. After

that, traces of slowing decaying plutonium, americium, curium, and, eventually, neptunium-237 will keep the activity relatively constant for the next several million years. The toxicity of the concentrated waste during that latter time will exceed natural pitchblende ore (which is 60% uranium) by a factor of ten or so. Eventually, after many thousands of years, the total toxicity of the wastes becomes less than that of the original uranium ore used up, but on that time scale the matter is somewhat academic.

According to present estimates, at an added fuel-cycle cost of perhaps 0.1 mill/kWhe, the long-lived wastes could probably be 99% removed from the 700-year ones and returned to the reactor itself. There they would eventually undergo fission, leaving only wastes of the 700-year variety. Thus what many conceive to be a million-year problem can become less than a thousand-year one. In any case, safe geologic structures—certain salt and hard-rock formations, for example—can be used for quite satisfactory permanent disposal, especially if the long-lived fractions have been substantially removed. As knowledge about ocean plate tectonics increases, placement in (not on) the ocean floor deserves reexamination.

All these schemes are still in the planning stage, and no high-level radioactive wastes from commercial nuclear reactors have yet been buried

---

Table 2. Summary of health effects of civilian nuclear light-water reactor power, per 1,000 MW(e) plant-year

| Activity | Fatalities | | | Injuries (days off) |
	Accidents (not radiation-related)	Radiation-related (cancers and genetic)	Total	
Uranium mining and milling	0.173	0.001	0.174	330.5
Fuel processing and reprocessing	0.048	0.040	0.088	5.6
Design and manufacture of reactors, instruments, etc.	0.040	—	0.040	24.4
Reactor operation and maintenance	0.037	0.117	0.154	158.0
Waste disposal	—	0.0003	0.0003	—
Transport of nuclear fuel	0.036	0.010	0.046	—
Totals	0.334	0.168	0.502	518.0

SOURCE: After P. Walsh, Ref. 30.

anywhere in the United States. The U.S. Atomic Energy Commission—and the Energy Research and Development Administration and the Nuclear Regulatory Commission—should be simultaneously commended for prudence and criticized for slowness in resolving the problem; matters will change dramatically in the near future, as ERDA intends to expand its activities in this area severalfold.

The commercial nuclear waste problem should clearly be distinguished from the unhappy situation which exists near, say, Richland, Washington, where about 70 million gallons of waste radioactive solutions from old weapons programs reside discontentedly in steel tanks. The technology is obsolete and will not be repeated, but that nuclear Augean stable awaits an expensive one-time cleanup.

India's recent development of nuclear explosives, the international sale of reactors and fuel enrichment facilities, and other events remind us poignantly of the Acheson-Lilienthal proposal of the early post-World War II period that all nuclear power and weapons everywhere be placed under strict and enforced international control. Many would now wish to retrieve that U.S.-sponsored possibility, which was not accepted elsewhere, but it is too late.

The decision for or against nuclear power in the United States will have negligible effect on international proliferation, since nuclear fuel enrichment and/or processing facilities are apparently for sale by other countries (though not now by the U.S.). New uranium enrichment technologies, such as the gas-dynamic centrifuge, can be built on a modest scale for a modest product yield and will soon be available to many small

countries. Thus it will soon not be necessary for a nation to purchase a nuclear reactor or an enrichment plant from abroad in order to develop nuclear weapons. Also, nothing prevents a country from signing the present weak Non-Proliferation Treaty, acquiring a nuclear capability, and then abrogating the treaty on 90-days notice.

Other possible forms of misuse of nuclear technology are the theft of fissionable materials for nefarious purposes and the sabotage of nuclear plants and/or fuel-reprocessing facilities. Willrich and Taylor (19) analyze a number of possibilities, including the (disputed) ease of building a nuclear weapon, given fissionable material. Most of the severe risks of theft could be eliminated by placing fuel reprocessing and fabrication plants on the same site. As for sabotage, it is possible, through an arbitrarily large effort, to provide arbitrarily good security around nuclear power or fuel-reprocessing plants; the problem is to judge the limits. While present U.S. security standards appear adequate to protect a plant and the public against a few saboteurs, it is not easy to prevent damage and, probably, severe injury to the public by a large, determined group. Grouping reactors in nuclear parks would substantially reduce the hazard.

The problems with coal

Compared to nuclear power, the picture for coal is more complex. Three main aspects require analysis: mining, direct burning, and the conversion of coal into one of the "clean" fuels through gasification or liquefaction.

Deaths and injuries fall into different categories than Table 2 showed for nuclear power; nevertheless, Table 3

forms some basis for quick comparison and serves to introduce the more detailed material (data come principally from Refs. 2, Sec. 3, and 20). The total coal-mining accidental deaths ($\approx$200/yr recently) and injuries have been allotted according to the amount of coal used per 1,000 MW electric power plant, similar to Table 2. Strikingly different here is the appearance of many "environmental deaths" (as we call them); they are total incapacitations owing to black-lung disease from past mining practices and, chiefly, heart and lung disease in the general public caused by breathing air polluted by power plant effluents. Most victims eventually die of their ailment. Data are inadequate—hence the query beside many entries.

Accidental mining fatalities per 1,000 MW power coal plant exceed those for nuclear plants by a factor of three, but much worse are the "environmental deaths," in this case coalworkers pneumoconiosis (black-lung disease); it has disabled tens of thousands of miners. Present U.S. welfare payments to incapacitated miners total well over one billion dollars per year. New, stricter dust standards of 2 mg/m^3 will substantially alleviate the problem in the future (2, Sec. 3). Note that nonfatal accidents in Table 3 are events, not man-days lost (which would be much higher).

Burning coal produces particulates, sulfur oxides, nitrogen oxides, and trace metals. Precipitators can remove as much as 99.8% of the particulates by weight, but they fail to remove the very small ones which penetrate deep into the lungs and are not filtered out by the body's natural mechanisms, and hence have the most serious health consequences.

We know relatively little about nitrogen oxides, although public health authorities become more worried about them as time passes. Sulfur oxides have been most intensively studied, but even here data are inadequate. About 63% of the coal mined is burned in electric power plants, and as most atmospheric sulfur oxides arise from coal combustion, about 60% of the health effects to be discussed can reasonably be attributed to the production of electric power via coal. (This percentage is somewhat uncertain because of the poorly

Table 3. Summary of present health effects of coal-generated electric power, per 1,000 MW(e) plant-year

| Activity | Accidents | | |
	Fatalities	No. of disabling injuries	"Environmental deaths"
Mining	0.5	60	10?
Coal transport	0.5?	?	
Coal burning (2.5% sulfur)	—	—	20–100?

measured effect of local use of coal for space heating.) Although air quality standards have been set on the basis of SO_2 concentration—80 $\mu g/m^3$ annual average—SO_2 now seems to be a precursor to worse offenders—acid sulfates, including sulfuric acid itself. These are formed when the SO_2 oxidizes in the atmosphere, at a rate that varies from 1% to 10% per hour, depending on moisture content, presence of fine particulates and of metal vapors deposited on the particulates, and so forth; these acid sulfates travel across the country with the prevailing wind until they pass over the Atlantic Ocean or are rained out.

Figure 1 shows the mortality data from chronic respiratory disease prepared by the Environmental Protection Agency and used by the National Academy of Sciences in their recent study of fossil fuels (20). The "best judgment" line neglects (probably properly) early London and Oslo data and forms the basis of the NAS and EPA mortality estimates. Acid sulfate levels in eastern U.S. urban areas are 16–19 $\mu g/m^3$, which are apparently safe *if* the best-judgment line applies strictly and *if* there is a real threshold at 25 $\mu g/m^3$, as shown. But few would feel satisfied to live so close to danger, especially in view of large uncertainties in the data and of environmental fluctuations. Using Figure 1, the EPA has estimated that, if the 1975 SO_2 standards were all met, the excess mortality in 1980 owing to SO_2 would be very small—in the order of one death, or less, per power-plant-year. This number is comparable to the nuclear-plant hazard.

If, however, the 1975 air standards are not met or are significantly relaxed, the numbers of deaths climb spectacularly—to about 4,500 in 1980 or some 20 deaths per 1,000 MW coal-fired plant per year. If the "mathematical best fit" of Figure 1 is assumed to apply instead (a pessimistic assumption), the 1980 deaths jump to about 100 per plant-year from air-quality deterioration alone. Such statistics overwhelm the nuclear risks of every kind.

As one more example of environmental risk, consider expected chronic respiratory disease as it changes with suspended sulfate level in Figure 2. Here, the "judgmental" and "mathematical" analyses give

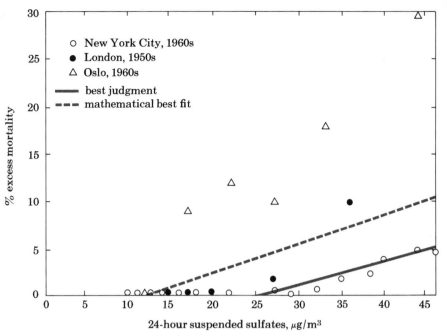

Figure 1. The percentage of expected excess mortality owing to acid sulfates in the air is estimated from suspended particulates and sulfur dioxide levels in three cities. The "mathematical best fit" line is currently considered to be a pessimistic assumption. (Data from EPA, summarized in Ref. 20.)

very similar results: if no sulfur restrictions are applied, the predictions are about 6 million cases annually in 1980; if all regulations plus conservation are enforced, this figure drops to undetectably few cases. The curve for cigarette smokers lies below that for nonsmokers, but no paradox exists: the smokers have already damaged themselves by their habit, and for them air pollution constitutes a smaller fractional degradation of their environment. The EPA and NAS also discuss aggravation of heart and lung disease in the elderly, aggravation of asthma, and excess acute lower respiratory disease in children, all with similarly trenchant prospects.

One fossil fuel problem common both to direct burning of coal and to the conversion technologies that has been generally overlooked may be the worst environmental danger of all: the

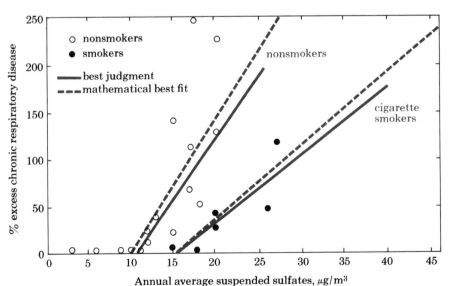

Figure 2. The excess chronic respiratory disease expected from acid sulfates has a threshold at 10 $\mu g/m^3$ for nonsmokers and 15 $\mu g/m^3$ for cigarette smokers. The data is based on studies in five areas and on the pooled results from the Community Health and Environmental Surveillance System program for 1970–71. (Data from EPA, summarized in Ref. 20.)

long-term effect caused by the increase of carbon dioxide in the atmosphere. The result is the "greenhouse" effect—a consequent long-term warming of the whole earth. Mitchell (21) estimates that the CO_2-driven warming will exceed 0.6°C before the year 2000, though it will possibly be moderated by a cooling effect from particulate pollutants, which itself is not a comfortable prospect. A global temperature increase of 1–2°C would almost certainly trigger a long-term global change, because sound meteorological theory predicts an amplification of the temperature rise at high latitudes (see 22).

Coal gasification and liquefaction technologies can produce three main fuel products: low-Btu gas, high-Btu gas, and a synthetic crude oil. Lurgi gasification and Fischer-Tropsch liquefaction plants (both dating basically from German World War II technology) are available, and a few Lurgi plants are being built in the U.S. today. The Wellman-Galusha, Koppers-Totzek, and Winkler gasification processes are also commercially available, but not as popular as the Lurgi. It is estimated that, with petroleum and coal prices at approximately $15/bbl and $20/ton, respectively, low-Btu fuel gas and liquid fuel would be attractive propositions to private industry, but high-Btu pipeline gas (for example, methane) probably would not. All these technologies are relatively expensive compared with the direct burning of coal, followed by stack-gas scrubbing. Also, the pollutants associated with them, and the rates of their emission, are not at present well understood. Leakages can be reduced to arbitrarily small amounts by arbitrarily large expenditures of money. Thus, no matter what the end product of the conversion process, as yet undetermined occupational hazards do exist. Coal chemistry yields more polycyclic organic compounds than does conventional petroleum chemistry, and these compounds tend to be more biologically active than the long-chain hydrocarbons in natural petroleum.

Because few pollutants from synthetic fuel processes have been measured directly, health effects must be mainly inferred. The following examples illustrate the state of the art and some data typically available. (1)

Table 4. Chemicals and fuels from a Fischer-Tropsch coal hydrogenation plant with a daily production capacity of 30,000 bbl*

Products	Production (bbl/day)	Weight (% of total product)
Tar acids		
Phenol	428	1.9
o-Cresol	48	0.2
m- and p-Cresol	530	2.4
Xylenols	374	1.6
Total	1,380	6.1
Aromatics		
Benzene	2,210	8.2
Toluene	3,770	13.9
Xylenes	4,190	15.4
Ethylbenzenes	750	2.8
Naphthalene	790	3.7
Mixed aromatics	1,780	6.8
Total	13,490	50.8
Liquefied petroleum gas	7,300	16.4
Gasoline		
Motor	5,260	15.6
Aviation	3,660	11.1
Total	8,920	26.7
Grand total	31,090	100.0

*This plant could also produce 450 tons of $(NH_4)_2SO_4$ and 89 tons of H_2SO_4 per day.
SOURCE: Ref. 31.

Many processes start with a washing and pretreatment heating of the coal to prevent caking, which produces much dust, SO_2, and perhaps also NO_x; control of these pollutants will be needed. (2) Many of these processes deal with coal fines, and the frothing agents used in their recovery are subject to release. (3) Particulate removal is not automatic in these processes, and tars and ash will contain sulfur and nitrogen unless post-treated. Also the tars and oils contain aromatic ring compounds. (4) To avoid gross pollution, water in many proposed locations must be cleaned before discharge.

Limiting our discussion to liquefaction plants only, if the Fischer-Tropsch process to make liquid fuel from coal is adopted extensively in the near future, very quick work will be needed to correct pollutant releases in the present technology. Table 4 gives a modest breakdown of the products of a hypothetical coal hydrogenation facility—a plant that adds enough additional hydrogen to the coal to make hydrocarbon liquids.

Note the preponderance of aromatic compounds and tars. From the point of view of pollutants, it should be noted that synthetic liquid fuel typically contains 0.02–0.5% sulfur (the actual content is determined by regulations and economics), and more polycyclic organics than conventional petroleum. The problems posed by these substances can be overcome at an additional, though at present undetermined, refining cost.

There are two types of coal gasification plants: those which produce a low-Btu product intended for use on site, while still hot, and those which produce a high-Btu product, essentially methane, which is a replacement for natural gas and is not a likely candidate for electric power generation. Intermediate-Btu gas is made by a similar process to low-Btu gas but uses oxygen or oxygen-enriched air instead of natural air in the gasification reaction. Thus, the intermediate-Btu gas lacks the large amount of nitrogen present in low-Btu gas and consequently contains fewer nitrogen-related pollutants.

The low-Btu gas option is now more economically attractive than other coal conversion processes, and because, in the case of large installations, it must be used locally, the health and environmental problems, which include H_2S, sludge, and waste water, become site-specific. The EPA is publishing a series of evaluations of the various gasification processes (23).

Synthetic fuels can be made clean but at still poorly determined costs, and the plants will almost certainly be more expensive to bring to the same degree of occupational and environmental safety that now obtains in petroleum refineries. This stands in sharp contrast to the problem of burning coal directly, where the cost might be low, but the environmental and health impact could be severe.

Economic comparisons

Nuclear power has been more expensive than fossil-fuel power in terms of capital costs, but the difference is likely to decrease. Many studies support the view that the nuclear-fuel cycle costs so much less than coal that nuclear power is much cheaper overall. Widely distributed comparative estimates have been made by the Arthur D. Little/S. M. Stoller consortium, e.g., for New England Electric in March 1975 (24). There they quote as the most probable capital costs for plants with 1983–85 startup: $863/kWe for nuclear, $565 for coal with no stack-gas scrubber, $697 for coal with scrubber, $512 for oil. Delivered coal costs were taken as $2.38/million Btu in 1983 for low sulfur, $2.31 for high sulfur. The two least expensive options turn out to be nuclear (3.4¢/kWhe in 1988, a typical year) and low sulfur coal with no scrubber (4.6¢/kWhe). The disparity is predicted to grow with time, especially if used western coalfields are properly reclaimed.

Davis (25) confirms and extends these economic opinions. Total investment for January 1985 completion, including fuel cycle allocation, varied from $945/kW(e) for coal without SO_2 cleanup, to $1,025 for coal with cleanup, to $1,060 for nuclear. Thus quotes for future clean plants can at present be figured at about $1,000/kW(e). (These are 1985 dollars; Ref. 24 uses 1975 dollars; figuring on inflation, the numbers are compatible.) Using $12/bbl oil and $25/ton coal, Davis computes total generating costs of 5.5¢/kWh for oil, 4.4¢ for coal, 3.4¢ for nuclear. The difference favoring nuclear power over coal could be overcome by a nuclear plant increment of $500/kW(e) at 14% interest, but no such cost penalty is foreseen. No believable uranium price increase could affect the balance; such an increase could only result from mining large volumes of very low-grade ore, which environmental damage would preclude.

Although these opinions are shared by almost all electric utility and other analysts, they are occasionally disputed. Bupp and Derian (26) predict a large and increasing divergence between nuclear and fossil costs, to the disadvantage of nuclear power. This is hard to envisage, because the entire nuclear steam-producing part of the plant accounts for no more than about 15% of the total cost. Why then the divergence? The difference arises through longer delays in obtaining licenses, finishing construction, etc. for nuclear plants, during which time interest and escalation charges accumulate. Such charges account for 40% or more of present nuclear plant costs. Reducing the time by two years between planning and operation would eliminate much of the nuclear capital cost disadvantage. Meanwhile, the high capital cost of nuclear plants tends to favor decisions for coal-burning ones, despite the certain long-term penalty.

One final economic issue needs putting to rest. The Federation of American Scientists' Public Interest Report of March 1975 disproves the statements of D. D. Comey (27) that nuclear reactors older than four years have poor performance.

Controlled fusion and solar power

For the long term, controlled fusion and solar power must be included in any discussion of the future energy picture. Fusion will cost a great deal to develop, but just how much will not be known for some years. Even with good fortune, it will not be generally available before the year 2000. It is expected to have advantages in security and general environmental quality; however, because no actual fusion reactor has yet been built, what follows is speculative.

A large fusion reactor, generating, say, 1,000 MW of electric power, is expected to contain an inventory of not more than 10 kg of tritium, one of its most important radioactive materials. Tritium, the fuel for the reactor, is radioactive (half-life, 12.3 years) and must be regenerated in the fusion reactor, using neutrons from the fusion reaction. The 10 kg of tritium would be mainly distributed in the moderating blanket of the reactor and in the tritium recovery equipment. Its activity is about 100 million curies—a large amount—but the relative biological hazard of tritium is very low compared to either fission products (strontium-90, cesium-137, etc.) or the transuranic elements (plutonium, curium, americium, etc.) produced in any fission reactor, whether it is a breeder or not. Best present estimates for fusion place the relative toxicity of the radioactive inventory at 1/1,000th or 1/10,000th that of an equivalent fission reactor (these numbers are guidelines only).

As with fission reactors, the relative safety of a fusion reactor depends on avoidance of a major accident, and experience with fission reactors indicates that safety against severe accidents can only be judged in relation to fairly specific designs. No such designs yet exist for fusion, but indications point to a relatively benign device.

The problems of theft and diversion of materials used in fusion technology arise obliquely and are probably not serious. To be sure, one can make an H-bomb with deuterium (readily available) and tritium; but an A-bomb is needed to trigger it, and there are easier ways of making H-bombs (given the trigger). Thus the incentive to steal tritium for clandestine purposes is very low. On the other hand, a small fraction of neutron yield from a fusion reactor could be diverted to illicit ^{239}Pu or ^{233}U production, hence to illicit weapons, but we state categorically that any nation achieving fusion on a large scale would find any fission technology easy, so this weapons scenario is not only improbable but illogical as well.

While no radioactive wastes arise from the fusion fuel cycle itself, there will be some items contaminated by tritium, and also the high energy (14 million eV) neutrons from the reaction will activate the reactor structure itself, with the degree depending on the materials used. Work by Steiner (28) at Oak Ridge, and others shows that the problem of old radioactive fusion reactor hulks should not be worse than that of old fission reactor hulks. This problem, in turn, is small compared to high-level radioactive wastes from the fission process itself.

Turning to solar power, it is necessary to distinguish between the heating and cooling of buildings, which shows great promise and will alleviate the drain on other resources, and bulk electric power generation, which is the target of our discussion. The goal of bulk electricity from solar power is to extract about 20 W/m^2 from a land surface covered with equipment, and it is this equipment which entails costs and certain environmental impacts.

At present the costs of the photovoltaic technologies are a factor of about 100 too high, but a substantial research program is underway to discover cheaper methods, for example, evaporated films. The various schemes to heat some working fluid using sunlight directly (e.g. boiling water for a steam turbine) suffer not so much from a lack of new science as from the difficulty of achieving substantial cost reductions using relatively conventional materials. For example, the solar "power tower" proposal must provide many sun-following mirrors which are accurate to a few milliradians and are 6 m on a side for a few hundred dollars each, including the foundation, drive, and control mechanisms. Hottel summarizes this and other proposals in a lively fashion (29).

It has been said that solar power is environmentally benign. This is true in certain applications and under certain conditions, but some qualifications should be noted. Presumably, large expanses of photovoltaic elements would be covered by glass, thus minimizing their erosion, but some of these substances, such as cadmium sulfide and gallium arsenide, are toxic.

Thermal conversion arrangements—for example, boiling a working fluid to run a turbine generator—are not so likely to entail chemical environmental difficulties, but they do produce thermal pollution, another difficulty which is common to most solar energy schemes. The system absorbs almost all the solar energy incident on the large array of conversion elements, then converts some of it at low efficiency into electricity. The remaining energy becomes waste heat rejected at the site. Calculations for typical solar power systems show that the waste-heat problem is comparable to or worse than that arising from a fossil-fuel or nuclear power plant, except that the heat is distributed over the large collecting area. It is possible, in principle, to overcome this difficulty by increasing the reflectivity of the surrounding and interstitial areas, thus preserving an unchanged overall heat balance. By this reasoning, thermal pollution per se is not a bar to using solar, fossil, or nuclear fuels, a point not generally recognized.

What are we to do?

The following seems clear: after conservation and more efficient energy use, the choice for new or replacement plants before the year 2000 must be between coal and nuclear power. The other options are uncertain, lie far off, or, in the case of solar electric power, are still uneconomic. In a stable soci-

ety, nuclear power is generally superior to coal unless (1) plutonium is as hazardous as its most severe critics claim (an issue that should be resolved as soon as possible); (2) emissions from coal technology are very strictly controlled (such as 90% or better removal of sulfur oxides and, probably, thorough treatment of fine particulates and nitrogen oxides); (3) the reactor safety studies, as amended, are still wrong by some very large factor. With this nuclear preference, it is possible to view some present activities and customs in a new light, not the least of which is that coal and what is left of the other fossil materials should start to assume their proper long-term roles as chemical resources rather than fuels.

The Price-Anderson Act, by which the electric utilities and the federal government provide no-fault insurance to the public against damage from civilian reactor operation up to $560 million per accident, should cost little on an actuarial basis. The present cash pool from utility contributions is now $125 million with no claims against it. The utilities could assume this light burden themselves via a consortium, perhaps arranged by the Electric Power Research Institute. How would the fossil-fuel industry respond to an equivalent act holding them responsible, on a no-fault basis, for ills deemed to result from fossil-fuel pollutants?

The Nuclear Regulatory Commission remains necessary, because the good nuclear record to date arises from the industry's diligence and care. What about establishing a "Fossil Regulatory Commission"? Its role is played by the EPA, but up to now it has been remarkably silent on these comparative issues, partly because of uncertainty of jurisdiction, partly because of overly rigid adherence to pro forma "criterion" pollutants, and partly because of underfunding. The Department of Health, Education, and Welfare, and especially its National Institutes of Health, also could have played a more active role in illuminating the problems arising from energy production. However, EPA and HEW *have* supported the small studies done up to now. Much more is needed.

On the legal side, federal environmental impact statements are required for all proposed nuclear power

plants, which often causes long delays. How long will it be before someone challenges a fossil-fuel plant on the grounds, say, that its effluents materially degrade the environment across state boundaries?

How did this nuclear–fossil debate get so far out of balance? First, the hazards of reactors and radiation have been perceived as "unknown" and, hence, very possibly large. Second, the public has accepted the health costs of polluted air, not realizing that much could be done, and not even aware that its understanding of the fossil-fuel hazard was inadequate. Probably a third reason dominates, however: more than one billion dollars was spent to illuminate and reduce nuclear hazards, but until recently almost nothing was done to measure similar hazards of fossil-fuel power. In retrospect, this is a scandalous omission. A vast amount of literature has appeared about nuclear hazards, providing material for a great public debate. The absence of any appreciable parallel assessment of fossil fuels has ensured that the debate would be unbalanced, and only now are semiquantitative social cost figures starting to appear. This profound issue can hardly fail to be resolved in the next few years as more data accumulate, especially on the effects of fossil fuels.

References

1. Energy Research and Development Administration. 1975. *A national plan for energy research, development and demonstration: Creating energy choices for the future.* Report ERDA-48, vol. 1, 28 June. Washington, D.C.: USGPO. See p. S–3.

2. National Academy of Sciences. 1975. *Mineral resources and the environment.* Report prepared by the Committee on Mineral Resources and the Environment, Commission on Natural Resources, Natural Resource Council (COMRATE study). Sec. 1, Estimation of mineral reserves and resources. Sec. 3, The implications of mineral production for health and the environment: The case of coal. Washington, D.C.: National Academy of Sciences.

3. M. King Hubbert. 1971. Energy resources of the earth. In *Energy and Power.* Scientific American Books. San Francisco: W. H. Freeman, pp. 31–40.

4. L. B. Lave and L. C. Freeburg. 1973. *Nuclear Safety* 14:409.

5. L. A. Sagan. 1972. *Science* 177:487.

6. L. D. Hamilton, ed. 1974. *The health and environmental effects of electricity generation: A preliminary report.* Biomedical and Environmental Assessment Group,

Brookhaven National Laboratory, Report BEAG-HE/EE 12/74 (20 July).

7. C. Starr, M. A. Greenfield, and D. F. Hausknecht. 1972. *Nuclear News* 15:37.

8. Atomic Energy Commission. 1974. *Draft reactor safety study*. Report WASH-1400 (Aug.). Rasmussen et al., eds. Washington, D.C.: USGPO.

9. H. W. Lewis (chmn.) et al. 1975. Report to the American Physical Society by the Study Group on Light Water Reactor Safety. *Rev. Mod. Phys.* 47 (summer): Suppl. No. 1.

10. H. W. Kendall. 1975. *Nuclear Power Risks*. Cambridge, Mass.: Union of Concerned Scientists, 18 June.

11. D. J. Rose. 1974. *Science* 184:351.

12. W. J. Bair and R. C. Thompson. 1974. *Science* 183:725.

13. A. R. Tamplin and T. B. Cochran. 1974. *Radiation Standards for Hot Particles*. Washington, D.C. Natural Resources Defense Council.

14. See, for example, "Hot Particles" in testimony presented by Chester R. Richmond, Oak Ridge National Laboratory, to the Subcommittee to Review the National Breeder Reactor Program, Joint Committee on Atomic Energy, U.S. Congress, 17 June 1975.

15. National Academy of Sciences–National Research Council. 1972. *The effects on populations of exposure to low levels of ionizing radiation*. Report prepared by the Advisory Committee on the Biological Effects of Ionizing Radiation (the "BEIR Report"). Washington, D.C., Division of Medical Sciences, National Academy of Sciences–National Research Council.

16. John W. Gofman. 1975. *The cancer hazard from inhaled plutonium*. Report CNR 1975-1-R (14 May). Dublin, Calif.: Committee for Nuclear Responsibility.

17. A. S. Kubo and D. J. Rose. 1973. Disposal of nuclear waste. *Science* 182: 1205-11.

18. U.S. Atomic Energy Commission. 1974. *High level radioactive waste management alternatives*. Report BNWL-1900 (May). Washington, D.C.: USAEC

19. Mason Willrich and Theodore B. Taylor. 1974. *Nuclear Theft: Risks and Safeguards*. Cambridge, Mass.: Ballinger Pub. Co.

20. National Academy of Sciences. 1975. *Air quality and stationary source emission control*. Prepared for the Senate Committee on Public Works. March 1975.

21. J. Murray Mitchell, Jr. 1975. A reassessment of atmospheric pollution as a cause of long-term changes of global temperature. In *The Changing Global Environment*, S. F. Singer, ed. Boston:Reidel.

22. W. S. Broecker. 1975. *Science* 189:460.

23. For example. H. Shaw and E. M. Magee. 1974. *Evaluation of pollution control in fuel conversion processes: Gasification. Section 1. Lurgi Processes*. EPA-650-12-74-009 c (July).

24. Arthur D. Little, Inc., and S. M. Stoller Corp. 1975. *Economic comparison of base-load generation alternatives for New England Electric*. Cambridge, Mass.: Arthur D. Little, Inc., and S. M. Stoller Corp. (March).

25. W. Kenneth Davis. 1975. *Combustion* 46:12 (June):25–31. (See particularly Exhibit 3.)

26. I. C. Bupp and J. C. Derian. 1975. *Technology Review* 77 (4, Feb.):15.

27. D. D. Comey. 1974. *Bulletin of the Atomic Scientists* 30(9):23–28.

28. D. Steiner and A. P. Fraas. 1972. *Nuclear Safety* 13 (5, Sept.–Oct.):361. See also D. Steiner. 1975. The technological requirements for power by fusion. *Nuclear Science and Engineering* 58:107.

29. H. C. Hottel. 1974. Solar energy. The 1974 Institute Lecture of the American Institute of Chemical Engineers, Washington, D.C., 2 Dec.

30. P. Walsh. 1974. Thesis, Nuclear Engineering Department, Massachusetts Institute of Technology.

31. Battelle Columbus Laboratories. 1974. *Liquefaction and chemical refining of coal: A Battelle energy program report*. Columbus, Ohio: Battelle Columbus Laboratories. See esp. p. 59.

R. Philip Hammond

Views

Nuclear Power Risks

A leading nuclear scientist takes a timely look at some of the hazards involved in the operation of atomic reactors

The energy crisis has come to public attention rather suddenly. The citizen who lately was urged to buy an all-electric home and to "see America first" by auto is now confronted with brownouts, reduced power voltage, and gasoline shortages. He is justified in complaining that someone should have foreseen this and have done something about it. Some did see what was happening and urged preventive action, though we know now that not enough was done. One far-reaching step *was* taken, however: Congress set up the Civilian Nuclear Power Program and instructed the AEC to find means of ensuring a plentiful supply of energy for the United States. The AEC fulfilled its instructions, and commercial nuclear power arrived on the scene in what seemed to be the nick of time.

But something went wrong. Instead of receiving the accolades of a grateful public, the AEC has become the target of impassioned censure for having produced a juggernaut, an inexorable monster

R. Philip Hammond has had over 30 years' experience with radioactive materials. He has worked with nuclear weapons, reactor fuels, fission wastes, and experimental reactors. He has first-hand knowledge of what can go wrong in the nuclear field and what has to be done to clean up after a spill. Safety is thus more than an academic matter to him. Dr. Hammond is well known for his developments in seawater desalting and for his studies of the application of nuclear energy to food production and industrial output in developing countries. He is author of the article "Atomic Energy" in the Encyclopedia Britannica, *has been an adjunct professor at UCLA, and is now a consultant in the energy field. Address: R & D Associates, P.O. Box 3580, Santa Monica, CA 90403.*

which could potentially destroy us if we become dependent on it for our power. The resulting dilemma is one of the most important social, political, and technical questions of our time. Energy is vital to our health, wealth, and safety, and yet the end of our gas and oil resources is in sight. Coal use could be expanded, but only at a very high cost in dollars and environmental effect, while oil imports pose economic and political problems which could become disastrous. It is not surprising that nuclear power has been welcomed by the utility industry as a clean, inexpensive, convenient, and inexhaustible source of energy. The AEC has assured us that reactors are safe, reliable, and economical, but public confidence in their assurances has been seriously undermined by events suggesting attempts to cover up the true status of reactor safety and waste handling.

The public has two big questions: What happens if a reactor breaks down and the fuel escapes? What are the risks and problems of shipping and storing nuclear waste for long periods? There is a third question, equally important, which is *not* asked: How do the risks for other energy sources compare with the nuclear risk? This question is not asked because most people have not realized there is a risk from using coal, oil, or gas. These questions are legitimate and urgent, and answers to them are available, but the AEC and the nuclear industry seem to increase suspicion, instead of confidence, with every reply they give.

To an observer who has worked with nuclear reactors and nuclear

wastes since the early days of the atomic project, the plight of the AEC is indeed ironic. Research in safety and attention to risks have been the watchwords throughout the years; the nuclear industry is without exception the safest in the world in which to be employed, and nuclear hazards are far better understood than are those of thousands of widely used chemical and biological agents, or of common energy sources such as coal. But it is difficult to judge these accomplishments, or the relative risks of various choices, without having information on these matters in clear and simple terms.

Our greatest need is for communication, and members of the scientific community can assist greatly in the process. Radioactivity and nuclear energy are complex subjects, and questions about safety tend to be answered by the experts in their accustomed language; i.e. in technical terms which are often meaningless jargon to the questioner. The basic facts must be put into plain, nontechnical language that people can relate to their own experience. As the President has noted, nuclear power can carry a major part of the energy burden if we can build reactors fast enough. We need to turn out power plants on a production-line basis, but public understanding and acceptance of the risks are essential. This article is an attempt to aid in this important communications step and to draw upon personal, first-hand experience for graphic illustrations of what nuclear fuels and waste materials are like, what amounts would be formed, and how they can be handled. In this effort I am acting as a spokesman for no

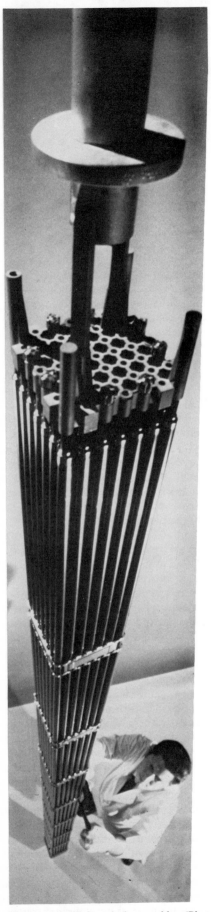

Figure 1. A nuclear fuel assembly. (Photo courtesy Exxon Corporation.)

one; my opinions and comments are not those of the AEC or any other institution, but are strictly my own.

Radiation

We live in a world which is by nature radioactive. Cosmic rays from space shower the earth steadily; all our food, every spadeful of earth in our gardens, every stone, the very rocks of the earth's crust, and the oceans all carry a small amount of radioactive material, and they always have. This background radiation we must assume is harmless to man, since the human race developed in its presence. Yet these radioactive ingredients of the rocks represent a fantastic amount of energy. A piece of New Hampshire granite contains 100 times as much energy as a piece of coal of equal weight, but in the form of uranium and thorium. Most of the earth's granite has only 10 times the energy of coal per pound, or about the same radioactive content as coal itself. The combustion energy of coal is thus greatly exceeded by the energy of the radium, thorium, and uranium it bears, which remain in the ash or go up the stack when it is burned.

Our ability to detect and measure radiation is millionsfold more sensitive than for other hazards. Present instruments can detect as little as one billionth of the amount of radiation that would be considered hazardous, while for many poisonous materials, such as mercury, our measurement ability is often not far below the threshold of noticeable injury. The great tuna-fish scare came only when new, more sensitive measurements were developed. If we are exposed to a nuclear hazard, at least we can tell it's there.

Two kinds of radiation are produced by nuclear fuels and wastes: short-range radiation and penetrating radiation. The fuel materials emit short-range radiations, called alpha particles. These particles can do severe damage to internal body tissues, but they cannot penetrate the skin or even a piece of thin paper or a coat of paint. Hence the fuel materials could be called radio-poisons because they must be actually eaten or deposited inside the body to be harmful. A famous

case in point is that of the radium watch dial-painters in the 1920s, who licked their brushes to point them.

The penetrating radiations, on the other hand, resemble X rays in that they can produce damage at a distance from their source, even though the emitting material is tightly sealed in cans that do not leak. To handle such materials safely one must work from behind a heavy shield which will absorb the radiation, such as 1,000 feet of air, 20 feet of water, 10 feet of earth or concrete, 3 feet of steel, or 1 foot of uranium metal. (These relative amounts of shielding are only approximately equivalent, but the heavier the shielding material, the less is needed.) Most substances are not affected by absorbing such radiation, but living organisms and photographic film are.

All radiation dies away with time, and it is fortunate that the penetrating radiation does so rapidly. In a mixture of reactor wastes, the intensity of radiation drops by 90% in a few hours and by another 90% in a few months. The residual level is quite low after 100 years, and, after 350 years, penetrating radiation has essentially disappeared as a major source of hazard. The radio-poisons, or short-range emitters, tend to be long-lived, however, lasting for tens of thousands of years, just as do the radium and uranium in the earth.

Identifying the hazards

Figure 1 shows a fuel element for a present-day power reactor, called a "light water" reactor, since it is cooled by ordinary water in the core. The element is an assembly of metallic tubes into which the actual fuel is sealed. The tubes permit the heat of the nuclear reaction to pass through the wall but prevent the escape of the radioactive materials. Other types of reactors have different fuel forms, but we will confine our discussion to the most common type.

The public rightly fears the escape of the reactor fuel, which is a radio-poison, and of the waste products, which emit penetrating radiation, into their living space. They also associate reactors with atom bombs and fear a nuclear explosion. Finally, they are concerned about having to store wastes in constantly increasing amounts in an uncertain and indefinite future. Let us try to deal with each of these problems in plain language.

How does a reactor differ from an atom bomb? As the scaremongers say, "A reactor contains enough fissionable material to make hundreds of atom bombs. Do you want *that* in your backyard?" From my experience at the nuclear weapons laboratory at Los Alamos, New Mexico, I can offer three reasons why a light water reactor cannot be a bomb: It has the wrong composition, the wrong surroundings, and the wrong timing. Imaginative people have done their best to think up improbable and hypothetical accident sequences that might produce a nuclear explosion, but they cannot get around the fact that the fuel contains substances which would prevent a bomb from igniting. Further, a bomb must be set off in "clean" surroundings, free from neutrons, or it will pre-ignite and shut itself off by thermal expansion. A reactor always has neutrons present and is thus the wrong surroundings. Finally, a bomb must be fired by pushing its parts together in a few millionths of a second, and there is nothing in a reactor to give such speeds, even if other conditions were met. Each of these three reasons is sufficient alone to prevent a nuclear explosion in a reactor. Thus one fear can be dismissed: A light water reactor may represent other hazards, but it cannot be a bomb.

What *can* happen to a reactor? In recent hearings on this subject, the AEC and the reactor manufacturers tried to defend their position on nuclear safety. They seemed to the public to be saying that, since all contingencies were provided for, nothing could possibly go wrong. In their defense, they refused to discuss what *would* happen if all the layers of prevention failed and the radioactive material escaped from the reactor. By their refusal, they tacitly agreed with the assumption of the uninformed that the consequences were unthinkably catastrophic. As a result, some writers have loosed their imaginations and have conjured up visions of a deadly, invisible miasma compressed inside a reactor, which, with the slightest failure, will escape and spread over the land, creating death, destruction, and instant blight, forever forbidding man's return to the region. Some typical statements are: "a damage potential beyond any other event that I can imagine. The hazard of fission products persists for a time that is longer than any I can conceive." "Where will your children live?" "Whole states may have to be evacuated." Thus in the absence of a clear statement of what *really* happens, people imagine a situation worse than if a full-scale atom bomb were released.

First, we must ask, is it possible to build a reactor so perfect that none of its components will ever fail? No! Humans being what they are, failures will occur despite the most stringent efforts in quality control and testing. A reactor is basically quite a simple device, compared to a boiler, for instance, but we must assume there will be failures. So far there have been about 2,000 reactor-years of operating experience and many kinds of failures. The majority of these are trivial—pumps to be replaced, bearings scored, control rods warped, various small leaks. There have been a few more serious defects and accidents: fuel failures, flow blockages, broken pipes, an unfastened lid, etc. Reactor operators and manufacturers point out that rarely are any of these failures in the nuclear portion; of those that are, most are trivial mechanical replacements. Of the few more serious failures, not one has yet caused even potential injury to the public. There was a near miss, though, when a British military reactor (of a type no longer used) caught fire while open for reloading and contaminated some pasture land. The reactor had no containment shell.

There is a good case for expecting rapidly diminishing levels of probability for really severe reactor damage and failure, not so much because of man-made safeguards, but because a reactor, of its own nature, tends to shut itself off if overheated. Yes, reactors can fail in dozens of ways, and the consequences can be costly delays and replacements. Very seldom is the fuel damaged, but if it is, there are some tedious and expensive clean-up jobs. Clean-ups within the plant are done by repeated flushing with detergents or other chemicals and rinsing of any spilled materials into special holding tanks. Everything must be done with long-handled tools or by remote control. None of these failures would be detectable outside the plant, and none would affect the public.

Thus far there has never been a reactor accident that ruptured or even damaged the main tank or vessel housing the fuel. For this to happen, all the previous defenses would have to fail, and a complete loss of all coolant would have to occur so fast that all the residual heat would still be on hand. Then the damaged fuel could in principle melt its way through the various structural levels and baffles within the reactor vessel and reach the heavy steel or concrete main vessel wall. No one really knows how likely the fuel is actually to escape in such a case. Most estimates say it would not penetrate the main vessel, but it might under some conditions. Almost any external cooling source would arrest the penetration, however. Some tests would seem to be in order, but no one has yet performed them.

Once through the main vessel, there is still another line of defense or, in some cases, two or three. All water-type power reactors except those in the USSR have a containment shell—an airtight dome or tank which is often the main visual feature of a nuclear power station. This containment shell encloses all the nuclear components—reactor, pumps, heat exchangers, coolant tanks, etc., and is intended to prevent any breakdowns or ruptures of the system from releasing anything to the outside. There is no doubt it can do this if intact, but a hot mass of fuel that has just melted its way through the thick reactor tank might be able to do the same to the containment. Some structural experts believe that the heavy concrete base of the reactor foundation, the presence of coolant that has escaped from the vessel, the diluting and cooling effect of the reactor internals, and the time delay involved in reaching the containment would essentially preclude further penetration. But until more tests are made, there remains a residual chance, however small, that the containment could be pierced.

Up to this point the interests of the public are not concerned. If the hapless reactor owner cannot prevent such serious damage to his half-billion dollar investment, that is his worry. We might reflect cynically that we will have to pay for it in higher electric bills, but there is no direct public risk. Once the containment is breached, however, the situation changes. It becomes very much our worry, and we have a right to know just what will happen. As mentioned above, the lack of official statements on this point has led to wild and imaginative speculations. The only official publication is the famous WASH 740 study *(1)*, made many years ago before it was completely clear that a reactor cannot function as a bomb. The authors of WASH 740 were asked to ignore all the improbabilities and assume that a small nuclear explosion *had* taken place and dispersed some of the reactor core in a "worst case" event. This study, often quoted by nuclear detractors, has never been replaced by a more modern, realistic appraisal, until recently when a new team of experts began work, led by Dr. Norman Rasmussen of M.I.T.

Assessing the risks

While we are waiting for their results, perhaps we can find some preliminary data to estimate the general scope of what we would be faced with and what could be done about it. The first thing that is apparent is that there are some seemingly deliberate attempts to mislead the public. For example, Ralph Nader and Friends of the Earth have recently asked for the shutdown of all operating nuclear power plants. They state that "the amount of radioactivity routinely present [in one of these plants] is equivalent to ten times the amount of radioactive fallout from detonation of the largest nuclear weapon in the United States defense arsenal" (complaint filed in U.S. District Court). In a fine example of misdirection, the reader of this statement is left with the implication that a reactor is ten times worse than an H-bomb, although

the statement does not say so directly. An H-bomb or an atom bomb does its damage, of course, primarily by blast, by scorching, intense heat from a fireball, and by a sudden burst of penetrating rays that are part of the fission process itself, not from the fission products or nuclear wastes. Fallout of wastes, though hazardous enough, is by comparison a trivial part of the effect of the bomb. Stating that these materials are present in a reactor, if there is no bomb to spread them over an area, is scaremongering. It is equivalent to saying that the chlorine gas stored at the city waterworks and swimming pools is sufficient to poison everyone in the city 8,726 times. Some of the AEC statements have been equally misleading in the other direction.

What facts of our own can we deduce? First, we know that none of the public is near the reactor, for each plant has an exclusion area, usually thousands of acres, which can be cleared in any emergency. Second, we know that there will be plenty of time for warning, evacuation, or preventive measures. The meltdown of a reactor core can occur only if the plant has recently been operated for several weeks at high power and if several kinds of safeguards have failed to provide a way to remove the 1% or so of the heat which is evolved for some hours after shutdown and continues at a diminishing rate for weeks. The rate at which the heat is produced is well known, and it is easy to calculate the maximum rate at which the reactor vessel internals and walls could be penetrated, which may be a matter of hours. Considerable additional time would then be needed to penetrate the containment vessel, so there is time to prepare.

As noted above, the consequences of a major meltdown which might penetrate the containment shell and come out into the ground have never received intensive study, seemingly because to discuss such an event was to admit the possibility of a bureaucratic failure. But let's try to imagine a blob of white-hot fuel emerging into the earth, about 30 or 40 feet under the surface, having penetrated the concrete foundation. There is a very good chance that the public would

have no measurable sign that anything had occurred. After several hours of sizzling away, all the easily vaporized materials in the fuel would necessarily have boiled out and probably would have solidified again somewhere inside the cool upper parts of the reactor containment or the concrete foundation. The emerging fuel material—heavy, white-hot, and semiliquid—is essentially inert, like so much molten steel or lava, except that it is emitting penetrating radiation and contains radio-poisons. If the containment is under pressure, the design provides either a means of venting it through a filter system which will remove radioactive particles, or else enough ice or water stored inside to keep everything condensed.

If I had to contend with such a material (and I have had some first-hand experience in cleaning up radioactive spills), I cannot think of a place where I would prefer to have it than far underground. It would be completely shielded by the overlying earth and concrete, it would be enclosed in a thick pocket of fused earth, and it would be completely dry, for it is known that heating of the earth would drive the soil moisture away for perhaps 20 feet or more. At a radius of about 20 feet or so, the heat flowing from the fuel mass would be spread out enough so that the soil could contain some water and so provide a rapid conduction of heat. Thus the system would stabilize and melt no further and would be completely safe until such time as salvage operations might begin. There would be no contamination of the water table because of the dry, heated zone. If the soil is dry, or if the foundation is rock, melting might continue somewhat further, but eventually a sufficiently large radius would be formed to carry away the heat steadily, and penetration would cease with debris encapsulated in a ball of fused earth or rock.

Where is the risk to the public in all this? If there is any, it is not from the melted fuel, but from the more easily vaporized materials driven off beforehand. If by some chance the containment cannot be vented, a blowhole might form through the soil, releasing some of the steam together with radioactive particulates. Such a release would contain only a small fraction of the fission products, but it could be a source of severe danger downwind, though far from a "catastrophe."

Hence it seems to me that all the controversy about whether or not the emergency core cooling system works is barking up the wrong tree. The owner of the reactor may very well want such a system to preserve his investment, but the safety of the public depends upon other factors, such as whether the containment vessel contains enough ice or water to assure condensation of the easily vaporized portion of the fission products, and whether there is a reliable way to vent excess pressure in the containment shell through an adequate filtering system. Such factors are much easier to determine, and the risks easier to guard against, than *proving* that a meltdown can never occur. If condensation can be assured, there is almost a negligible public hazard, in my opinion, from a meltthrough of the containment shell. I would be glad to tackle the job of drilling into the spilled fuel and bringing it up in small bits for recovery. This could be done safely and completely.

Some experts have worried about whether there could be a so-called steam explosion inside the main vessel, in which molten fuel would suddenly become dispersed in fine droplets in water, and thus generate a volume of steam sufficient to blow off the vessel lid, rupture the containment, and disperse the core over the surrounding area. There is indeed sufficient energy in a melted core to do this, and the consequences, although not in any way resembling the havoc of a nuclear bomb, could be injurious or fatal to persons in the vicinity. However, such an explosion is, in my opinion, incredible because the conditions required for its occurrence are incredible. Such an explosion could not happen if the reactor vessel were full of water, as it normally is, or if the vessel were empty, as it would be if ruptured. (Water stays liquid at the temperature in a reactor only because it is pressurized—if the vessel has even a small leak, all the water inside flashes to steam and escapes.) Thus it is very difficult to have it both ways—dry

enough inside to permit an uncooled blob of melted fuel to form, and yet wet enough to provide a pool of water into which it could fall. The scenarios which try to arrange such conditions certainly need to be studied thoroughly, but so far these seem more remote than other kinds of hazards.

Regardless of the precautions we take and the safeguards we install, there will always be a residual hazard from a nuclear power station. The best we can ask is that such a station be at least as safe to the general public as an alternative power source. For the near future, the only practical domestic alternative to nuclear power is coal. Although the coal industry has not yet been required to produce an environmental impact statement, much very disturbing information is now available about the consequences of using coal. One careful 1964 study (2) showed that about 19,000 deaths per year in the U.S. could be attributed, directly or indirectly, to the use of coal and oil, which contain carcinogenic, radioactive, and acid-forming materials. In addition, we are paying a heavy environmental cost for coal in mining areas and in continuous damages in corrosion and cleaning losses. A large nuclear power plant could displace nearly 1% of the 1964 U.S. coal consumption and could thus be looked upon as saving 1% of 19,000, or 190 lives per year. Over the 30-year life of the plant, 5,700 lives would weigh in the balance, plus untold property damage. Even the most pessimistic estimates of nuclear plant failures predict a smaller toll than that, with a probability of occurrence not once in 30 years but once in hundreds or thousands of years. As noted above, in a total of 2,000 reactor years of experience, there has been thus far no failure which was a significant hazard to the public.

The above example does not exhaust the list of "What if?" questions about reactors. There are also other types of reactors than the water type considered here. But for all cases I have seen, similar conclusions can be drawn, and the same safety standards will have to apply, so that failures, even rare ones, could hardly be described as catastrophes. On balance, it is hard

to escape the conclusion, after 25 years of experience, that reactors, failure-beset as they are, are already much safer than the alternative of using coal, and that they have indeed arrived in the nick of time. There is no doubt that they will be improved as experience is gained.

Storage of waste

The storage of nuclear waste is the other big question that concerns the public. This is not a technical question so much as a social problem, for it involves taking on a responsibility which we cannot discharge completely ourselves but must hand on in some form or other to our successors. What is lacking in the public's concept of this problem is an appreciation of the real size, scope, and cost of the commitment and the nature of the hazard involved. People are confused by the apparently careless leakages of stored liquid wastes at the Hanford works of the AEC. The soil conditions at Hanford may be such, as the AEC claims, that no risk has resulted. But a continuing series of unintentional spills is not the way to gain the public's confidence.

First, what is the nature of the waste material? When spent nuclear fuel elements are removed from a reactor, they still look like Figure 1, and both waste products and unburned fuel are completely contained. The penetrating radiation emitted from the wastes, however, means that the elements must be transported in a thick-walled cask or shield to the reprocessing plant. At the plant, in a sealed chamber behind heavy walls, machines chop up the fuel rods and dissolve the material. Chemical treatment then separates the valuable unburned fuel, which is recovered for reuse, from the wastes, which remain in a highly purified liquid form. (There are some gaseous waste products, which must be absorbed on charcoal, pumped into storage tubes, and held for decay; but these are a relatively minor problem, since both the amount and the costs are low.)

After some intermediate storage period, the liquid waste can be boiled down in an electric pot furnace and melted to form a glassy solid. When cooled, the solid waste is a black, ceramic clinker resembling obsidian or lava. It is inert; it does not dissolve in water or react with air. The procedures for converting a batch of waste to this solid form and sealing it into a metal tube 12 inches in diameter and 10 feet long are completely worked out and the costs are known. Once sealed, the tube can be safely moved in a thick-walled shipping cask (designed to survive all conceivable shipping accidents) to a final storage place. If the contents somehow escaped, it would not be dispersed but would lie where it fell until scraped up by remote control vehicles. The public hazard from this material consists only in proximity—it must be kept behind a heavy wall or shield for about 350 years. After that, it could be given more routine storage but, since it contains traces of fuel, which is a radio-poison, it must be kept isolated from food or water, like so much pitchblende or other radioactive ore. Alternatively, further processing could remove the fuel traces.

The part of the waste problem that is hardest to grasp is the extremely small volume produced. A single aspirin tablet has the same volume as the waste produced by 7,000 kw hours, or one person's annual share of U.S. electric output. If the entire electrical capacity of the U.S. were nuclear and ran at the present rate for 350 years, the total waste produced could be stored in a single pit or vault 200 feet long by 200 feet wide and 200 feet deep. (In an engineered waste storage facility, one would allow some extra space for accessibility and cooling passages.) The oldest cans could then be removed as new ones were added, making a perpetual capacity. This volume of waste is quite small compared to the corresponding volume of ore needed, which would occupy a space about 200 feet by 200 feet by 5 miles long (assuming 0.25% ore and breeders available in 10 years or so). This amount in turn is dwarfed by the environmental impact of producing the same energy from coal: 33 cubic miles would be needed, or the equivalent of a pit 200 feet wide and 200 feet deep extending clear around the earth! In either case, we are perforce handing over a prob-

lem to our successors. The nuclear one is much the lesser of the two.

Care in handling nuclear waste is obviously important, especially before the inert solid form is reached. Thus the public should scrutinize the safety measures at fuel-processing plants. But, compared to the large quantities of other lethal materials necessary to our society, the minuscule volume of the nuclear waste reduces the problem, since it cannot add measurably to the overall risk, and the cost of treatment, transport, and storage is only 1/1000 of the cost of electricity.

Whether we want to store the wastes in a retrievable form or effect some permanent disposal, as in an ice cap or salt mine, is a decision of little importance now. Since we have very little waste so far and we may find a use for it later, I think we should keep our options open for the present by retrievable storage, say for the next 30 years or so, until the need and best means of final disposal become apparent.

The issue of sabotage and terrorism using nuclear materials is raised by those who have imagined a deadly, compressed gas which would disperse itself over the countryside. In reality, these heavy, inert solids are less of a threat than so much dynamite. A fanatic could cause trouble with them, mostly to himself. In the long run, just as with reactors, there are going to be spills, casualties, or even fatalities from nuclear waste. But it is quite clear that no new, unique avenue is offered to terrorists, nor is the total risk measurably increased. As for do-it-yourself atom bombs, we noted above that nuclear fuels have the wrong composition, so that the materials that move in industrial nuclear power channels are of little use. Whether or not a highly sophisticated, heavily financed organization could acquire the array of special talent needed to seize, separate, and purify plutonium and produce a bomb is a separate question which needs intensive study. It is also a question which does not much affect the choice of a civilian energy source, for such an organization could also intercept weapons materials or finished bombs. Internal disruption of this type is a social question the whole world must

face. The terrorist has many types of threat to choose from, no matter what type of power station we have.

Making the choice

The existence of a constantly expanding human population on this hostile ball of clay is fraught with hazard at every turn, and there are *no* completely safe alternatives. All we can hope to do is choose wisely from the paths that are available. Daily experience shows that the public accepts risk when necessary, provided the nature of the risk and the alternatives are understood. Legitimate questions are being asked about the risks of using nuclear power, and they must be answered. The public is not served by those who exaggerate the risks or by those who claim there is no risk. Much of the outcry against nuclear energy is from those who fear it because it is new to them—forgetting that it is 25 years old and that its hazards have been studied from the first—or from those who, hearing of these hazards, have not stopped to compare the dangers of choosing another path.

What I have tried to do in this article is to dispel the vision of unthinkable catastrophe if there *is* an ultimate nuclear failure. The hazard is real—somewhere along the line in a nuclear economy there will be some lives lost, some injuries, and some nasty messes to clean up and decontaminate. But there will be no catastrophe, and we know from experience that radioactive spills *can* be cleaned up. It seems clear that each of the other available paths will have an even higher cost in lives, in dollars, and in damage to the environment. The *real* friends of the earth can assist the public in such balanced assessment.

References

1. USAEC Division of Civilian Applications. 1957. *Theoretical Possibilities and Consequences of Major Accidents in Large Nuclear Power Plants.* WASH 740 (March 1957).
2. C. Starr. 1964. Radiation in perspective. *Nuclear Safety* 5(4):325–35.

PART 4 *Solar and Other Energy Sources*

Farrington Daniels

Direct Use of the Sun's Energy

Abundant and continuing supplies of energy are necessary for modern civilization and they will become ever more important as the number of people on the earth increases and the poorer nations push toward a higher standard of living. Only when men could multiply the work of their muscles many fold by the use of power machines could they have a chance for a successful economy which assured sufficient food, adequate education, and modern conveniences of living. Our present dependence on mechanical energy is astonishing. In agriculture, by way of example, for every calorie of food produced in the United States another calorie of fuel has been consumed in preparation of that food—with tractor ploughing, mechanical harvesting, food processing, motor transportation, and delivery. Thus, under current practices, when we run out of fuel we will run out of food also.

At present the world has an ample supply of fossil fuels (coal, oil, and gas) but they are not distributed evenly and they will not last indefinitely. Our engineering is geared to these fossil fuels which were produced by sunlight long ago and preserved for us through geological accidents. We are fortunate to have had these transportable, concentrated sources of energy, but, looking ahead, we are trying through research and engineering to develop new sources of energy. Atomic energy through fission came as a surprise a quarter of a century ago and its commercial use is now here. A future, but still uncertain, source of energy may be the nuclear fusion of the oceans' hydrogen.

Scientists and engineers should now give more serious attention to finding practical ways to use the sun's energy directly. There is an ample supply of solar radiation falling on the earth to do all the work that will conceivably be needed, and a new supply will always be available

Farrington Daniels was Emeritus Professor of Physical Chemistry at the University of Wisconsin, President of the American Chemical Society, and for the years 1965 and 1966 President of Sigma Xi. Willard Gibbs and Priestley Medalist of the American Chemical Society, he was the author of the most popular textbook of physical chemistry in a preceding generation. His early researches in chemical kinetics were highlighted by his demonstration of the existence of first order—or unimolecular—reactions. Later researches centered on the direct use of the sun's energy. The Sigma Xi-RESA National Lecture printed here deals particularly with the many aspects of utilization of solar energy.

every sunny day—as long as there are people on the earth. But the intensity of solar radiation is low and large areas are required for collecting it; large areas of any material are expensive. Moreover, the solar radiation is intermittent due to regular interruption by night and irregular interruption by clouds, so that costly devices are usually needed for storing the energy obtained when the sun is shining.

In the future, atomic energy and solar energy should be complementary rather than competitive—atomic energy will be produced in large units requiring high capital investment in temperate latitudes and cloudy areas, and near cities, where there is not enough sunlight to supply the concentrated energy needs. Solar energy will be used in the sun belt, perhaps 30° north to 30° south latitude in small inexpensive units widely distributed in rural communities. The sunlight is already distributed in rural communities, while the electricity generated at large atomic energy plants must be carried through transmission lines which are costly if they service many small users separated by large distances.

All our food and all our fuel come from solar radiation through the marvelous process of photosynthesis in nature, which we are just beginning to understand. But the natural chemical storage of solar energy by photosynthesis in agriculture is inefficient. Ordinarily, only about one-fifth of one per cent of the annual supply of solar radiation falling on agricultural land is utilized in plant growth. But in devices, with the direct use of the solar radiation, higher efficiencies are possible—perhaps 40% for the distillation of water, 25% for solar refrigeration, and 10% for conversion into electricity with silicon photovoltaic cells. We can use the sun effectively for cooking, for heating water and heating houses, for cooling and refrigeration, for distilling water, and for producing mechanical and electrical power with heat engines, thermoelectric converters, and photovoltaic cells. All these things can be done, and in fact they have been done.

Yet solar energy cannot now compete economically with cheap fuel and electricity as we have it in industrialized countries. But there are 2 billion people in the world who do not now have any electricity, and most of these people live in the sun belt near the equator. In many of these areas there are few deposits of fossil fuel and no opportunities for hydroelectric power. Solar radiation is often the most important, potentially available, natural resource.

There is a great challenge to scientists and engineers to develop new materials and new ways of doing things so as to make practical use of the sun in raising economic standards. It is not easy to develop new techniques which will compete with well-established techniques and it is difficult to obtain their social acceptance. Electricity produced by atomic energy is now competitive in some areas with electricity generated from coal, but it has taken twenty years of intensive government-supported research and many billions of dollars to achieve this success. According to one philosophy, new developments will come automatically as the technological ceiling rises. But the ceiling does not rise spontaneously; it rises only because individual scientists, engineers, and entrepreneurs like to pioneer—and their efforts make it rise.

The future use of solar energy will be an interesting development to watch, not only in practical results achieved but also in a study of the efforts which are being made to accelerate its use. Very little support for solar energy research has been available from government or industry. The U.S. government has financed some research on the solar distillation of salt water; the Rockefeller Foundation has made a substantial grant to the University of Wisconsin for solar energy research with particular emphasis on trying to help the rapidly developing countries; the United Nations in 1961 organized a large symposium in Rome on new sources of energy. A few countries are supporting solar energy research, important among which are the USSR, the USA, France, Australia, and Israel.

The Solar Energy Society, with headquarters on the campus of Arizona State University at Tempe, Arizona, has 1000 members in 87 different countries who are endeavoring to increase the use of solar energy. The Society publishes a professional journal and a news bulletin, and it holds annual international meetings. Is it possible for a worldwide group of scientists and engineers and solar enthusiasts to advance significantly a new field of science and technology, and to influence its impact on human affairs?

Solar energy applications are certain to be important in time. If those who advocate a new development are only five or ten years ahead of the time of general acceptance, their work is probably considered praiseworthy. If they are fifty years ahead of their time, their efforts may probably be considered misplaced. Possibly the development of solar energy will be considered satisfactory if major practical applications come in twenty years—the time it took for atomic power. The span of twenty years, starting arbitrarily with the founding of the Solar Energy Society in 1955, is now more than half gone, but already real progress is beginning to appear in some areas. After several pilot demonstrations, solar distillation of ocean water is now producing 2 million gallons of fresh water a year. Solar hot water heaters are sold in many tropical countries. Active research is going on with solar heat engines and solar refrigerators. A recent report from the UN states that in Africa, south of the Sahara, fractional horsepower solar water pumps should have the highest priority. Unexpectedly, the direct production of electricity from sunlight through silicon photovoltaic cells has become a multimillion dollar solar business for supplying auxiliary power to rockets and vehicles being used for the exploration of space. A very substantial amount of work on solar energy applications has been published since 1949, when the author participated in an early Sigma Xi lecture tour on "Atomic and Solar Energy." [2]

Solar energy research is unique in several ways. It calls for interdisciplinary attacks by physicists, chemists, chemical engineers, mechanical engineers, electrical engineers, astronomers, meteorologists, and biologists. Much important work can be done with standard, inexpensive equipment and it has an appeal to some workers in that it can be useful now in some parts of the world.

The use of solar energy is not new. A solar still constructed of many wooden trays with glass covers occupying over an acre was built at a nitrate mine in north Chile in 1872, and it operated over a period of 40 years. A solar steam engine was demonstrated in Paris in 1878, and Ericcson built a small solar hot air engine in 1870. Several other solar operated devices have been tried, but they failed to achieve continuing economic success. The opportunities should be re-examined now, however, because we have new ideas in science, accumulated experience in engineering, and a broader knowledge of worldwide opportunities and needs. We have new materials to work with, such as sunlight-resistant plastics and new semiconductors of high purity. Moreover, we realize that, although a solar operation may not be able to compete now in our economic system where fuel is cheap, labor costs are high, and capital is easy to obtain, it may be able to compete in some other parts of the world where fuel is scarce, labor is abundant, and capital and foreign exchange are difficult to obtain.

The purpose of this lecture is to call attention to the interesting research opportunities which are available in both basic and applied solar science, and to point out important unsolved problems in solar science and engineering.

For those who may wish to undertake research problems in the field of solar energy, there are several books and reports of symposia with ample references [1, 3, 4-10].

Current researches on solar energy are reported quarterly in *Solar Energy* [11], and the first ten years of this publication are available in a single bound volume [12].

Solar radiation

On a sunny day, the solar radiation intensity varies from zero in the early morning and late afternoon to a maximum at noon of perhaps 0.5 to 1.5 calories per cm² per minute depending on the latitude and the season. Figure 1 [13] gives a typical record of solar radiation throughout the day in Wisconsin at different times of the year, and Figure 2 [14] gives daily radiation in calories per sq. cm. in July throughout the world. This reference [14] gives a world survey of solar radiation by months. Solar radiation is greatly affected by clouds and by local weather conditions. For example, the solar radiation received along a foggy coast may be markedly less than that a few miles inland.

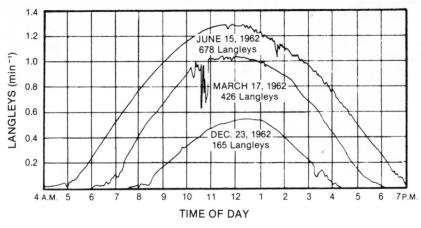

Figure 1. Typical solar radiation on clear days at Latitude 43 N.

In sunny climates one can make a rough estimate of solar radiation as averaging around 1 calorie per cm^2 per minute (i.e., one "Langley" per minute) or 500 calories per cm^2 per day or nearly 500,000 calories per $ft.^2$ per day. Fifteen square feet of this radiation give per ft^2 15 kilocalories per minute or a heat rate of 1 kilowatt. The roof of a house (1000 ft^2) receives 70 kilowatts while the sun is shining or 560 kilowatt hr of heat in an 8-hr day. This is equivalent to 150 lb of coal or 14 gal of gasoline. If converted directly into electricity with a conversion efficiency of 10% (and such conversion is now technologically possible but extremely expensive), the sunlight striking the roof of a house could average 7 kilowatts of electricity while the sun is shining, or 56 kilowatt hr of electricity on a sunny day. This is enough to operate the electrical equipment of a house, but the solar batteries and the storage of electricity are prohibitive in cost. An acre of sunlight at 10% conversion could give 280 kilowatts of electricity and a square mile 180,000 kilowatts. These are potentially large quantities of power, but they are bound to be expensive. To cover an acre of land with any collecting material is very costly.

For the long range future, it is comforting to realize that theoretically it may be possible to make portable fuels in sunny, wasteland deserts. For example, in north Chile, there is a desert 180 miles long and 100 miles wide which recieves nearly as much heat from the sun in a year as is produced by the burning of all the coal, gas, and oil used in the whole world in a year. Here is a tremendous challenge (which will be discussed later) to develop methods for producing concentrated fuels for transportation to substitute for our present fossil fuels.

Measuring instruments

Instruments are available for measuring solar radiation, most of which depend on measurements of electrical currents produced by thermocouples, or the arms of a Wheatstone bridge, attached to a reflecting and a heat-absorbing surface. Simple photovoltaic cells are also satisfactory when calibrated against the standard heat-measuring instruments. Chemical actinometers have not been used much, although they have some advantages for long integrated exposures to sunlight. For focused sunlight of high intensity, small calorimeter cans holding water and a thermometer are simple and often accurate enough.

Collectors

Solar collectors are of two types. In flat plate collectors, black, radiation-absorbing plates are covered with transparent covers of glass or plastic which trap the heat and reduce the loss by infrared radiation. In the more expensive focusing collectors, curved, reflecting surfaces concentrate onto a small target the solar radiation which strikes the whole collector. The flat plate collectors commonly give temperatures up to about 100 °C, while the focusing collectors can give localized heating in a small area from hundreds of degrees up to 3000 °C and higher, depending on the optical precision of the curvature and the reflectivity of the surface. The focusing collectors must be moved to keep them always turned toward the sun, while the flat plate collectors are usually stationary and, accordingly, can be cheaper and much larger in size. The flat plate collectors are mounted horizontally or are tilted to the south at an angle approximately equal to the latitude. The flat plate, nonfocusing collectors respond to all the sunlight, diffuse and ground reflected as well as direct, but the focusing collectors can use only the direct rays of the sun and are not satisfactory with cloudy or hazy skies. With long cylindrical, focusing collectors, it is possible to mount larger areas than with circular collectors, but the temperatures obtainable are much less, though still well above the temperatures obtainable with flat plate collectors.

The reflecting surface may be of aluminized plastic, with the reflecting surface on the back, or it may be composed of a mosaic of many small glass mirrors. The glass or plastic reflectors are attached with a plastic adhesive to a parabolic shell made on a master mold and constructed of fiber glass cloth and liquid plastic or burlap cloth and plaster of Paris. The high capital cost of solar collectors is the economic bottleneck in many plans to use the sun directly. Details of construction and careful analyses made of the efficiency and reliability of many different types of collectors are available [15].

Devices used for turning the collectors to follow the sun include hand operations of rotating and tilting, clockwork mechanisms, electric clocks, steady leakage of a reservoir of liquid, and bimetallic strips or photo cells and electrical relays which respond when the focused sunlight moves beyond the range of the sensing target.

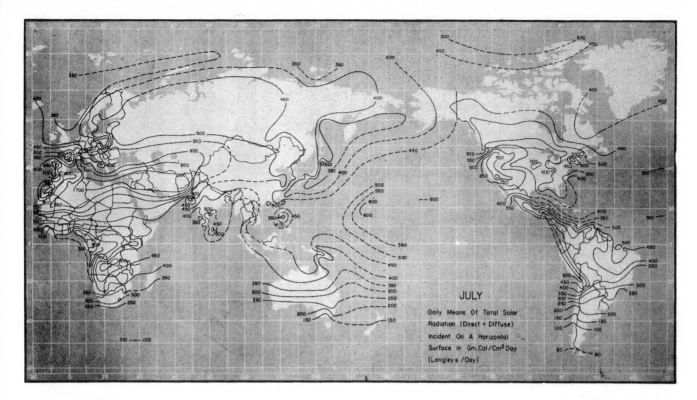

Figure 2. Average daily total (direct and diffuse) radiation on a horizontal surface throughout the world in July.

Efforts have been made to develop inexpensive focusing collectors [16] which can be made by hand, locally, without power machines. It is easy to build a focusing collector 4 ft in diameter which will deliver half a kilowatt of heat to a target 8 inches in diameter, but it has been difficult to perfect the techniques so that 70% of the solar radiation hitting the collector is transferred to a target 3 or 4 inches in diameter. Results have been obtained with shells of burlap cloth costing 6 cents per ft², impregnated with plaster of Paris and mounted in square frames of thin-walled electrical conduit tubing which are bent and bolted. The shells are set on parabolic mounds of wet sand which are shaped by a rotating parabolic metal edge attached to plywood. The parabolic mounds are covered with a half inch of Portland cement and sand smoothed by the shaping tool. After sandpapering and shellacking, the shells, cast on the parabolic mound, are lined with aluminized plastic. The total cost of materials for these reflecting shells can be less than $1 per ft² of collector area. The shells may be lined also with a mosaic of glass mirrors 1 inch square, at somewhat higher cost.

Eventually, when the demand for small focusing collectors becomes large enough to justify expensive machinery for mass production, it should be a simple matter to stamp out cheaply sheet iron parabolic shells similar to automobile fenders. Apparently, focusing collectors of 4, 5, and 6 ft will be practical. Larger ones are more difficult to transport and operate.

As in many applications of solar energy, minor problems such as the removal of dust in some areas can be serious.

Cooking

Ten years ago, solar cooking appeared to be a simple way of using the sun to substitute for cow dung and other scarce fuels in certain semi-desert regions. Plastic solar cookers operating with about 500 watts of solar heat were developed and tested in Mexico with the cooperation of anthropologists [17]. The field tests disclosed structural weaknesses, particularly in heavy winds, which were corrected. There was no difficulty in showing the women how to adjust the cooker so as to follow the sun. The cookers were technologically satisfactory, but they were not socially accepted. They can be considered only in areas where bright sunlight is reliable, wood is not available close by, and petroleum fuels are expensive. The technical problems involved in solar cooking have been carefully analyzed [18].

One of the important research areas to be explored is that of heat storage, which will be discussed later. The social problems might well be solved if, instead of heating food with the sun at noon time, the housewife could expose a pot of chemicals or a block of high heat capacity material for any two or three hours during the day while the sun is shining, and store a kilowatt-hour of heat at 200 °C in a package weighing no more than 25 lb for cooking in the kitchen in the evening. Such heat storage is difficult, but it should not be impossible. At present, it must be concluded that, in spite of its simplicity and obvious advantages, solar cooking has not been successfully introduced in Mexico.

Heating water

Solar hot water heaters are manufactured in Israel and sold in many countries around the world. It has been reported that there are 300,000 solar water heaters in Japan. Formerly there were many in use in Florida and California, but they have nearly been driven out by cheap natural gas.

One common type consists of sheet metal painted black with copper pipe soldered to the back. A reservoir tank is placed above the solar heater and the hot water circulates and is stored in it. An auxiliary electric heater in the tank is a great convenience in cloudy weather. A tube-in-sheet collector has been developed in which two sheets of metal are seam-welded in strips and alternate channels blown out with hydrostatic pressure into pipes for circulating water. In freezing weather, the bursting of metal pipes is a hazard.

Black plastic pillows for heating water on south-facing roofs are manufactured in Japan for about $10. The great increase in the number of swimming pools in the US may lead eventually to a potentially large market for solar water heating.

Heating houses

Progress in the solar heating of buildings has been slow. From the point of view of conserving the world's supply of fossil fuels this is unfortunate, because over one-fifth of our annual consumption of fuel goes to space heating and the requirements are only for low temperatures which can be met with solar heating. One reason for the lack of interest in the solar heating of houses is the fact that it is so easy to heat buildings with the burning of fuel in simple, inexpensive equipment. There are only a few solar heated houses with heat storage in the world now, and nine of these were described in detail at the UN symposium in Rome in 1961 [19]. The whole field of house heating was critically reviewed by Löf. In some houses, water was circulated through the solar heat collector and stored in tanks; in others, the solar-heated air was circulated through a heat-storage pebble bed of cheap gravel. In earlier work by Telkes, the heat was stored in cans of sodium sulfate and water, which has a transition temperature of 32 °C. All the houses were fairly expensive, multi-room houses and all the houses used electrical power for circulating the air or water.

It would be desirable to direct research attention to the solar heating of inexpensive one-room houses, perhaps with plastic collecting channels on the south roof or wall, or preferably along a south-facing hill below the house. Air circulation through underground pebble beds of gravel could be the simplest way of storing heat, particularly if a substitute for a motor driven fan can be developed where electricity is not available. Effective, hand-operated fans have been produced for civil defense shelters.

One serious difficulty with the solar heating of houses arises because large collecting areas of several hundred square feet are architecturally unacceptable. On a flat roof, they can be located out of sight.

An attractive feature of solar heating is the fact that the equipment can be used for cooling as well as heating. A black solar collecting surface will radiate heat to the night sky and also the surface will be cooled by the circulation of the cooler night air. In summer, the "cold" can be stored in the same storage units that the solar heat is stored in in the winter. Moreover, the heat storage units can also be used in connection with heat pumps, which

are becoming more widely used because they produce heat more efficiently from electric power than do ordinary electric resistance heaters.

Drying

Solar drying of crops in the field is standard practice in agriculture. If hay is stored before drying, spontaneous combustion may occur, and if grains are stored without drying they undergo a serious loss of quality. The taste of tobacco is very sensitive to the temperatures and times of drying of the leaves stored in sheds. To insure an earlier harvest and proper quality, fuel-heated or solar-heated air may be blown through the stored agriculture crop. Solar-heated air may also tend to eliminate dirt, overheating, and damage from long continued fungus or bacterial growth in unheated crops. The factors involved in the drying of crops have been carefully studied and reports are available on the construction and tests of solar heated air [20]. In one of the simplest heaters, ducts for blowing air are placed directly under a black-painted metal roof. Other types include channels of transparent plastic with bottoms of black plastic laid on nearby ground. Talwalkar and Duffie [21] have made a thorough study of the complicated flow of heat into a porous bed from radiant heat striking the top surface and from a stream of warm air with low relative humidity flowing through the bed. The simultaneous vaporization of water in the upper region of the bed and the diffusion of water vapor from deeper in the bed are also analyzed.

Solar radiation could well be used in sunny climates for industrial drying, remembering that one square foot receives about 500,000 calories per day and about 580 calories are required to vaporize 1 gram of water. The evaporation of water to obtain salt from the ocean has been used, of course, for a very long time.

Solar furnaces

Solar furnaces are used for producing very high temperatures for scientific investigation [22, 23]. In 1774, Lavoisier used a solar furnace with a glass lens as tall as a man for obtaining high temperatures in chemical studies. In 1957, a survey was made of 21 solar furnaces operating in the US [24]. Many laboratories are using old army reflecting searchlights, 5 ft (150 cm) in diameter, which give temperatures over 3000° in a focal spot about 0.6 cm in diameter. In most experimental work, it is inconvenient to have the heated target moving continuously with the sun, and so a moving heliostat of flat mirrors reflects the solar radiation onto the stationary parabolic focusing collector.

Trombe and his associates at the Laboratory of Mont Louis, France, have been leaders in the development of large solar furnaces [7]. A heliostat 30 ft on a side with 3500 plane glass mirrors produces at the focus 200 calories per second or 0.8 kilowatt per cm². A much larger solar furnace will soon be in operation.

Large parabolic reflectors and their heavy mountings are very expensive. They can be used only on clear, sunny days, and the hot zone is confined to the front of a very small area. They have the great advantage, however, of freedom from electric and magnetic fields and freedom from contamination from fuel gas, electric heaters, and crucible material. The mass of unmelted material is the only material that contacts the pool of liquid.

Solar furnaces provide an excellent way of studying physical and chemical properties of refractory materials at very high temperatures. They are used in melting and sintering ceramics such as zirconia and in purifying materials by zone refining and by driving out relatively more volatile materials. They have been used also for the high temperature production directly from their ores of pure chemicals such as the oxides of zirconium, beryllium, and tungsten.

Solar furnaces can be used also in carrying out at room temperatures photochemical reactions in which the solvent is transparent to the solar radiation [25].

Solar furnaces can perhaps be used in basic scientific studies of chemical kinetics using flash pyrolysis with instantaneous heating of small black particles followed by quick chilling, thus possibly permitting the production of intermediate, unstable compounds which are not observed with ordinary heating and cooling. If it should become possible to make parabolic focusing collectors of sufficient optical precision and low enough cost (perhaps less than $1 per ft²), it might be possible to use them in some special locations for ceramic and metallurgical manufacturing at intermediate temperatures of perhaps 500 to 1500 °C.

Already solar furnaces are being studied in connection with the exploration of outer space, such as, for example, the heating of rocks on the moon to release water and the operation of heat engines on space vehicles, using plastic parabolic shapes released after the vehicles are beyond the region of frictional resistance caused by the atmosphere.

Solar distillation of water

The production of fresh water from sea water has become a subject of great popular interest and concern. As the world's population grows and the pressure for living space increases, the semi-desert and even the desert lands are being given attention. The supply of water is the limiting factor for growth in many parts of the world and there are worries for the future in some areas where pollution and increasing salt concentration may threaten a water supply that was formerly complacently considered adequate. Intense research is going on along many different lines—distillation, freezing, solvent extraction, vapor compression, centrifugation, and electrodialysis. Reverse osmosis is a promising approach in which a plastic membrane with pores of just the right size permit pure water vapor to pass through when high pressure is applied to the salt water. Distillation with atomic heat is another solution which appears to be very cheap (20 cents per 1000 gallons), if the operation can be carried out on a very large scale with capacities of hundreds of millions of gallons per day, and very large amounts of electricity can be sold as a by-product.

In most parts of the world, an abundant supply of good water is taken for granted and bought for a few cents a thousand gallons. What other high purity commodity weighing four tons can be obtained so cheaply? Irrigation water from rivers can often be obtained in Arizona on agricultural land without piping for around $4 per acre ft which amounts to about 0.5 cent per 1000 gal. Irrigation water pumped from wells may cost about four times as much. The present tentative goal in producing fresh water from sea water in moderate sized installations is less than $1 per 1000 gal, and it is being met.

Solar distillation is probably the cheapest method for small units of a few gallons per day up to perhaps 10,000 gal per day. The other methods require trained technical operators and special equipment such as multiple stills, compression engines or refrigerators so that a reasonably large production is necessary to meet the minimum cost of the investment.

In the solar distillation of water with simple equipment, the maximum amount of water distilled is limited by the heat of vaporization of water and the total solar heat received. To heat 1 gram or 1 cc of water to 50 °C and evaporate it requires over 500 calories of heat. Assuming 500 calories of solar radiation per day per cm² and a 100% efficiency still, with no heat losses, the theoretical maximum yield would be less than 1 cc per cm² or 1 gal for an area of 4 ft². If a solar still has an efficiency of 33%, an area of 12 ft² would give a capacity of 1 gal per day. Many stills have an efficiency somewhat higher than this in clear sunshine in warm climates. Fuel-fired stills have temperatures high enough to permit multiple effect distillation so that some of the heat of condensation can be reused to evaporate additional water, but the capital investments are higher.

A considerable amount of research has been carried out on the solar distillation of water and progress is being made in the construction of practical solar stills [26, 27]. At the meeting of the Solar Energy Society in March 1966, reports were given on stills up to capacities of several thousand gallons per day on Greek islands in the Mediterranean, and in Australia, Spain, the US and the USSR. Löf [28] has made a special study of the heat balance in solar stills, including the losses of heat from the cover and the salt water, from radiation and convection, and from conduction into the ground.

The simplest solar still consists of a black tray of water tightly covered with a transparent roof which allows the solar radiation to pass through and heat up the water. The water vaporizes, condenses on the air-cooled roof, and runs down on the inside and is collected in a trough which drains into a storage vessel. In the research project described here, the emphasis has been on the development of a family-sized still, which is simple to build and operate and which involves the lowest possible capital investment. Figure 3 shows plastic solar stills which have been operating successfully at Tempe, Arizona, for a year and a half with an efficiency of about 25% in winter and 40% in summer. Many of these stills are about 12 ft long and 2½ ft to 5 ft wide with a tray and side troughs of concrete, a wooden ridge pole, and a cover of weatherable Mylar stretched over the central pole and held firmly by long plastic sand bags squeezed into the drainage troughs. They produced 2–3 gal per day on bright summer days and the cost of materials (ex-

Figure 3. Small plastic solar water stills.

cluding labor costs) was about $15, giving a capital investment as low as $5 for a capacity of 1 gal per day. Earlier models were tested in Wisconsin, Florida, and on the islands of Rangiroa and Guam. It was found essential to keep the stills tightly sealed with the sand bags because any air escaping through leaks carries moisture which is lost. Another difficulty was due to the condensation of water on the plastic roof in the form of drops which scatter the light and tend to fall back into the salt water tray. This problem was solved by scratching the plastic with waterproof sandpaper to give a water-wettable surface, as suggested by Frank E. Edlin. Heat losses under the tray were lessened by using a layer of sawdust set with Portland cement. Rain drainage was effected by introducing an over-flow at the end of each trough. In some stills, a rubber floor was used, giving a higher temperature than the concrete floor because its heat capacity is much less. The daytime distillation was appreciably greater. However, the concrete stills gave a higher night-time distillation, and on hot summer days the two types of stills gave about the same efficiencies. The rubber-floored stills are considerably more expensive and their drainage pipes are more apt to leak.

The efficiencies obtained in summer, up to 40% and higher, involved water-tray temperatures of 60–65 °C. The plastic covers appeared to resist deterioration for at least a year and a half, and longer tests are continuing.

There are many problems in the practical development of solar stills to be solved by further research. Horizontal stills are easy to operate, but stills tilted toward the sun give higher radiation intensities and greater efficiencies (except when located near the equator). Flat land is usually needed for other purposes, but cheaper sloping, hilly land with southern slopes is often available, as for example in the neighborhood of Hong Kong.

Plastic stills are potentially cheaper than glass-covered stills and easier to construct, but are less permanent and less transparent. The glass stills are heavier and breakable and more difficult to transport. Great progress has been made in the industrial development of suitable plastics which will withstand years of exposure to ultraviolet sunlight, and to high temperatures while bathed con-

tinuously with a film of water. The plastics must also withstand vibrations caused by the wind. Further longtime testing under operating conditions is needed. Inflated stills and stills with covers mounted over a framework have been tried.

Stills with deep water layers are easier to operate, but the shallow water stills reach a higher temperature and give higher efficiencies. At the higher temperatures, the vapor pressure of water is higher and, since the total pressure is one atmosphere, the ratio of water vapor to air is greater. The heat transferred by water from the tray to the cover is necessary for condensation of the water, but the heat transferred by the air fulfills no useful purpose and increases the heat losses and lowers the total efficiency. In the deeper stills, much of the heat is conserved and used for distillation at night, but the distillation is then at a lower temperature and is less efficient.

Solar distillation can be used for brackish water or polluted or dirty water as well as for ocean water. A very inexpensive emergency solar still has been developed for use by travelers in arid lands [30]. Long after the last rain, the ground still contains an appreciable amount of moisture. A hole is dug about 3 ft in diameter and 2 ft deep, and a piece of clear plastic is laid across the top and held in position with the sand piled around the edges. A rock is placed in the center to form an inverted cylindrical cone and an open can is placed below it. The solar radiation heats up the ground and drives out the water which condenses on the plastic and drips down into the collecting vessel. If the soil is too dry, cactus or other plant material may be cut up and placed in the hole. It is usually possible to obtain a quart or two of drinking water by this technique.

It is possible also to obtain drinking water from the moisture in the air. A water-absorbing material such as calcium chloride, silica gel, or ethylene glycol is exposed to the air, and it takes up moisture, particularly at night. The material is then heated with solar radiation in a modified solar still, or with focused solar radiation if a higher temperature is required for the release of the water.

Cooling

Cooling may well be one of the first widespread applications of solar energy. The demand is great for refrigeration and air conditioning, and the present methods are expensive. In many isolated tropical communities, the cost of electricity is too high. According to one statement, solar refrigerators making ten pounds of ice per day without operating costs would be important in Burma [31]. Such a refrigerator has been demonstrated [32].

The principles of absorption-desorption solar cooling are well established [33] and fuel-operated refrigerators have long been on the market. In electrically operated refrigerators, a vapor such as ammonia is condensed to a liquid with a motor-driven pump, and the heat evolved is removed with circulating air or water at room temperature. The liquid is then vaporized in an insulated box and heat is removed by the vaporization to give the cooling effect. In solar refrigeration, the cycle is similar except that the pressure is built up by heating a concentrated solution of ammonia to give a high vapor pressure, instead of compressing the vapor mechanically. There are two connecting, gas-tight vessels, one of which contains liquid ammonia and the other a very concentrated solution of salt in liquid ammonia. The salt solution has a much lower vapor pressure, and the liquid ammonia vaporizes in its compartment, thereby cooling it, and dissolves in the salt solution contained in the other compartment. The system is regenerated by using focused solar radiation to raise the temperature of the salt solution to such a high temperature that the vapor pressure of ammonia in the solution exceeds the vapor pressure of the pure liquid ammonia in the second compartment. In this way, the operating cycle produces cooling by evaporating ammonia as it goes into the concentrated solution of salt, making it more dilute; and the solar regeneration drives out the ammonia from the diluted salt solution to produce pure ammonia and leaves a more concentrated solution. If the operation is intermittent, the apparatus can be simple and inexpensive, but the hand operation is troublesome. With more complicated apparatus, the operation can be made continuous and simpler.

A good salt solution for this solar refrigeration is 70% sodium thiocyanate in 30% liquid ammonia, which has a vapor pressure less than one atmosphere at room temperature, whereas the vapor pressure of pure liquid ammonia is about 10 atmospheres. Ammonia is chosen because of its high heat of vaporization, 300 calories per gram (second only to water), and its low boiling point, $-40°C$. The solution of sodium thiocyanate in water is cheap, nontoxic and nonexplosive, nonviscous, a good conductor of heat, and noncorrosive to iron. Its physical chemistry properties have been studied [34]. This very concentrated salt solution has interesting properties which are more like those of a fused salt than those of an ordinary solution. This solution is being tested in a laboratory study of solar refrigeration and further studies are being made on the coefficient of performance (COP) which involves both the performance of the refrigerator and the solar collector. Thus,

$$COP = \frac{\text{heat absorbed by vaporizing coolant}}{\text{solar heat incident on collector}}$$

Experiments with a solution of ammonia in water [35] gave approximate heat losses: reflectivity 30%, absorptivity 5%, thermal conductance 3%. About 40% of the total solar radiation intercepted by the collector was picked up by the ammonia-water solution and 24% of the heat was lost to the surroundings by radiation and convection. The total COP was 0.16.

Solar air conditioning is simpler than refrigeration in that the temperature differences are not so great and flat plate collectors can be used rather than focusing collectors, and there are more cooling liquids to choose from because the temperature does not have to be so low. A cycle has been studied [36] in which a concentrated solution of lithium bromide absorbs water vapor and causes liquid water in another compartment to vaporize and produce a cooling effect. The lithium bromide solution is concentrated again by heating the diluted solution with solar radiation and the system is operated on a continuous basis. A laboratory was partly air conditioned by the sun for a while during these tests. For ordinary use, the solar collector needs to be nearly as large in area as the floor space which is to be air conditioned.

There should be an interest in developing solar cooling for vegetable cellars in tropical climates, where a temperature only as low as 60°F and a humidity of 60% could save much of the vegetable crop such as tomatoes and potatoes. In some tropical areas, as much as 25% of a crop is frequently now wasted because of spoilage.

Another approach to the problem of solar cooling is by reduction of the humidity of the air by adsorption of moisture followed by evaporation of water into the dehumidified air with cooling. Absorbing agents, silica gel or dehydrated salts may be used. Löf [37] used a spray tower with ethylene glycol to absorb moisture from the air and regenerated the ethylene glycol with an air stream heated by the sun.

Heat engines

Solar heat engines of 0.1 kw could be very useful in some areas for irrigation, for communications with radio or television, and for village industries. A considerable effort has gone into the development of solar engines and some progress has been made [38].

Tabor [39] demonstrated a 5 horsepower solar engine operated by reflectors at the bottom of inflated cylindrical tubes 5 ft in diameter and 20 ft long. The solar radiation was focused onto a central pipe under conditions such that the collector did not need to be moved through the day. The turbine was operated on the vapor of monochlorobenzene with an efficiency of 10 to 15%.

There seems to be considerable hope in hot air engines, using the Stirling cycle [40]. They can be operated by relatively unskilled labor because there are no boiler hazards and no necessary treatment of the water for the boiler. In small sizes they can be considerably more efficient than steam engines. One of the difficulties lies in getting sufficiently rapid transfer of heat from the sun across the cylinder-head into the air contained in the cylinder. According to one suggestion, an improvement

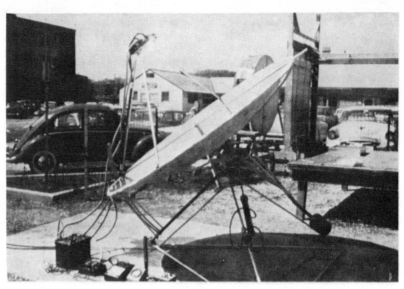

Figure 4. Thermoelectric solar converter.

can be made by using a transparent window of quartz or pyrex which allows the solar radiation to be focussed directly inside the engine onto the moving piston. Farber [41] has constructed and tested a 1/5 horsepower solar Stirling hot air engine.

The production of electricity with small heat engines is expensive and cannot compete with conventional electricity from large power plants. But it may eventually become competitive with electricity at 5 to 15 cents per kilowatt hr produced by small diesel engines in isolated communities where fuel transportation is expensive, repairs are costly and chargeable to only a few customers, and transmission lines cost $1500 per mile.

A promising attack has been made on the solar production of larger scale power using power ponds [42]. Water ponds about 6 ft in depth and an acre or more in area with black bottoms can be heated, under special conditions, to give temperatures up to 90 °C and operate low pressure steam turbines. The heated water is prevented from rising to the surface and vaporizing with large heat losses by using a layer of a very concentrated salt solution at the bottom. Even when heated, this salt solution has such a high density that it remains stratified. Such power ponds have the advantage that large areas of glass or plastic are not required for the collector and the heat is stored for continuous operation at all hours of the day and night.

All heat engines and other devices that produce work from heat are limited by the Carnot relation according to which the maximum amount of the heat, Q, which can be converted into useful work, W_{max}, is equal to the heat supplied at the higher temperature multiplied by difference in temperature of the hot part T_2 and the cold part T_1 divided by the temperature of the hot part, thus

$$W_{max} = Q \frac{(T_2 - T_1)}{T_2}$$

Through high-precision focusing, the sun's energy can be delivered at high temperatures, thus giving a high efficiency for conversion of the heat into work.

Thermoelectric converters

The sun's radiant heat can be converted directly into electricity by heating one end of a thermocouple and cooling the other. The thermocouple is made very simply by joining two unlike metals or semiconductors. The production of electrical work is limited by the Carnot equation just discussed. To be effective, the thermocouple should give a large electromotive force per degree difference in temperature and it should have good electrical conductance but poor thermal conductance. Metals and metal alloys give reliable thermocouples but they have low voltages. About 15 years ago, it was difficult to exceed 3% in the conversion of heat into electrical work. Advances in solid state physics have led to new thermocouples of considerably higher efficiency using semiconductors which have higher voltages, and a better chance for obtaining good electrical conductance and poor thermal conductance because the conducting mechanisms are different. Highly purified elements such as antimony, bismuth, germanium, and tin combined with silicon, sulfur, or tellurium are doped with impurities of different valence to give positive and negative electrodes. There are problems in mounting so that good electrical contacts and mechanical structures will be maintained in spite of the extreme conditions of thermal expansion caused by heating one face of the thermocouple box to several hundred degrees with focussed sunlight. The front surface is of thick copper or other thermally conducting material to make the whole area uniform in temperature, and the surface is blackened with a selective radiation coating, discussed later. The operation temperature is made as high as possible in order to obtain high thermodynamic efficiency but it must be kept low enough so that the impurities will not diffuse on long, continued use from the region of high temperature to the region of low temperature. This deterioration with time at high temperatures is an important problem to be solved.

The present status of the production of electricity through thermoelectric conversion of solar radiation was discussed at the UN Conference on New Sources of Energy [43] with an introduction by Baum.

One thermoelectric solar converter, shown in Figure 4,

was reported [44] in which a 6 ft diameter plastic focusing collector intercepted 2 kilowatts of heat and focused 1 kilowatt of heat onto a 5-inch square thermoelectric generator of germanium, bismuth and lead tellurides developed experimentally by the Westinghouse Electric Corporation. The temperature was brought up to 500 °C. The rear of the box, 1 inch thick, was water-cooled. Forty watts of electricity were produced at 4 volts, which pumped water and charged a storage battery. Since the voltage was low, the storage battery was charged in parallel and discharged in series for operating a motor. The 40 watts of electricity from the 2 kilowatts of intercepted solar heat is an efficiency of only 2% and the efficiency became less after continued operation at 500 °C.

Thermionic converters of solar radiation are also worthy of further research. Electrons are driven across a very narrow gap in a high vacuum by heating a specially prepared surface to 1500 °C or higher. The high temperature is attainable with sunlight only with expensive, high precision collectors. Efficiencies comparable with those of thermocouple conversion have been obtained.

Photovoltaic converters

The most spectacular development in the use of solar energy has been the silicon photoelectric converter or "solar battery." These barrier layer cells have a P-N junction between parts of a highly purified semiconductor containing impurities in which the positive charges or holes are free to move in the P-layer and the electrons are free to move in the N-layer. When light enters the crystal, electrons are released which migrate to an electrode, pass through the connecting wire, operate a motor, and join up with the positive holes at the other electrode. No material is used up and the cells continue to operate indefinitely. These photovoltaic cells have been widely used for photographic exposure meters. In 1953, it was stated that an efficiency of 1% in the conversion of sunlight into electricity was all that could then be attained. Two years later the Bell Telephone Company laboratory announced the silicon "solar battery" with a conversion of 11% and it is now used for auxiliary power and communications on most rockets and space vehicles. Efficiencies appreciably higher than this have been reported in special laboratory experiments.

Only the solar radiation below 11,000 A is usable in these silicon cells. They do not depend on heat and the Carnot limitation does not apply. These cells are light in weight, have no moving parts and will operate directly, without a focussing or special collector. They are still extremely expensive, costing around $2 per square inch. They are bound to be expensive because, though abundant in the earth's crust, the element silicon has to be chemically prepared, highly purified, and grown as a single crystal, cut with a diamond saw and then exposed to vapors of a pentavalent compound like arsenic and a trivalent compound like boron. Intensive research has been devoted to trying to find cheaper ways of making these photovoltaic cells. If polycrystals made by vapor plating or electroplating could be used instead of the single crystal and its cutting with a diamond saw, the cost could be reduced greatly. Cadmium sulfide has been used in polycrystalline form but its efficiency is considerably less. Silicon is especially advantageous because it responds to all the visible light and a considerable part of the short infrared light.

Photovoltaic generators for converting sunlight were discussed at the UN conference in Rome in 1961 [45].

Photochemistry

Instead of using the sun's energy as heat, it is possible to use it as *light* in bringing about photochemical reactions to give useful products which can be stored and transported and used as fuel. Photochemical reactions are not subject to the limitation of the Carnot relation, and a temperature difference is not required. However, photochemical reactions respond only to the ultraviolet and visible portion of the solar radiation. The infrared portion, which comprises about half of the total energy, is too low in energy intensity to activate molecules for chemical reaction. The basic laws of photochemistry are well understood but very little research effort has been directed toward the photochemical use of sunlight [47]. A committee of the National Research Council in 1957 formulated criteria for selecting photochemical reactions of possible usefulness in obtaining energy directly from sunlight [48]. Perhaps the most important quantity in photochemical investigations is the quantum yield Φ which is defined as follows:

$$\Phi = \frac{\text{number of molecules reacting}}{\text{number of photons absorbed}}$$

The amount of photochemical reaction depends on the intensity of light, the thickness of the absorbing layer, and the concentration of the absorbing material, but when Φ has been determined experimentally, it is possible to calculate the amount of chemical product from the amount of light under a variety of conditions.

For every photon of light absorbed, one molecule becomes excited and this excited molecule may undergo chemical reaction, but it is more probable that it will lose its energy by collisions with other molecules with a resulting increase in temperature. A quantum yield less than unity may be due not only to losses as heat but to a reverse reaction, or to other secondary reactions, or to absorption of light by other absorbers. On the other hand, if a long series of secondary reaction occurs, giving a chain reaction, the quantum yield may be very large—up to a million molecules per photon in the case of hydrogen and chlorine exposed to light.

The number of photons in light may be calculated by dividing the total energy of the light by the energy of one photon, called the quantum of energy ϵ. It depends on the frequency of light v and Planck's constant h as given by the relation

$$\epsilon = h\nu$$

The number of photons available in an acre of sunlight and the number of molecules of photochemical product can be very large. For example, it is calculated [47] that if a photochemical reaction with a quantum yield of unity and a molecular weight of 100 is assumed, and one-tenth of the sunlight of 500 calories per cm² per day is absorbed and is chemically active, and the effective wavelength is 5000 A, then the photochemical product of the acre should be 2×10^{28} quanta or 2×10^{28} molecules or 3.6 tons of material per day. Nothing approaching this photochemical productivity has ever been achieved, and yet the assumptions are not unreasonable. A great deal of basic research in photochemistry will be necessary in order to find suitable photochemical reactions for using sunlight.

For the storage of the sun's energy in photochemical products, several special requirements must be met. The photochemical reaction must be endothermic with a quantum yield of unity or only slightly less than unity and the reverse reaction must take place spontaneously in the dark with the evolution of energy. Most photochemical reactions are exothermic reactions which give the products spontaneously, and the activation by light simply makes them go faster. The thermodynamics is such that the products will not restore the original reactants. Such nonreversible photoreactions are valueless for storing energy. Even if there is a suitable reverse reaction in the dark, it must proceed fast enough for practical use and yet not so fast that the reaction is completed during the exposure to the sunlight. Other requirements for a suitable solar energy-storing photochemical reaction are that the photoreacting materials must absorb light throughout most of the visible spectrum in order to use as much of the solar radiation as possible, the energy stored should be at least 50 to 100 calories per gram and the materials must be inexpensive. Liquids and solids are preferred because of their small volume, but reactions involving gases are not ruled out.

The photochemical products may be separated and stored indefinitely for use at a later time, or they may remain inactive in contact with each other at room temperature until a catalyst is introduced, or they are heated to a higher temperature at which the reaction proceeds rapidly enough and remains at the high temperature due to the evolution of heat.

The wavelength range of photochemical absorption can be extended by the introduction of photosensitizers or by changing the molecular structure of the absorbing material.

Photoexcited molecules or unstable fragments or free radicals often have too short a lifetime to study, but sometimes they may be preserved in rigid glasses. Again, molecules excited in the singlet state lose their energy rapidly, with a half-life of about 10^8 seconds, whereas some photoexcited molecules with electrons in the triplet state may have half lives of the order of 10^3 seconds and thus have a better opportunity for bringing about chemical reactions which can store solar energy for longer periods of time.

All these criteria for storing the energy of sunlight are met in the marvelous natural process of photosynthesis which we are just beginning to understand.

Carbon dioxide and water combine through the action of sunlight on green chlorophyll in the living plant. The products are a carbohydrate and oxygen at room temper-

ature. The carbohydrates and other materials produced in the plant are stable in air at room temperature, but at high temperatures they burn with the evolution of heat, or in the presence of certain enzymes at room temperatures they combine to supply the physiological energy for living organisms. The thermodynamic energy requirement for the reaction is 112,000 calories, which corresponds to light energy in the ultraviolet. Moreover, the reactants are transparent throughout the whole spectrum of sunlight, and hence photochemically inactive, but the chlorophyll absorbs visible sunlight below 6800 A and becomes activated and releases hydrogen atoms from the water which then react with carbon dioxide. Red light is effective in growing plants, and in it the energy intensity corresponds to only 40,000 calories per mole so that at least three photons of light must be required for each molecule.

The efficiency of photosynthesis in algae under optimum laboratory conditions was found to be about 1/8 molecule per photon [49, 50], corresponding to a chemical storage of about 30% of the light actually absorbed by the growing algae. This is a very high efficiency. There are many complicated reactions which follow the primary activation by light and some of these use up energy. The primary photoreactions as a rule are uninfluenced by temperature, but the secondary thermal reactions increase in rate as the temperature is increased. There are many unique features in photosynthesis including structural features in the plant which permit quick diffusion of carbon dioxide gas and which keep some of the intermediate products separate so that the back reactions are retarded.

Intensive research should be directed toward finding photochemical reactions other than photosynthesis, which can store the energy of sunlight.

The photolysis of nitrosyl chloride in carbon tetrachloride

$$2 \text{ NOCl} \underset{\text{dark}}{\overset{\text{light}}{\rightleftarrows}} 2 \text{ NO} + \text{Cl}_2$$

meets many of the requirements [51]. The nitrosyl chloride absorbs almost all of the visible sunlight and dissociates with a quantum efficiency of about unity. The products of the photoreaction are separated automatically, before they can recombine. The nitric oxide is insoluble and escapes from the carbon tetrachloride and is stored in tanks, whereas the chlorine remains dissolved in the solution. When the nitric oxide is bubbled back through the solution of chlorine, the nitrosyl chloride is reformed with the evolution of heat for another cycle of reuse, but, unfortunately, the photochemical dissociation produces chlorine as atoms which combine immediately to give molecules with the loss of a considerable part of the heat of recombination.

Many other photochemical reactions have been explored, but the search for new ones should be continued. Several photopolymerizations and photoisomerizations might be considered except for the fact that they require ultraviolet light, and the amount of ultraviolet light in solar radiation is only 2 to 3%.

Special agricultural possibilities

The world's population is expanding faster than the world's supply of food and serious difficulties for the future have been predicted. Temporary and local surpluses of food in the industrialized, temperate zones have led to a complacency regarding the good supply. Remarkable advances have been made in greatly increasing agricultural productivity through the use of fertilizers, irrigation, pesticides and weed killers, and plant breeding, but these will not be sufficient for the future. Radically new ways of obtaining more food from sunlight should be explored.

In spite of the very high efficiency of modern agriculture, the efficiency with reference to the utilization of sunshine in ordinary agriculture is very low—about 2 tons of dry organic material per acre per year corresponding to the storage of only about 0.2% of the annual sunshine. This is low in comparison with the 2 tons per acre per day illustrated in the hypothetical example just calculated and with the 30% storage of light absorbed in growing algae in the laboratory under optimum conditions. But agriculture can't utilize the infrared half of the solar radiation, it cannot grow crops in the winter time, it must use the 0.03% carbon dioxide in air as against the 3 per cent used in the laboratory, and the acre of farm land is covered with photosynthesizing green leaves for only a short period of year. Moreover, the sunlight is too intense for maximum efficiency. Perhaps research should be directed toward the genetic breeding of plants for high energy efficiency as well as for disease resistance, size, and other features.

The mass culture of algae has been studied and it appears likely that dry algae can be produced at a cost of from 25 cents to $1 per pound. The yield can be as much as 20 tons per acre per year, in contrast to 2 tons of ordinary agriculture. Tamiya [52] has grown algae in ponds up to an acre in size with the addition of extra amounts of carbon dioxide. Fisher [53] and others have grown algae in large plastic tubes. Expensive engineering equipment was used with water pumps, heat transferring units, and centrifuges for harvesting the algae. It is difficult to keep the water in the closed plastic containers from becoming too hot, and it has been proposed to use algae that have evolved in hot springs such as those of Yellowstone Park which can tolerate the high temperatures. Another difficulty has been contamination with larger organisms which eat the algae and do not themselves contribute to the products of photosynthesis. Among the advantages of algae production is the fact that the percentage of carbohydrates, fats, and proteins in the algae can be varied over wide limits and most of the amino acids necessary for nutrition are present.

Another solution to the need for more food is to squeeze the juices out of higher plants for proteins and other items of food and use the remaining fibrous material for fuel. M. A. Stahman [54] has found that certain types of alfalfa can give, by extraction, 20% of edible proteins of high nutritive value. The standard way of obtaining proteins for food is by feeding plant material to animals for the production of meat. But over four-fifths of the food energy of the plants is lost in this method of feeding, and, in areas where the food supply is marginal, the luxury of meat cannot be afforded.

Research should be directed toward the growing of vegetables and food crops in arid lands making use of solar-distilled water. The water obtained from salt water is much too expensive to use with ordinary agricultural practices. Most of the costly, distilled water evaporates from the ground and is lost. It should be possible, however, to grow special vegetable crops in gardens which are entirely enclosed with sheets of plastic to reduce the water vaporization losses. But, without the cooling produced by evaporation of water from the leaves, the plants become overheated and die. The heat load can be cut nearly in half by using a filter of 1 inch of water containing 1% of cupric salts. Such a solution absorbs most of the infrared radiation which heats up the plants but does not contribute to the photosynthesis; but it does not absorb any appreciable amount of the visible light which produces photosynthesis. Some cooling by air and the introduction of carbon dioxide into the growth chamber can be effected by circulating, over the plants, air which has been saturated with the salt water, so that there will be very little loss of the valuable distilled water.

Photochemical production of electricity

It is possible to convert the energy of some chemical reactions directly into electrical work without going through the intermediate stage of producing heat and operating a heat engine. In ordinary thermal reactions, the displacement of electrons leads to the formation of energy-rich molecules which on collision with other molecules give rise to higher velocities and the evolution of heat. In electrochemical reactions, electrons are transmitted to one electrode and pass through a wire to the second electrode, where they are absorbed by the other reacting material. The electrochemical process is not limited by the Carnot relation, and the fraction of chemical energy convertible into work is much higher. The thermal reaction is in competition with the electrochemical reaction, and it is necessary for the reactants to collide with the electrodes quickly before they collide with each other.

Rabinowitch [55, 56] has described one photogalvanic cell which indicates an interesting possibility for the direct conversion of sunlight into electricity. The reaction is

$$\text{Thionine} + 2\,Fe^{++} \underset{\text{dark}}{\overset{\text{light}}{\rightleftharpoons}} \text{leukothionine} + 2\,Fe^{+++}$$

When an aqueous solution of thionine and ferrous ion is illuminated by light, the ferrous ion reduces the purple-colored dye thionine by transferring an electron to it, thus producing ferric ion and colorless leukothionine. When the light is removed, the reaction reverses itself over a period of several minutes. The operation can be repeated for an indefinite number of cycles. If two electrodes are placed in the solution and one electrode is illuminated, a voltage of 0.4 volt is produced because of the difference in concentration of oxidizing and reducing materials. When the two electrodes are connected with a wire, a current is generated as long as one of the electrodes is illuminated. Another similar reaction is the photoreduction of mercuric ion by ferrous ion. Other photogalvanic systems similar to these should be explored.

Another possibility lies in the use of a concentration cell in which ferric ions surround one electrode and ferrous ions surround the other. The two electrodes are connected with a wire. Electrons flow through it from the ferrous to the ferric ions, and a mixture of ferrous and ferric ions is produced throughout the whole cell. The cell is then regenerated by oxidizing half of the solution all to ferric ions with air and reducing the other half to ferrous ions with waste organic material, such as grass or garbage produced by photosynthesis. The cell is then ready to operate again and the operations can be made continuous. This waste organic fuel cell has been studied in the laboratory [57]. The voltages are low and the chemical regenerations are slow. Thus far it has been necessary to use a platinum catalyst and to heat the solutions to boiling.

Storage of heat and power

The storage of heat and power is necessary for many uses of solar energy, except for the pumping of irrigation water and the distillation of sea water. Several methods of storage have already been referred to, but it is important to emphasize the need of solving the problem of storage facilities. There is a considerable literature on storage [58, 59].

Rock piles or pebble beds of gravel have large heat capacities, excellent heat exchange with an air stream flowing through them, and are cheap. Metal tanks are used for holding water heated by the sun. A liter of water stores 1 kilocalorie of heat per degree C, a liter of iron stores 0.89 kcal, and a liter of aluminum 0.63 kcal. The metals are good heat conductors but they are expensive. It is necessary to insulate the heat storage units.

Considerably greater amounts of heat can be stored in systems which undergo chemical changes or physical changes such as melting or vaporizing.

The reaction between unhydrated sodium sulfate and water

$$Na_2SO_4 \cdot 10H_2O \rightleftharpoons Na_2SO_4 + 10H_2O$$

with a transition temperature at 32.3° has been used by Telkes [60] for storing heat. The heat evolved per liter is more than four times as much as that obtained by cooling a liter of water through 20 °C. A variety of salt hydrates is available for storing heat at different temperatures. Goldstein [61] discusses chemical systems capable of storing heat, together with the technical and economic considerations.

Experiments with J. Kruger on the system $AlCl_3$-KCl looked promising for heat storage in solar cooking, until an explosion occurred due to moisture adsorbed from the air during the time when the material was being packed in tight iron containers. This system melts at 250 °C and has a large heat of fusion, 60 calories per gram. The heat of fusion of water is only 80 calories per gram. At the temperature of fusion, the aluminum chloride hydrolyzes rapidly with the water and builds up a high pressure of HCl. The system deserves further study.

One problem in using the heat of fusion of crystals for heat storage is the tendency to supercool so that solid crystals do not form on cooling. After many cycles of freezing and melting, the supercooling is apt to be accentuated. Basic research on the nucleation of freezing salts is needed. One solution [60] is to introduce a few crystals of a similar salt of higher melting point which have the same crystal lattice form and remain unmelted, acting as nucleating reagents to start the solidification.

The heats of vaporization are larger than the heats of fusion, and with a two-vessel system it is possible to vaporize the solvent out of a concentrated solution and then recover the heat as the solvent diffuses back into the solution. Such a system, with sodium thiocyanate and ammonia, was described in the section on solar cooling. There are many systems of this type which should be studied such as phosphoric acid and water, $Ca(OH)_2$ to give CaO and H_2O, $CaCl_2 \cdot 6H_2$, $ONiCl_2 \cdot 4NH_3$, and silica gel with adsorbed vapors.

The storage of power is much more difficult than the storage of heat. One of the simplest power-storage systems involves the pumping of water to a reservoir at a higher level and operation of a water turbine when the water returns to its original level. This type of storage has been used on a large scale in a few installations at river dams where coal-burning, steam-driven pumps are also available. If there is no opportunity for a natural reservoir on a hill, the hydraulic storage of power is not practical because elevated tanks are too expensive. Possibly, in some areas, a turbine pump and water wheel could be placed at the bottom of an abandoned mine. Power storage of this type is not practical for small fractional horsepower units because the pumps and water wheels in small sizes are too inefficient. The use of springs and falling weights on pulleys is not very practical. For example, it would require over 2500 sacks of sand, each weighing 100 lb, falling through 10 ft to produce 1 kilowatt hr of mechanical work.

Storage batteries are highly developed and very efficient. The lead storage battery used in automobiles is the cheapest. The nickel cadmium cell is more suitable for general use. In considering storage batteries for solar energy storage, it should be pointed out that it is not necessary to design the batteries to take a high amperage drain as automobiles do, but that it is necessary to discharge the cells each day down to a low voltage, and this deep discharge is apt to shorten the life of the battery. The cost of storage of electrical power in secondary batteries depends to a large extent on the number of cycles of charging and discharging which they can tolerate. The life of most such batteries is now usually limited to several hundred or a few thousand cycles.

Fuel cells in which new, chemically reacting materials are fed into a permanent battery offer promise for the efficient storage of power. Industrial laboratories are spending considerable effort in developing fuel cells. The hydrogen-oxygen fuel cell is one of the most advanced types. Inexpensive catalytic surfaces to accelerate the electrode reactions, particularly when gases are involved, constitute one of the important areas of research. In another type of thermally-regeneratable fuel cell, the materials which react electrochemically at room temperature are regenerated by heating with focused sunlight [62]. The chemical materials for operating the fuel cell produced by solar energy may be stored indefinitely for later use.

One way of storing and transporting fuel made by the sun is to electrolyze water with a solar engine and dynamo or with photovoltaic cells and use the hydrogen in a fuel cell or an internal combustion engine. The hydrogen could be transported in pipe lines, or steel tanks or as liquid ammonia (after combination with nitrogen by the Haber process). The best way would be by loose chemical combination with a solid or liquid. Palladium would be suitable, but it is too expensive. Research should be directed toward finding a practical reversible hydrogenation of a suitable organic or inorganic liquid or solid. The production of methanol from hydrogen and the carbon dioxide of the air is another possibility of obtaining a portable fuel from solar radiation.

The cost of stored electrical or mechanical energy is high because of the inefficiency of the solar production of power combined with the inefficiency involved in its later use. The economics of the situation have been stated [63] as follows:

$$T = \frac{P}{e_s e_d} + \frac{M}{n} + \frac{I}{n} + \frac{C}{ny}$$

Where T is the cost of 1 KWH of stored electricity, $P =$ cost of 1 KWH of electricity produced initially by the sun; $e_s =$ efficiency of storage; $e_d =$ efficiency of delivery; $M =$ cost of maintaining and operating storage per year; $n =$ number of KWH delivered from storage per year; $I =$ annual interest charges on capital investment; $C =$ capital investment for storage and delivery of energy; and $y =$ number of years of life of storage unit.

Selective radiation surfaces

Much can be done to increase or decrease solar heating by using specially prepared surfaces. The temperature of a surface exposed to sunlight rises until the heat losses offset the heat gains. Much of the heat loss is due to radiation emitted by the heated target, and it increases directly as the area exposed and as the fourth power of the absolute temperature. This infrared radiation has a maximum at about 10μ when the temperature is somewhat above room temperature and at about 5μ when the temperature is $300°C$. To achieve high temperatures, the target should absorb as much of the solar radiation as possible, but it should emit as little infrared radiation as possible. Usually, any black surface which is a good absorber of radiation will also be a good emitter of radiation, but because the solar radiation falls between 0.4 and 2.5μ and the radiation emitted by the target falls between perhaps 10μ and 5μ, it is possible to have selective radiation surfaces with high absorptivity but low emissivity [64]. Tabor [65] has pioneered in the development of these special coatings which consist of a bright, reflective under-surface of metal (which emits very little radiation in the infrared) covered with a very thin layer of a black absorbing material. The thickness of this outer black layer is about the same as that of the wavelengths of sunlight and so the solar radiation is absorbed. However, it is only a fraction of the wavelengths of radiation at 10μ, and so the emission of the infrared radiation is determined by the material beneath the thin black layer. Another factor in some selective radiation coatings is the microstructure of the surface. The micropores can be of such a size that the shortwavelengths are absorbed

but the long wavelengths are not emitted by these microcavities, which are smaller than the wavelength of the emitted radiation.

In one research [66], about 85% of the sunlight was absorbed but the radiation emitted was only about 10–15% as great as that of an ideal black body. The selective surface was made by electroplating silver or nickel on a metal plate and then electroplating a thin layer of copper 5×10^{-5}cm thick on it. When heated in air, the copper turns to black cupric oxide and adheres tightly to the metal in spite of violent temperature changes. Black paints containing organic binders may be burned off at sufficiently elevated temperatures, but a surface of thin black cupric oxide adhering to a shining metal surface is permanent at elevated temperatures.

A selective radiation coating is useful in conserving heat and attaining higher temperatures with flat plate collectors which have large areas and also with focusing collectors which produce high temperatures.

In cooling buildings in the hot sun, the problem is to absorb as little as possible of the solar radiation, by reflecting much of it, while at the same time emitting as much infrared radiation as possible. Good progress has been made along these lines with a paint containing white potassium titanate [67]. Another area of needed research lies in treatment of window glass to reduce the solar heating in summer without reducing the transmission of the visible light [68].

There may be a possibility of improving the moisture content of the soil in some special semi-arid regions by research on the selective radiation of soils and rocks [69]. Light-colored soils, such as those composed of volcanic ash, reflect part of the sunlight and do not attain a very high temperature through the day. If the nights are clear, the rocks radiate to the night sky and may cool down to a temperature low enough to condense moisture from air of high humidity as it comes in from the sea in the early morning.

Conclusion

We have seen that solar radiation can be used (and has been used) for cooking, heating water, heating buildings, cooling buildings, refrigeration, drying, and distilling drinking water from salt water or polluted water. It can be used (and has been used) for generating mechanical power or electrical power using heat engines, thermoelectricity and photovoltaic cells. In principle, it can be used through photochemical reactions for making portable fuels but no suitable reaction has yet been found, except the all-important natural process of photosynthesis on which we have always been completely dependent.

Many uses depend for their economic success on further research in basic science, applied science, and engineering. Several specific research problems have been pointed out.

Only in a few cases can solar energy be competitive *now* [1967] with conventional fuel and power using coal, oil, and gas. Special opportunities exist in some of the rapidly developing countries situated not too far from the equator where sunlight is abundant and fossil fuel is expensive. These countries need technological and financial help from the industrialized countries to get started. Soon they can carry on their own research and development of solar energy will be readily available in some of the industrialized countries when inevitably, in the future, it will be needed.

Several applications of solar energy, such as water heating and distilling water, are ready now for evaluation as to general public acceptance, and many more will soon be ready. There are social problems connected with their introduction. Will the introduction be supported first by governments or the United Nations or by private industry or by philanthropic foundations? How does one obtain a mass market to achieve a low manufacturing cost when the low cost is necessary in order to have a mass market? Should solar devices such as water stills, ice machines, and electric generators be introduced on the family scale or the village scale? The family scale has the advantage of independence and low capital investment; and in the case of electric power, the elimination of transmission power lines. However, the village scale has the higher efficiency of larger units and the requirement for fewer technologically trained workers in a society where technologists are limited in number.

It is hoped that many solar applications will be tried in places where local conditions are different, and that many workers in the fields of science and engineering, and in economics and foreign affairs, will become interested in helping to use the sun's energy directly. The author wishes to acknowledge the support of his research by the Rockefeller Foundation.

References

1. Part of the material presented here is covered in greater detail, and with complete references, in a paperback book, F. Daniels, *Direct Use of the Sun's Energy*, published in 1964 by the Yale University Press, New Haven, Connecticut. 374 pages.
2. F. Daniels. Atomic and Solar Energy, *American Scientist*, 38, 521–48, (1950).
3. United Nations Conference on New Sources of Energy, Rome 1961. UN Publications, New York, 1963 and 1964, Vol. 1, 63 I 2 Vol. 4, 63 I 38. Vol. 5, 63 I 39. Vol. 6, 63 I 40.
4. Proceedings of the World Symposium on Applied Solar Energy, Phoenix, Arizona. Stanford Research Institute, Menlo Park, Calif., 1956.
5. Transactions of the Conference on the Use of Solar Energy, The Scientific Basis, 1958. University of Arizona Press, Tucson, Arizona, 1958.
6. F. Daniels and J. A. Duffie, Eds., *Solar Energy Research*, Univ. of Wis. Press, Madison, Wisconsin, 1955.
7. Applications Thermiques de L'Energie Solaire dans le Domaine de la Recherche et de l'Industrie, Symposium Mont Louis, France, Centre National de la Recherche Scientifique, Paris (1961).
8. Research Frontiers in the Utilization of Solar Energy, Proc. Nat. Acad. Sci., 47, 1245–1306 (1961); Solar Energy, Special Issue, Sept. 1961.
9. A. M. Zarem and D. D. Enway, *Introduction to the Utilization of Solar Energy*, McGraw-Hill Book Co., New York, 1963.
10. A. G. Spanides and A. D.Hatzikakidis, Proc. of Advanced Study Institute for Solar and Aeolian Energy, Sounion, Greece, Plenum Press, New York, 1964.
11. *Solar Energy* (previously J. Solar Energy Sci. and Eng.) published by the Solar Energy Society, Campus, Arizona State University, Tempe, Arizona.
12. Solar Energy (1957–1966), Solar Energy Society, Arizona State University, Tempe, Arizona—$65.
13. Reference 1, page 26.
14. G. O. G. Löf, J. A. Duffie and C. O. Smith. World Distribution of Solar Radiation, Solar Energy, 10, 27–37 (1966); Engineering Experiment Station Bulletin, University of Wisconsin (1966).
15. Summarized in Reference 1, Chapters 4 and 5.
16. F. Daniels, Construction and Tests of Small Focusing Collectors. Forthcoming publication.
17. J. A. Duffie, G. O. G. Löf, and B. Beck. Laboratory and Field Studies of Plastic Reflector Solar Cookers, Reference 3, 5, 339–346. *Solar Energy*, 6, 94–98 (1962).
18. G. O. G. Löf, Reference 3, 5, 304–334. *Solar Energy*, 6, 94–98 (1962).
19. Reference 3, 5, 114–247.
21. A. T. Talwalkar and J. A. Duffie, Reference 3, 5, 284–293.
22. Reference 1, Chapter 11. Reference 3, 6, 307–454, Reference 5, 146–186, Reference 7, 83–376.
23. Proc. of the Solar Furnace Symposium, Phoenix, Arizona, *Solar Energy, 1* (2–3) 3–115 (1957).
24. R. K. Cohen and N. K. Hiester, A Survey of Solar Furnaces in the United States, *Solar Energy, 1* (2–3), 115–17, 1957.
25. R. J. Marcus and H. C. Wohlers, Chemical conversion and storage of concentrated solar energy, Reference 3, 1, 187–194; Chemical synthesis in solar furnaces, Reference 3, 1, 187–194; Chemical synthesis in solar furnaces, Reference 3, 6, 393–397.
26. Reference 1, chapter 10.
27. Reference 3, 6, 139–306.
28. G. O. G. Löf, Fundamental Problems in Solar Distillation, *Proc. Nat. Acad. Sci.*, 47, 1279–89 (1961); *Solar Energy 5*, Special Issue Sept., 35–45 (1961).
29. F. Daniels, Construction and Tests of Small Solar Water Stills. Forthcoming publication.
30. R. D. Jackson and C. H. Van Bavel, Solar distillation of water from soil and plant material. A simple desert survival technique, *Science, 149*, 1377 (1965).
31. Anon. A Case for a Solar Ice Maker, *Solar Energy, 7*, 1–2 (1963).
32. R. Chung, *Solar Energy, 7*, 187–188 (1963).
33. Reference 1, Chapter 13, Reference 3, 6, 3–139.
34. G. C. Blytas and F. Daniels, Concentrated Solutions of NaSCN in Liquid Ammonia, *J. Am. Chem. Soc., 84*, 1075–83 (1962).
35. R. Chung and J. A. Duffie, Cooling with Solar Energy, Ref. 3, 6, 20–28.
36. R. Chung, G. O. G. Löf, and J. A. Duffie, Experimental Study of a LiBrH$_2$O Absorption Air Conditioner for Solar Operation, *Am. Soc. of Mech. Engrs.*, Paper 62 WA347, (1962).
37. G. O. G. Löf, Reference 6, pages 43–45.
38. Reference 1, Chapter 14. Reference 3, 6, 3–138.
39. H. Tabor and J. L. Broncki. Small turbine for solar energy power package. Reference 3, 4, 68–78.
40. Reference 1, 256–265.
41. E. A. Farber. Closed-cycle solar hot-air engines. *Solar Energy, 9,* 170–82 (1965).
42. H. Tabor. Large-area solar collectors (solar ponds) for power production. Reference 3, 4, 59–66.

43. Reference 3, *4*, 115–228.

44. K. Katz. Thermoelectric generators for the conversion of solar energy to produce electrical and mechanical power, Reference 3, *4*, 153–166. F. Daniels, J. A. Duffie, G. O. G. Lof and R. R. Breihan, Preliminary tests of a solar heated thermoelectric converter, Reference 3, *4*, 145–146.

45. Reference 3, *4*, 229–285.

46. W. A. Beckman, P. Schoffer, W. R. Hartman, Jr., and G. O. G. Löf, Design Considerations for a 50-watt Photovoltaic Power System using Concentrated Solar Energy. *Solar Energy, 10*, 132–136 (1966).

47. Reference 1, Chapter 17.

48. L. J. Heidt, *et al.*, *Photochemistry in the Liquid and Solid States*, Wiley and Sons, New York (1960).

49. F. Daniels, Energy Efficiency in Photosynthesis, in A. Hollaender, *Radiation Biology*, McGraw-Hill Book Co., New York, 1956.

50. E. L. Yuan, R. W. Evans and F. Daniels, Energy Efficiency of Photosynthesis by *Chlorella, Biochem. Biophys. Acta, 17*, 187–93 (1955).

51. O. S. Neuwirth, The Photolysis of Nitrosyl Chloride and the Storage of Solar Energy, *J. Phys. Chem., 63*, 17 (1959).

52. Reference 4, H. Tamiya, 231–42.

53. Reference 4, A. W. Fisher, Jr., 243–54.

54. Private communication.

55. E. Rabinowitch, Reference 6, 195–201.

56. E. Rabinowitch, Reference 48, page 84.

57. Reference 1, page 339–40.

58. Reference 1, chapters 8 and 18.

59. Reference 3, Vol. 1, 156–216. Vol 5, 401–408.

60. Reference 6, M. Telkes, Solar Heat Storage. 57–62.

61. M. Goldstein, Some Physical Chemical Aspects of Heat Storage. Reference 3, *5*, 411–417.

62. Reference 1, page 341.

63. F. Daniels, Energy Storage Problems, Reference 1, page 347 and Reference 3, *1*, 156–173; *Solar Energy, 6*, 78–83 (1962).

64. Reference 1, chapter 12.

65. H. Tabor, Solar collectors, selective surfaces and heat engines. *Solar Energy* Special Issue Sept., 27–30 (1961).

66. P. Kokoropoulos, E. Salam, and F. Daniels, Selective Radiation Coatings, Preparation and High Temperature Stability, *Solar Energy, 8*, 69–73 (1964).

67. F. E. Edlin. Selective Radiation Reflection Properties of the Fibrous Titanates, *Solar Energy Symposium*, U. of Florida, Gainesville, Fla. (1964).

68. J. L. Yellott, Calculation of Solar Heat Gain Through Single Glass, *Solar Energy, 7*, 167–75 (1963).

69. Reference 1, 192.

W. D. Johnston, Jr.

The Prospects for Photovoltaic Conversion

Solar cells convert sunlight directly to electricity. Can they be made efficient and cheap enough to contribute significantly to a solar energy economy?

Beyond considerations of engineering practicality or economic necessity solar energy has an underlying psychological appeal based on long familiarity. Wind energy has been harnessed for more than a thousand years, solar hot-water systems were used by the Romans, and sunlight has served man as the primary source of warmth, creature comfort, and illumination since before the dawn of history. Most of the general public understand the basic principles of solar hot-water systems or windmills, although the design of modern, economically practical wind and solar thermal systems is highly sophisticated. The solar cell (see Figs. 1 and 2), a true product of the electronic age, is unique among solar technologies in its lack of historical precursors, and the principles governing its operation are not familiar to the general public—or even to many scientists.

When the potential of photovoltaic conversion is compared to various solar and nonsolar energy technologies, the conclusions reached depend of course on assumptions of efficiency and cost for the solar cells, but in general they are not very encouraging for the development of photovoltaic devices (*1*). The question is whether "modern technology" can save the day by bringing forth more efficient,

less expensive cells. In this article I will discuss the basic principles underlying the operation of solar cells, review the current state of the art in terms of demonstrated performance and cost for established technologies, and discuss the motivation for and apparent potential of present research for which the technology is not yet well established.

Basic principles

Briefly, in photovoltaic conversion two key steps produce an electrical current: *creation* of pairs of positive and negative charges in the cell by absorbed sunlight and *separation* of the positive and negative charges by an electric field within the cell (Fig. 2). The current produced is collected by a metal grid attached to the cell (Fig. 1). Although one great virtue of the process is its simplicity, various problems—both technical and economic—are encountered with the materials and procedures used to manufacture the cells. To understand what the problems, the potentialities, and the limitations of direct conversion of solar energy to electricity really are we must look more closely at the basic principles that underlie the creation and separation of the electric charges in the cell.

As the name implies, a photovoltaic device—or solar cell—produces an electrical voltage when it is illuminated. Two criteria, corresponding to the key steps in photovoltaic conversion, must be met for the effect to take place: the device must be made of a material that can absorb light of the wavelength being provided, and it must have an inherent electrical asymmetry so that the charge carriers excited by the light are separated. Although a great variety of organic,

inorganic, and even biological materials can be used to give a measurable effect, only semiconductors are known to generate enough electrical power to be useful on a practical scale. Among the semiconductors typically used are silicon, germanium, cadmium sulfide, and gallium arsenide.

In a semiconductor the electrical charge carriers, or electrons, occupy one of two bands. The lower energy *valence band* is filled or nearly filled with electrons, and the higher energy *conduction band* is only slightly occupied or empty. These two bands are separated by an energy gap in which there are no electrons. The semiconductor is relatively transparent to light of wavelengths longer than that corresponding to the energy of this gap—the *bandgap energy*. Light of shorter wavelengths is absorbed and excites some of the electrons, which move from the valence band across the energy gap to the conduction band, leaving "holes" in the valence band (see Fig. 2). These holes behave like charge carriers—with a positive charge—and are analogous to the negatively charged electrons in the conduction band. The creation of the electron–hole pair is the first of the two key steps in the photovoltaic process.

Both the hole in the valence band and the electron in the conduction band are mobile and can be separated by an electrical asymmetry, or *potential gradient*—the second essential step in producing photovoltaic power. If they are not physically separated they will eventually recombine, emitting either light again or heat. Such gradients exist near the interface between two semiconductors or between a metallic conductor and a semiconductor; they arise because some of the

Dexter Johnston received the B. S. degree from Yale University and the Ph. D. from the Massachusetts Institute of Technology, both in physics. In 1966 he joined the technical staff of Bell Telephone Laboratories at Holmdel, NJ, where he has worked on a variety of research topics dealing with the interaction of light with matter. Address: Bell Telephone Laboratories, Holmdel, NJ 07733.

electrons in one of the materials are in higher energy states than empty, lower levels in the other. When the two materials are joined, these electrons flow across the interface to fill the lower-lying states. This process tends to give one material a positive charge and the other a negative charge and establishes an electric field at the interface that acts to resist further charge transfer and causes the energy bands to appear "bent" at the interface.

Such an interface between two semiconductors is called a *junction,* between a metal and a semiconductor, a *Schottky barrier.* A common example of a junction is the interface between two semiconductors that are the same chemically (both silicon, for instance) save for small additions of different impurity elements. Some impurity elements (boron, for exam-

ple) "accept" electrons from the silicon because they contain fewer electrons than the silicon. These *acceptor* impurities leave a less than completely filled valence band, giving rise to positive, or *p*-type, material with free holes. Conversely, a *donor* impurity (such as phosphorous) donates electrons to silicon, giving rise to negative, or *n*-type, material with free electrons in the conduction band. Introducing impurity atoms to obtain a desired concentration of carriers is called *doping.* The union of two complementarily doped semiconductors is called a *pn junction.* The two semiconductors can be chemically different—germanium and gallium arsenide, for instance—in addition to being doped with oppositely acting impurities, and this union is called a *heterojunction.*

When the region of the semiconduc-

tor adjacent to the *pn* junction or Schottky barrier is illuminated, absorption of the light produces an electron–hole pair. These electrons and holes are then separated by the electric field associated with the band-bending, or potential gradient, at the interface. If the light is absorbed in a *p*-type region, the electrons excited into the conduction band are *minority* carriers, whereas in *n*-type material, the electrons in the conduction band are the *majority* carriers and the holes created by the light are the minority carriers.

The maximum voltage that can be produced by a solar cell is a temperature-dependent fraction of the bandgap energy. The maximum current possible corresponds to one electron transferred for each light quantum (photon) absorbed. In practice, both Schottky barrier cells

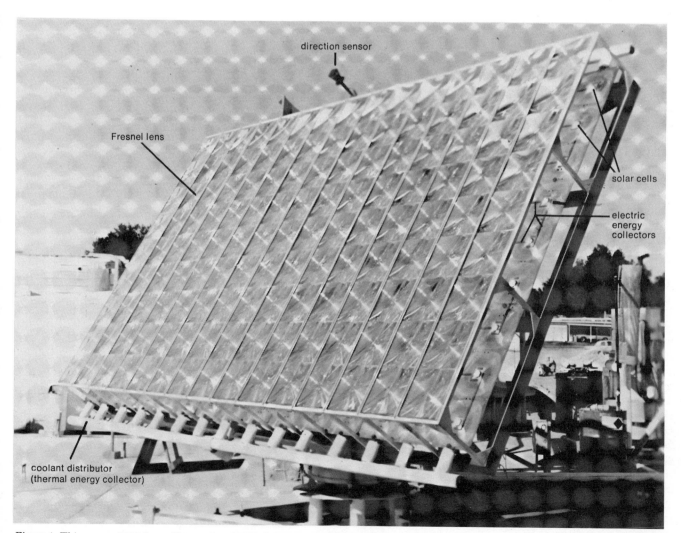

Figure 1. This array of 135 8-cm silicon solar cells mounted behind 30-cm-square plastic Fresnel lenses produces a *peak* electric power of 1 kW, together with about 5 kW (18,000 BTU/hr) of peak thermal output (~80°C). The approximate cost, which includes the tracking mount and lenses, of $3,500 for the 3-by-5 m array must be reduced by a factor of six to eight before such arrays will be economically attractive. (Photograph courtesy of E. L. Burgess and M. W. Edenburn of the Sandia Corporation.)

and *pn* junction cells are efficient in producing current, with the ratio of electrons collected to photons absorbed approaching unity. However, *pn* junction cells are generally more efficient than Schottky barrier cells in maximum voltage output, with values ranging from two-thirds (for silicon) to three-fourths (for gallium arsenide) of the bandgap energy available at room temperature. Schottky barrier cells are normally illuminated through the metal layer, which absorbs some light even if the layer is quite thin (∼5 nanometers), and thus *pn* junction cells enjoy at least a small advantage in potential current efficiency as well.

For maximum power conversion the right material for the cell must be chosen. The voltage increases as the bandgap energy of the semiconductor (the substance with the smaller bandgap in the case of a heterojunction) increases. However, since light of wavelengths longer than that corresponding to the bandgap energy is not absorbed, the current available from the solar spectrum diminishes as the bandgap energy increases, and there is a point at which optimum bandgap energy is reached. For terrestrial sunshine, optimum bandgap energy is about 1.45 electron volts— very close to the value for gallium arsenide (1.43 eV at room temperature) and somewhat greater than that for silicon (1.2 eV).

A solar cell is a diode (an electrical rectifier with two terminals), and the characteristics that separate the photoexcited charges also cause current–voltage rectification, with the current an exponentially dependent function of applied voltage (see Fig. 3). Under illumination, the current–voltage plot shifts downward along the current axis, and the useful power available is shown by the area of the largest rectangle that can be formed by the current and voltage axes and the current–voltage curve. This area is always less than the product of open-circuit (zero-current) voltage and short-circuit (zero-voltage) current by a ratio called the *curve factor,* or *filling factor,* which depends on the perfection of the junction, the temperature, and the bandgap energy. Curve factors of different types of solar cells range from 0.7 to 0.85.

When these intrinsic characteristics of solar photovoltaic cells are taken

into account, the maximum conversion efficiency that can be obtained from a single-junction cell under terrestrial conditions is about 28%—corresponding to a high-quality *pn* junction in a semiconductor with a bandgap of about 1.45 eV. Thus gallium arsenide cells are expected to yield about 27%, and silicon cells about 22%, under ideal conditions. Of course a variety of extrinsic considerations, which do not directly relate to the physics of solar cells but which are essential for efficient power generation, contribute to small reductions in the final efficiency. For example, a pattern of narrow metal contact fingers must be applied so that the charge carriers can be collected efficiently from the front surface of the cells, perhaps in conjunction with a transparent conductor such as indium-doped tin oxide. The cells must be coated with thin layers of transparent dielectric materials to minimize the reflection of visible light; they must be mounted and in-

terconnected into arrays; and they have to be encapsulated against atmospheric contaminants.

One other aspect of semiconductors bears directly on the design of solar cells. Light may be absorbed in a semiconductor either *directly* with the creation of an electron–hole pair or *indirectly* with the creation of an electron, a hole, and the emission or absorption of a phonon—a quantum of lattice vibrational energy. Direct absorption is simpler and hence more probable if the semiconductor is the type that *can* absorb light directly— which is not always the case, since momentum as well as energy must be conserved in the absorption process, and in some semiconductors (called *indirect-gap* materials) this requires emission or absorption of phonons because of the nature of the electronic states. Silicon is an indirect-gap material, while gallium arsenide, for instance, is a *direct-gap* material. The practical consequence of this is that

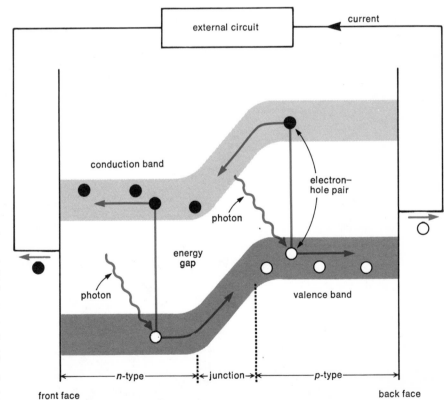

Figure 2. In photovoltaic conversion, a device known as a solar cell produces an electrical voltage when it is illuminated. In this simplified diagram, the energy of a charge carrier is plotted vertically and distance into the cell horizontally. When the cell absorbs a photon (a quantum of sunlight), an electron in the valence band is excited. The electron moves from the valence band to the conduction band, leaving a hole in the valence band. This creates

an electron–hole pair: the hole has a positive charge, and the electron in the conduction band has a negative charge. An electrical asymmetry in the cell (*at the junction represented by the bend in the diagram*) causes the positive and negative charges to separate, propelling positive charges in one direction and negative charges in the other. A collection grid attached to the cell collects the current produced by the charges.

light with energy exceeding the bandgap of the semiconductor is absorbed rapidly, very near the surface, in a direct-gap material, but propagates considerably further into an indirect-gap material. Thus a layer of gallium arsenide only 5 μm thick suffices for essentially complete visible-light absorption, while a layer of silicon more than 100 μm thick is required to achieve that end.

In an indirect-gap material the photogenerated carriers must diffuse over a greater distance without suffering recombination along the way, before being collected at the junction. Since recombination is highly probable if the carriers collide with crystal defects or certain kinds of impurity atoms, purity and crystalline perfection for indirect-gap material is even more essential than for direct-gap materials. Also, since a greater thickness is needed, more material is necessary for indirect-gap semiconductors.

Established technologies

The principal breakthrough in solar cell technology came in 1954 with the disclosure of the silicon solar battery at Bell Laboratories (2). With improvements over the years, this device has provided power for interplanetary probes and terrestrial communication and reconnaissance satellites. The space environment places a large premium on reliability, resistance to radiation, and high power per unit of weight; cost of cells is not a primary consideration. Today's advanced "space-qualified" silicon solar cells represent the culmination of extensive engineering design and testing: cells with operational efficiencies of about 15% in the near-Earth space environment (equivalent to 18% terrestrial efficiency) are now manufactured (3).

Since silicon is an indirect-gap semiconductor, the single-crystal cells of the silicon solar battery absorb the solar spectrum through a relatively thick (tens of micrometers) layer. High-quality silicon, characterized by a minority-carrier diffusion length comparable to this absorption distance, is required for the most efficient utilization of the longer wavelengths, which penetrate deepest into the structure. Cells with back contacts that are optically reflective, reflective for minority carriers, or both

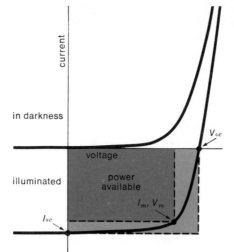

Figure 3. Current and voltage characteristics of a solar cell are plotted for darkness and illumination. Under illumination the current–voltage plot shifts downward along the current axis. The short-circuit current (I_{sc}) increases (to larger negative values) proportionally to light intensity; the open-circuit voltage (V_{oc}) increases more slowly, approximately as the logarithm of the light intensity. The "curve factor" is the proportion of the maximum useful power available ($I_m \times V_m$) to the "ideal power" $I_{sc} \times V_{oc}$. This equals the ratio of the area of the dark blue rectangle to that of the larger pale blue rectangle.

are advantageous. The front surface of the cell is usually constructed for the most efficient absorption of ultraviolet rays and includes a transparent cover, with antireflection coating, to provide protection from particle radiation (the solar wind).

There are three essential differences between the requirements for space and for terrestrial applications. The terrestrial solar spectrum lacks the ultraviolet component and some portions of the infrared, owing to atmospheric absorption and scattering. For terrestrial solar cells, resistance to water vapor, atmospheric contamination, and abrasion by airborne particles, rather than protection from ionizing radiation, is the primary duty of the encapsulant. Most important, power per dollar, rather than per unit of weight, is the single overriding concern in evaluation of the practicality of a terrestrial solar power facility. This last condition implies that cell efficiency as well as cell cost is important. Since the cost of land and interconnection increases with total area, the use of cheaper, lower-performance cells (which require more area for the same output) can be pursued only to a limited extent. Indeed, what we now know seems to

suggest that, for a cell to be useful, minimum efficiency will have to be somewhat *above* 10% regardless of the total capacity of the facility envisaged—be it a residential rooftop installation in the kilowatt range or a central power station at the gigawatt level—and regardless of how inexpensive the cell turns out to be (4).

During the past several years manufacturers of silicon solar cells have turned their attention to the terrestrial market, to see if by relaxing somewhat the performance standards for space cells a cheaper but still useful terrestrial cell can be made. Arrays with efficiencies (based on total area) above 12%, and made up of rectangular cells with individual efficiencies up to 15%, are now available (5). The price of similar arrays but composed of round cells (which give lower areal efficiency) has recently come down to about $15 per peak watt (6). Such arrays would be quite satisfactory for many applications—if the cost were 50 to 100 times lower. Efforts along traditional lines—e.g. using cheaper, less highly purified silicon; streamlining and automating production; increasing volume to mass-production levels—will undoubtedly continue, and the price of these conventional silicon cell arrays can be expected to decline, in all probability with the same "learning curve" behavior characteristic of other mass-produced high-technology products. As the price comes down, the civilian market, now restricted to such special applications as powering communications equipment for forest fire lookout towers, will expand.

The only other material currently used for commercial production of solar cells is the n-cadmium sulfide/p-copper sulfide heterojunction system, which has a theoretical efficiency of 16%. This system is highly attractive in terms of cost and manufacture, since the cadmium sulfide can be deposited by a variety of inexpensive processes, such as spray pyrolysis, onto metallized polymer or conductive glass substrates. The copper-sulfide layer is formed by an ion-exchange dipping and low-temperature heat treatment. In contrast to silicon, however, where laboratory cells with 19% efficiency compare well to the theoretical expectation of 22%, the best cadmium sulfide/copper sulfide cells have reached only about 8%, with 6% apparently more representative of production units. In addition, the quality of the copper-sulfide portion of the heterojunction depends strongly on a slight departure from stoichiometry. Since copper diffuses rather easily in this structure, the junctions degrade with use, and short lifetimes pose a serious problem.

The effort to develop this system has been second only to that for the silicon cell, and the persistence for more than a decade of lifetime problems and low efficiencies must be taken as a strong indication that the problems with copper sulfide are unlikely to be overcome. In spite of this, some research on the system continues, motivated by the very attractive economies of the manufacturing technology, which leave little doubt that cost goals would be met. However, achieving efficiencies above 10% concurrently with projected service lifetimes of more than 20 years can easily be regarded as the long-odds dark horse of the photovoltaic field.

The other system that can be considered to be in some sense established is based on heterojunction cells of aluminum arsenide or an aluminum-gallium-arsenide alloy deposited epitaxially on single-crystal gallium arsenide substrates. The theoretical efficiency of this system is very high: the bandgap of gallium arsenide is an almost ideal match with the solar spectrum for maximum power conversion with a single-junction device, and the aluminum arsenide provides a transparent window for most of the visible spectrum. Gallium arsenide is a well-understood and relatively well-behaved semiconductor material, for which an extensive device technology—second only to that for silicon—already exists. Light-emitting diodes, semiconductor lasers, and microwave field-effect transistors are among the gallium-arsenide devices manufactured more or less routinely by processes using epitaxial growth on single-crystal substrates from vapor or solution phases. In these processes thin layers of semiconductor material are grown with the same crystallographic orientation and lattice spacing as is characteristic of the substrate material.

No solar cells based on this technology are available commercially, but impressive results have been obtained by several independent research groups, with the highest reported terrestrial efficiency over 23% (7). The theoretical efficiency is 26–28%, depending on details of the cell design—primarily thickness and composition of the aluminum-gallium-arsenide layer, which influence series resistance as well as blue-light efficiency. Most of these results have been obtained for small cells produced by liquid-phase epitaxy, which is generally unattractive as a manufacturing process since it is not readily scaled to producing larger areas and since large volumes (at least in relation to the volume of the epitaxial layer grown) of expensive gallium must be used. However, similar performance has recently been obtained for large cells produced by vapor deposition (8), for which scale-up problems should be much alleviated.

In the modification of the cells produced by vapor-phase epitaxy, no gallium beyond that in the gallium-arsenide substrate is used, and growth rates up to 1 m^2 per hour per reactor appear possible. The cells are still very costly, as gallium-arsenide substrate wafers cost well over 10 times as much as silicon wafers. These cells do provide higher efficiencies than silicon (both in theory and in practice), as well as improved resistance to ionizing radiation and significantly better performance at elevated temperatures. These qualities, which result from the wider direct bandgap in gallium arsenide, suggest that such cells can be used advantageously in space, and limited commercial production for space applications can be anticipated in the near future.

Concentrators

One way to reduce the cost of photovoltaic cells is to use solar collectors to concentrate the sunlight. Since this technique means that the semiconductor area can be smaller, high-performance cells can be used, in principle at least, in spite of their cost. Then the expense of the focusing lenses or mirrors and the sun-tracking mount they require becomes the limiting factor. This method will be cost effective if this equipment can be made to cost significantly less per unit of area than do the solar cells.

Optical constraints limit focusing concentration to several thousand, and the thermal and electrical properties of the solar cells and their mounting and interconnection systems add further limitations. For silicon cells the highest practical concentration is limited to several hundred, owing to series resistance effects in the relatively thick, lightly doped structures necessitated by the indirect bandgap. An artificial cooling system must be used because efficiency decreases rapidly at elevated temperatures. Sandia Laboratories has recently demonstrated a silicon cell/Fresnel lens concentrator array that provides 1 kW of electric power and 4–5 kW of low-grade thermal power ($\sim$80°C) suitable for hot-water or space-heating use, at an estimated manufacturing cost of less than $3,500, including plastic lenses, tracking mount, active heat sink, and cells (9) (Fig. 1). This is cheaper by a factor of about four than the least expensive nonconcentrator 1-kW silicon array and provides significant thermal energy in addition.

Gallium-arsenide cells can be used at much higher concentrations than silicon cells: efficiencies over 19% at concentration ratios of nearly 2,000 have been reported (7). Since current densities of tens of amperes/cm^2 are generated at these high concentrations, problems with the collection grid and interconnection become severe if cells as large as even 1 cm^2 are used. (Separate smaller cells can be put in series electrically, permitting operation at high voltages and with lower currents.) Absolute heat-sink temperatures can be about 20% higher for gallium arsenide than for silicon. These cells can thus be used in a hybrid electric-thermal system requiring higher temperature heat input. For example, absorption air-conditioning equipment requires working fluid input temperatures of 120–140°C, and gallium-arsenide cells used in that temperature range are about twice as efficient as silicon cells. Meaningful cost estimates for gallium-arsenide concentrator systems are not yet available, since neither the cells nor high-ratio optical concentrators are being manufactured commercially.

Research directions

The practical limitations on the technologies based on silicon and gallium arsenide stem largely from the costs of traditional single-crystal cell fabrication. For silicon, for example, metallurgical-grade elemental material is purified by conversion to chlorosilane, distillation, and reduction back to elemental polycrystalline silicon. Single-crystal boules are pulled from a melt of this material, sliced into wafers, and polished.

These steps consume the lion's share of time, energy, and dollars, before the actual device processing—junction diffusion, contacting, and array bonding and encapsulation—can begin. Similar steps are involved with the manufacture of gallium-arsenide and other single-crystal cells. Typically, the wafers as cut are not of the appropriate rectangular or hexagonal shape to permit assembly into closely packed arrays, and thus waste is incurred from trimming as well as from slicing and polishing.

Because minority-carrier diffusion lengths in the 100-μm range are required for efficient silicon cells, the material must be free from defects, such as grain boundaries or impurities, which promote minority-carrier recombination. Most efforts to obtain good cell performance in polycrystalline material, even with millimeter-size grains, or in metallurgical-grade silicon without purification have been disappointing. Recently, however, H. Fischer and co-workers at AEG-Telefunken reported efficiencies over 10% for 120-cm^2 cells, and over 14% for smaller 4-cm^2 sections, made from a directionally solidified, cast block of silicon in which an oriented, fibrous polycrystalline structure was produced (10). Although purification and sawing are still necessary, the process itself produces rectangular slices, and dip-etching is substituted for polishing. It is possible that thin sheets, rather than blocks, can be cast in a similar way. In any case, the development is important, because it shows that efficiencies of around 12% can be achieved with a non-single-crystal material.

Progress has also been reported recently in finding alternatives to the traditional purification process (11). The goal is to provide "solar-grade" silicon at less than $10/kg in production quantities of several thousand metric tonnes per year. Single-crystal cells made from material purified by directional freezing of metallurgical silicon from a slightly upgraded smelter have shown terrestrial ef-

ficiencies of 12–14%, and it seems that these techniques would increase the present $1/kg cost of metallurgical silicon by a factor of only four or five. Successful use of such solar-grade material in polycrystalline cells, however, remains to be demonstrated.

A variety of other approaches for nonconventional silicon are being pursued. Notable among them is growing silicon ribbon by pulling the silicon from a melt through an edge-defining slot-shaped aperture. Cells up to 2.5×10 cm with efficiencies as high as 9.5% have been made from ribbon 2.5 cm wide, although contamination with impurities from the ribbon-defining growth apparatus appears to be the limiting problem (12). Other techniques that may hold promise involve dip-coating ceramic substrates in molten silicon, or growth on inexpensive substrates by decomposition or reaction of volatile silicon compounds from the vapor phase. These last techniques produce polycrystalline material, and because methods to control the critical grain size and orientation have not yet been developed, lower efficiencies of about 5–6% are typical.

Carlson and co-workers at RCA have described cells made from a different type of silicon material (13). The material—amorphous silicon—is formed by introducing silane into a DC discharge plasma. It can be deposited on substrates such as metals or conductively coated glass. The films produced have very different optical properties from crystalline silicon, with the absorption coefficient in the visible region about ten times higher. Amorphous silicon acts more like a direct bandgap material, and material with diffusion lengths of a few micrometers can be used for solar cells. Because the amorphous nature of the material is, of course, detrimental to the transport properties, series resistance effects are troublesome. The collection efficiency for red and near-infrared light is low, since the optical absorption drops below that of single-crystal silicon in this range. A potential efficiency of approximately 15% for Schottky barrier cells on n-type amorphous silicon can be estimated, but this material is not yet well understood, and the practical problems that stand in the way of this goal are still largely unknown.

Thin-film approaches

Since silicon constitutes somewhat more than 25% of the earth's crust, there is little worry about its availability. However, this is not the case for other semiconductor materials of potential interest for solar photovoltaic conversion. There is 10^4 times less gallium than silicon, 10^5 less arsenic, and 10^6 less cadmium or indium, for example (14). Consequently, serious efforts to develop nonsilicon solar cell materials for terrestrial applications have concentrated on thin-film approaches using direct-gap semiconductors, which require a minimum amount of material.

Gallium arsenide is one of the most promising thin-film materials, since it would provide, in theory, the highest efficiency and hence could tolerate the greatest loss as a consequence of thin-film imperfections while still exceeding the practical efficiency requirements. Furthermore, it ranks second to silicon in availability. Films with the desired grain size, which need be only about 5 μm, owing to the high optical absorption coefficient, can be prepared on a variety of substrates by pyrolytic decomposition of arsenic hydride and trimethyl gallium or by chemical vapor-deposition processes similar to those used to produce single-crystal gallium-arsenide cells.

Several other conditions must be met for a successful thin-film device. The film must stick to the substrate without chipping or flaking, which requires a good match of thermal-expansion coefficients, and the film-substrate interface must have the desired electrical characteristic—usually a low-resistance, nonrectifying contact. The upper film surface must be such that a junction-forming layer or barrier-forming semitransparent metal film can be applied. Cells intended to provide a thin-film version of the efficient single-crystal aluminum-arsenide/gallium-arsenide heterojunction cells have been prepared, for which good internal quantum efficiency may be inferred (15). The external efficiency, which is what really counts, of course, appears to be limited by the effects of grain boundaries in restricting majority-carrier flow in the junction-forming aluminum-arsenide layer. Again, efficiencies of about 15% can be projected, assuming the difficulty with grain boundaries can be overcome.

Other thin-film gallium-arsenide devices that use Schottky barrier contacts rather than heterojunctions show efficiencies in the 4–5% range. (The first polycrystalline gallium-arsenide Schottky barrier cells were described over a decade ago, with up to 5% efficiency, *16,* and no substantial advance in efficiency has been reported since.) Problems with interface states and effects of grain boundaries on the metal–semiconductor or metal-oxide–semiconductor interface appear to dominate here and may turn out to be of a more fundamental nature than those for the heterojunction thin-film cell, where the problem is really one of developing a transparent ohmic top contact. Thin-film gallium-arsenide cells, either of the Schottky barrier or heterojunction type, could be combined with concentrators for solar thermal systems provided that the series resistance arising in the top layer can be reduced to acceptable levels.

The difficulties in lifetime and performance with the cadmium-sulfide/copper-sulfide cells are thought to arise from the relative instability of copper sulfide, which would not in other contexts be considered a device-quality material with today's technology. Indium phosphide can be used in place of copper sulfide. Although its cubic crystal structure differs from the hexagonal structure of cadmium sulfide, there is a good coincidence between atomic spacings when the (111) planes of indium phosphide—perpendicular to the cube-body diagonal—are matched to the (0001) planes of cadmium sulfide—those perpendicular to the hexagonal axis.

Such cadmium-sulfide/indium-phosphide cells have been made on single-crystal indium-phosphide substrates. They have terrestrial efficiencies up to 15% and do not appear to have any stability problems (*17*). Although indium phosphide does not provide quite as good a match with the solar spectrum as gallium arsenide, it is better than silicon; and the blue cut-off due to the cadmium-sulfide "window" would permit utilization of approximately 45% of the terrestrial solar spectrum. The junction characteristics for sin-

gle-crystal cells have not permitted realization of the full expected open-circuit voltage, presumably because the interface between the two semiconductors is not as perfect as that for aluminum arsenide/gallium arsenide. Thin-film cells of this material do not have a spreading resistance problem in the top layer, because grain boundaries do not appear to interfere with majority-carrier flow in the cadmium-sulfide layer. When prepared on graphite substrates, the normally weakly rectifying *p*-indium-phosphide/graphite interface can be made ohmic and low in resistance by interspersing a layer of polycrystalline *p*-gallium arsenide. Terrestrial efficiencies up to 5.7% have been reported for such composite thin-film cells (*17*).

Liquid/solid cells

A qualitatively different approach to photovoltaic conversion has recently been explored, in which the junction-forming layer of the cell is replaced by an aqueous electrolyte (*18*). This structure, called a photoelectrolytic cell, may be considered a form of Schottky barrier cell, with the electrolyte taking the place of the thin metal film. An *n*-type semiconductor is used as the anode. In fact, when current flows, redox reactions will take place at the cathode as well as at the anode, with the specific reactions dependent on the ionic species and on the composition of the cathode and anode.

Cells of this type were first investigated for their potential to convert water to hydrogen. When the semiconductor anode is illuminated, minority carriers (holes) are formed and diffuse to the electrolyte interface, where they may combine with electrons from anions in solution. For example

$$4OH^- + 4h^+ \rightarrow O_2 + 2H_2O$$

while the electrons flow around an external circuit to the cathode and reduce cations

$$2H_3O^+ + 2e^- \rightarrow H_2 + 2H_2O$$

Charges are separated near the semiconductor surface because of the band-bending caused by the difference in chemical potential of carriers in the electrolyte and in the semiconductor. In general the chemical

potential in the electrolyte is pH-dependent and can be adjusted somewhat. The minimum bandgap for the semiconductor is set by the requirement of 1.49 eV for each charge transfer in the $H_2O \rightarrow H_2 + \frac{1}{2} O_2$ reaction. Additionally, there are "electrode potentials" to be overcome as well as circuit resistance losses. Moreover, most suitable semiconductors (such as gallium phosphide or cadmium telluride, with bandgaps from 1.6 to 1.8 eV) are unstable in water under anodic conditions. Gallium phosphide, for instance, forms an anodic oxide that inhibits further reaction; cadmium telluride is dissociated since $CdTe + 2h^+ \rightarrow Cd^{++} + Te$ proceeds rather than the oxidation of the hydroxide ion. Although some metal-oxide semiconductors *are* stable—titanium oxide and strontium titanate, for example—they all have bandgaps in the near-ultraviolet and thus do not efficiently match the solar spectrum. In practice it is also usually necessary to provide an additional external bias source—either a battery or a conventional solar cell—to allow for the fact that band-bending potentials of only a fraction (typically half) of the semiconductor bandgap are actually realized.

The best use of the photoelectrolytic cell may be to supply electric power rather than to produce hydrogen. Adding suitable ionic and neutral species to the electrolyte semiconductor can stabilize anodes that otherwise erode. For example, adding potassium hydroxide, potassium selenide, and selenium to water produces a bath that stabilizes cadmium selenide. The reactions

$$2Se^{--} + 2h^+ \rightarrow Se_2^{--}$$
$$\text{(at CdSe anode)}$$
$$Se_2^{--} + 2e^- \rightarrow 2Se^{--} \text{ (at cathode)}$$

proceed with no net change in electrolyte content and without anode attack. Efficiencies of 7–9% have been reported for cells with cadmium-telluride, cadmium-selenide, or gallium-arsenide anodes stabilized with polysulfide or polyselenide redox systems (*19*). The gallium-arsenide cell is still subject to corrosion at a low rate, and undetermined factors cause deterioration of the outputs of the other cells mentioned also, with the result that their useful projected lifetimes are only a few years or less, which poses a problem. In terms of manufacturability, the liquid cell would appear to offer advantages and disadvantages, and it seems premature to suggest that this approach will lead to important practical applications.

Prospects and prognosis

A discussion of the degree to which photovoltaic cells might eventually contribute to a solar energy economy must assume one of several systems concepts and should consider the overall environmental and economic viability of photovoltaic conversion in competition with other solar and nonsolar energy technologies. In comparison with several other solar technologies (see ref. 20), it appears that photovoltaic conversion—even after the cost and performance goals discussed here are reached—will make a significant but not a dominant contribution.

The recent progress in silicon technology makes it appear reasonable to project that 10% of our total energy needs in the year 2020 could be met with solar cells—possibly of the concentration type, which provides easier integration with a thermal recovery system, or more probably with the simpler rooftop arrays of low-cost, 12% efficient, oriented polycrystalline silicon cells. Either possibility assumes decentralized photoelectric generation interconnected to conventional central electric-generating stations, perhaps fired with char-oil from biomass or driven in the near term by coal or fission power. The likelihood that solar cells would be used for central station electric generation appears remote.

It is also reasonable to expect that gallium arsenide cells will be developed for space applications, and, given their superior performance and radiation resistance, they may come to dominate that market. A thin-film version suitable for terrestrial applications might well result as a spin-off, in a parallel to the history of the development of the silicon cell. There is as yet no reason to expect that any of the other solar cell materials discussed here are as likely to see wide practical application.

The fundamental limits set by physical laws and the practical limits set by economic constraints combine to indicate that solar cells alone will probably never carry the full burden of a solar energy society. It *does* appear possible that solar cells could meet about 30% of our predicted total electric needs (that is, 10% of total energy needs). Although the development of practical cells that can achieve this goal appears imminent, a large manufacturing industry, with a product volume of approximately 5 $\times 10^9$ dollars per year, will have to be established to provide the necessary quantities. For this reason extensive implementation of photovoltaic conversion appears further in the future than other solar technologies, and the time-scale of thirty to fifty years envisioned by the United States national solar energy program seems realistic.

References

1. See, for example, W. G. Pollard. 1976. The long-range prospects for solar energy. *Am. Sci.* 64:424–29.

2. D. M. Chapin, C. S. Fuller, and G. L. Pearson. 1954. A new silicon *pn* junction photocell for converting solar radiation into electrical power. *J. Appl. Phys.* 25: 676–77.

3. M. Wolf. 1975. Outlook for Si photovoltaic devices for terrestrial solar-energy utilization. *J. Vac. Sci. Technol.* 12:984–99.

4. P. A. Rappaport. Progress in low cost solar cells. Paper 1.3, Int. Electron Devices Meeting, Washington, D.C., Dec. 1976.

5. Solarex Corp. Beltville, Md.

6. L. M. Magid. 1976. The current status of the U.S. photovoltaic conversion program. In *Proc. 12th IEEE Photovoltaic Specialists Conf.,* p. 607. N.Y.:IEEE.

7. L. W. James and R. L. Moon. 1975. GaAs concentrator solar cells. In *Proc. 11th IEEE Photovoltaic Specialists Conf.,* pp. 402–08. N.Y.:IEEE.

8. W. D. Johnston, Jr. 1976. Vapor phase epitaxial growth, processing and performance of AlAs/GaAs heterojunction solar cells. In ref. *6*, pp. 934–38.

9. E. L. Burgess and M. W. Edenburn. 1976. One kilowatt photovoltaic subsystem using Fresnel lens concentrators. In ref. *6*, pp. 774–80.

10. H. Fischer and W. Pschunder. 1976. Low cost solar cells based on large area unconventional silicon. In ref. *6*, pp. 86–92.

11. L. P. Hunt, V. D. Dosaj, J. R. McCormick, and L. D. Crossman. 1976. Production of solar grade silicon from purified metallurgical silicon. In ref. *6*, pp. 125–29.

12. K. V. Ravi, F. V. Wald, R. Gonsiorski, H. Rao, L. C. Garone, J. C. T. Ho, and R. O. Bell. 1976. An analysis of factors influencing the efficiency of EFG silicon ribbon solar cells. In ref. *6*, pp. 182–90.

13. D. E. Carlson, C. R. Wronski, A. R. Triano, and R. E. Daniel. 1976. Solar cells using Schottky barriers on amorphous silicon. In ref. *6*, pp. 893–95.

14. W. S. Fyfe. 1974. *Geochemistry,* p. 9. Oxford: Clarendon Press.

15. W. D. Johnston, Jr. 1977. Vapor phase epitaxial growth of n-AlAs/p-GaAs solar cells. *J. Cryst. Gr.* 37:117–27.

16. P. Vohl, D. M. Perkins, S. G. Ellis, R. R. Addis, W. Hui, and G. Noel. 1966. GaAs thin film solar cells. *IEEE Trans.* ED-14:26–30.

17. S. Wagner, J. L. Shay, K. J. Bachmann, E. Buehler, and M. Bettini. 1977. Chemistry and preparation of InP/CdS solar cells. *J. Cryst. Gr.* 37:128–36.

18. M. S. Wrighton. 1977. The chemical conversion of sunlight. *Tech. Rev.* 79:31–37.

19. K. C. Chang, A. Heller, B. Schwartz, S. Menezes, and B. Miller. 1977. Stable semiconductor liquid junction cell with 9% solar-to-electrical conversion efficiency. *Science* 196:1097–99.

20. In addition to the article by Pollard cited in ref. *1*, see also W. G. Pollard, 1976, The long-range prospect for solar-derived fuels, *Am. Sci.* 64:509–13, and M. A. Duguay, 1977, Solar electricity: The hybrid system approach, *Am. Sci.* 65:422–27.

"Our problem, once solar energy is in operation, is to find a way to have the citizens whose homes are heated by the sun continue to pay *us* every month."

M. A. Duguay

Solar Electricity: The Hybrid System Approach

Generation of solar electricity may first be achieved most economically in systems that combine it with other functions, such as heating and lighting

Part of the worldwide activity in solar energy research is aimed at deriving electricity economically from sunlight. Photovoltaic solar cells, which convert sunlight directly into electricity, are already profitably used in some portable electronic equipment and in remote locations not connected to national power grids. The present challenge is to generate solar electricity, even if only in small amounts at first, at a cost that is competitive with electricity put on the network by conventional central power plants.

There is universal agreement that the economic generation of solar electricity would be an achievement of major importance. As the supplies of economically recoverable oil, natural gas, and uranium approach exhaustion in the 1990s, new sources of energy will be needed. The United States has much coal to fall back on, but the mining and burning of coal impose a tremendous burden on the environment and present serious health hazards to the general population. The controversial nuclear breeder reactor will not have demonstrated its technical feasibility, economic viability, and societal acceptability for yet a decade or two and

Michel Duguay received his B.S. from the University of Montreal and his Ph.D. from Yale University, both in physics. Since 1966 he has been employed by Bell Telephone Laboratories, where he has done work in the field of ultrashort laser pulses. From August 1974 to February of this year, he was on leave of absence to Sandia Laboratories in Albuquerque, New Mexico, where he worked in the field of solar energy. In March of this year he returned to Bell Labs to work on lasers and semiconductors. The author acknowledges the help and support of Everet H. Beckner, Robert P. Stromberg, and Robert M. Edgar of Sandia Labs. Address: Bell Labs 4B-423, Holmdel, NJ 07733.

therefore cannot now be relied upon with confidence. For many nations of the world that have neither coal nor the inclination to run nuclear reactors, solar electricity may be the most desirable option for the future.

A wide variety of schemes have been proposed for deriving electricity from sunlight (*1–3*). Many of the solar systems that are undergoing active development at the present time utilize either of two schemes: conversion of sunlight to heat and heat to electricity via a heat engine (e.g. a steam turbine) or direct conversion of sunlight to electricity by the photovoltaic effect in semiconductors (*4,5*).

Solar central power plants based on these two schemes have been analyzed (*6,7*), and initial experimental efforts are under way. Some writers have expressed pessimism about the future economic viability of such solar central power plants (*8,9*), and the central receiver or "solar tower" project, in particular, has been the subject of much criticism. In an experimental installation now under construction at Sandia Laboratories in Albuquerque, New Mexico, thousands of sun-tracking mirrors will be used to reflect and focus sunlight into boilers placed at the top of a central tower. Steam generated in the boilers would later be used to drive a turbogenerator.

A recent calculation by W. G. Pollard finds that *base load* electricity (that generated steadily 24 hours a day) from the solar tower would be about four times more expensive than electricity produced at present by coal-fired power plants (*9*). In that calculation, the cost of solar electricity was doubled by the assumed need to store

power for five days of cloudiness. However, economics are more favorable to solar electricity if it is applied directly, with little or no storage, to the *intermediate load,* i.e. to the rise in electric demand that accompanies daylight hours (*10*). Much of daytime electricity in the United States is generated by burning precious oil and gas. Because of the high fuel costs and relatively lower efficiency of electrical generation, intermediate load electricity is more expensive and therefore constitutes a better market for early penetration by solar devices.

Another way of improving the economics of solar electricity, and the one which will be discussed here, is to combine electrical generation with other services, such as heating and lighting. Since sunlight is present everywhere, *central* generation of solar electricity is not a necessary constraint, and we can envisage multiple-use or hybrid systems installed at the point of use. A focusing mirror mounted on the roof of a building, for example, can concentrate sunlight onto a photovoltaic device and thereby provide the building with both electricity and heat (*11*). Commercial systems that use sun-tracking concentrators for the single purpose of heating water are already finding a market that is sure to expand with the passage of every cold winter. The incorporation of solar cells into such systems would result in a situation where the economics of solar electricity and solar heating are helping one another.

Sunlight is most valuable in its pristine form, i.e. as light, a fact which was recognized early (Genesis 1:1). A simple calculation shows that the present monetary value of sunlight, delivered as light for illuminating interior spaces, is about ten times its value when used for heating. This simple fact speaks well for the prospective economics of the second hybrid system described in this article, namely, a system that uses the visible part of sunlight for lighting and converts the infrared to heat and electricity in solar cells.

Solar heat plus electricity

Figure 2 shows a house in Albuquerque, New Mexico, which has been equipped with commercial suntracking concentrators for heating. Aluminized mylar mirrors, shaped to a parabolic trough, rotate about a north-south axis as they track the sun. The reflected solar flux is concentrated by about a factor of 5 by the time it impinges on a water-filled blackened copper pipe that coincides with the mirror rotation axis. Only the direct or collimated part of sunlight is utilized in concentrating solar collectors. However, a recent study of insolation data for the U.S. (12) shows that a tracking concentrator "sees" over a year as much and often more solar energy than a fixed flatplate collector tilted south at the optimum angle (the latter collects both the diffuse, or scattered, and the direct sunlight). Tracking concentrators of the type shown in Figure 2 are now priced at about \$100/m² and have an efficiency (ratio of collected to incident solar energy) on the order of 60%.

Sun-tracking concentrators used for heating offer at their focus a place where photovoltaic solar cells could begin to generate electricity economically. Single-crystal silicon cells were selling in 1976 for about \$15 per peak Watt. At 10% electric efficiency and 1 kW/m² peak insolation, this translates to 15¢/cm² of cell area. In present applications, solar cells are simply put in the sun in some optimal fixed position. Electricity thus generated by single-crystal silicon cells is about 40 times more expensive than that available from the network. But if similar cells are placed at the focus of a concentrator, where the solar flux has been increased 40 times or more, the cost of the solar cell itself ceases to be a deterrent to generating solar electricity economically.

Photovoltaic solar cells could be incorporated into a concentrator heating system, as shown in Figure 3, by

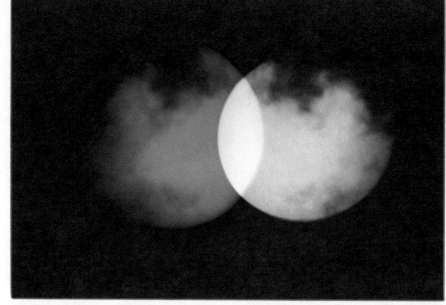

Figure 1. The parts of the walls and ceiling where the projected beams of sunlight impact become the "light fixtures." The overlapping blue and yellow circles—separated by a color filter to provide an aesthetic lighting effect— are images of the sun and make up a "solar light bulb." The projected circles can serve as a local weather indicator; here, they are partially obscured by moving clouds passing overhead.

covering one side of the water-filled pipe with close-packed solar cells. At present, between 10% and 20% of the concentrated solar flux can be converted to DC electricity in various types of semiconductor devices. The major part of the collected energy is in the form of heat which can be profitably recovered by the water-filled pipe. Photovoltaic devices have been operated at temperatures of up to 100°C without seriously degrading their electrical conversion efficiency.

To give an idea of where the economics of such a hybrid system now stand, let us consider a location on the East Coast, where 6 Gigajoules (GJ) of direct solar radiation might typically be available each year to a tracking concentrator that presents to the sun a 1 m² cross-section. (For the American Southwest, a typical value is 9 GJ/m²/year, i.e. a factor of 1.5 times more.)

Let us assume a geometrical concentration ratio (ratio of the area over which sunrays are intercepted to the area onto which they are focused) of 33, a collection efficiency of 70%, and an overall efficiency of 10% for converting collected sunlight to AC electricity. Assuming a unity load factor (i.e. all the output is consumed as generated throughout the year), the system would deliver 3.8 GJ$_t$ of

heat and 0.42 GJ$_e$ or 120 kWh of AC electricity.

At present, the following approximate cost figures appear reasonable: \$100/m² for the concentrator, \$50 for 300 cm² of silicon solar cells (per 1 m² of concentrator), and \$15/m² for the DC to AC inverter. If the heat recovered is used to pay for the concentrator, and the electricity to pay for the solar cells and inverter, and if a fixed charge rate of 15% on invested capital is assumed, then (neglecting small expected costs for operation and maintenance) the levelized solar heat cost will be

$$\frac{0.15 \times 100}{6 \times 0.7 \times 0.9} = \$4.0/GJ_t$$

and the solar electricity cost will be

$$\frac{0.15 \times (50 + 15)}{6 \times 0.7 \times 0.1} = \$23/GJ_e$$

$$= 8.3¢/kWh$$

With heating oil selling at 42¢/gal (11¢/liter), the cost of heat obtained by burning oil at 70% efficiency is \$4.2/GJ$_t$, which makes solar heat from this system look very good for year-round applications like domestic hot water heating. When oil purchased at \$13/barrel is burned during the day to generate electricity at an overall efficiency of 25%, the fuel cost of that electricity is 3.2¢/kWh, and the price it can command is higher yet. At 8¢/kWh, solar electricity would not be so far off. Of course, a much more detailed analysis than this is required to determine how solar electricity would interface with the electric utilities and how much it would be worth to them (7). But this simplified analysis indicates that the economic gap that solar electricity has to bridge is not very large.

It was not so long ago that single crystal silicon solar cells were ten times more expensive than they are today. As experience and cumulative production increase, due in large part to a vigorous development effort on the part of the U.S. government, the cost of these remarkably simple devices is sure to move down the learning curve. Funds devoted to research in photovoltaics could be expected to lead to breakthroughs in key manufacturing processes or in new materials and structures for photovoltaic conversion. One ambitious goal that has been set by both government and private enterprise is to reduce the cost of photovoltaic solar cells from 15¢/cm^2 to about 0.5¢/cm^2 or less. At that price (which incidentally is about the price of a U.S. copper penny) solar cells without concentrators would become economically competitive. For single-crystal silicon cells, this requires a cost reduction by a factor of about 30. In the example discussed here, with concentration, a cost reduction by a factor of 3 will suffice.

As we will see, another way of reducing the relative cost of the solar cells is to increase the concentration ratio by using concentrators that focus sunlight in two dimensions. When the concentration ratio falls in the range of 100–1,000, the cost of the solar cells becomes a small fraction of the concentrator cost, and the problem is reduced to making a sufficiently cheap concentrator. This is the approach followed by one project at Sandia Laboratories (11).

Light and electricity from the sun

Sunlight is undoubtedly the best light. The gradual exclusion of natural light from the work environment is not to the credit of the modern industrial era. Much evidence has accumulated on the inadequacy of artificial light sources at maintaining physiological and psychological well-being (13–15). When the sun shines outside, why not bring its

Figure 2. Sun-tracking concentrators for space and hot water heating are constructed on the roof of a house in Albuquerque, New Mexico. The concentrators are cylindrical concave mirrors (shown in cross-section in Fig. 3) which rotate about an axis approximately parallel to the Earth's polar axis and focus sunlight onto a blackened water-filled pipe. (Photo courtesy Albuquerque Western Industries.)

cheerful rays inside for lighting? Here again a hybrid system can be devised, where sun-tracking concentrators coupled into an optical distribution network could be used not only to satisfy the lighting needs of a building but also to generate electricity and heat in solar cells (16). The combined economics of such a system appear to be quite good.

Figure 4 illustrates one way in which sun-tracking mirrors could be used

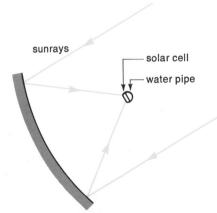

sunrays

solar cell

water pipe

Figure 3. A cross-sectional view of a cylindrical mirror-type concentrator that might be used in a hybrid system for heating water and generating electricity shows the photovoltaic solar cell that absorbs the focused sunlight and can transform 10–20% of it to electricity. The remaining energy appears as heat which is recovered via the water pipe.

partially to light, electrify, and heat a multistory building. A flat mirror, M_1, tracks the sun in both azimuth and elevation in order to reflect a sunlight beam in a fixed direction, toward a curved mirror, M_2 (Fig. 3). This mirror focuses the beam through a hole made in the roof and ceiling. The sun-tracking optics are placed under a transparent cubicle for protection against various weather conditions. After emerging from the hole, the beam is recollimated by a lens, L_1. A so-called "cold" mirror, or dielectric filter, F, reflects the visible portion of the solar spectrum (wavelengths under 7300 Å) and transmits the infrared, which is focused by a mirror into photovoltaic solar cells optimized for the near infrared, from which electricity and heat are recovered.

Small mirrors, like M_3 and M_4, intercept light from the visible beam and spread it around the room as needed for illumination. The remaining part of the beam is relayed by lenses, L_2 and L_3, to lower stories. Light reflected or scattered from these lenses is not lost since it serves to illuminate the rooms in which the lenses are installed.

Controlled sunlighting creates a very pleasant lighting effect: one corner of my windowless office in Sandia Laboratories was brightly lit by controlled beams of sunlight (16). In a small-scale prototype experiment, two sun-tracking mirrors were used in an arrangement similar to that shown in the upper part of Figure 4, with a total effective collection area of 0.33 m². The office, which was 3 × 5 m, was located in the middle of a one-story building and had no conventional access to natural light.

Following a conservation ethic, the first task is to light the work area, in this case a desk top, at the adequately high level of 1,000 lumens/m². This is achieved by broadcasting a beam of sunlight onto a sheet of translucent diffusing plastic placed about one meter above the desk, which is thus uniformly illuminated with diffuse shadowfree light.

The remainder of the office is lit by projecting beams of sunlight onto the walls and ceiling, which ideally should have matte white surfaces where the beams strike so that the sunlight will be diffused all around the room. The parts of the walls and ceiling where the beams impact become the "light fixtures." A beautiful colored lighting effect is obtained without any energy waste by placing a dielectric color separation filter in one of the white sunlight beams. What the filter does not reflect is transmitted, and the complementary colors blue and yellow recombine in white in the room (Fig. 1).

One potentially important aspect of a sunlighting system is that it gives the worker visual contact with the outside, thereby fulfilling what appears to be a basic psychological need. The sunlighting system can operate through certain types of cloud covers. In fact, clouds seen running across the projected suns, as in Figure 1, give a strong feeling of being in contact with nature. Electronically controlled interfacing with a conventional electric system would of course maintain an adequate lighting level at all times. In a practical system, solar and electric lighting would be completely integrated.

Sun-tracking mirrors (or heliostats) can bring light into buildings in other ways as well. A precedent in controlled sunlighting was established in 1974 at the Hyatt Regency Hotel in Chicago, where three heliostats were used to send diverging beams of sunlight vertically down through the glass roof of the hotel lobby, thereby

creating a pleasing ornamental lighting effect (17). Where building geometry permits, beams reflected in this simple manner from flat mirrors may be used for sunlighting (18). The advantage of combining focusing optics with a heliostat is that it makes the fitting of existing buildings much easier, and the infrared can be separated out economically by using a small "cold" mirror (the cost of large "cold" mirrors would be prohibitive).

Economics of sunlighting

The anticipated economics of this triple use system are made more attractive by the high cost of artificial light as a form of consumed energy.

Fluorescent lamps transform 20% of the electricity they consume to visible light energy, the rest appearing as heat or invisible radiation. Incandescent lamps, used for their directionality (reflector lamps) or color rendition, are about 10% efficient. The cost of visible light energy is therefore 5 to 10 times the cost of electricity. Assuming a cost of 4¢/kWh for daytime electricity, this implies an operating expense of $55/GJ_i (i stands for illumination) for light from fluorescent lamps, and $110/GJ_i for incandescent light. When the cost of lamp replacement is added, the figures are increased to about $65/GJ_i and $130/GJ_i.

About 50% of the solar spectrum is in

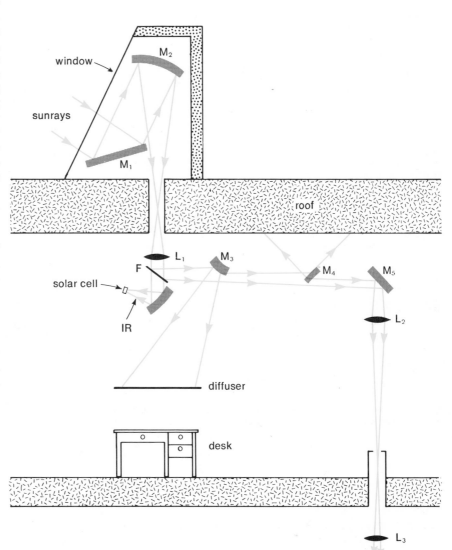

Figure 4. This proposed system distributes direct sunlight throughout a building for lighting the rooms. A "cold" mirror, F, is a dielectric filter which reflects the visible part of the solar spectrum (4000–7300 Å) and trans-mits the infrared. The visible is used for lighting and the infrared is converted to electricity and heat in a solar cell. Visible beams can be relayed to various parts of the building by means of lenses (L) and mirrors (M).

the infrared, at wavelengths above 7300 Å, and 50% in the visible. We lump into the "visible" the solar UV, which is an important biologically active band extending from 4000 to 2700 Å; at sea level, about 4% of the available energy falls there. Some studies suggest that the solar UV may affect vision in some beneficial way (13,15). The solar UV could be transmitted in the sunlighting system by UV transparent plastic windows and lenses and aluminum mirrors.

For a concentrator with an effective collection area of 1 m^2 and an efficiency of 65%, we have, in the 6 GJ/m^2/year of incident direct sunshine at a typical East Coast site, a light energy value (or savings value) of $6 \times 0.5 \times 0.65 \times 65 = \$127/m^2/year$. Where sunlight would compete with incandescent light, the value is much higher, $253/m^2/year. By comparison, at the present cost of oil heat, the heating value of the solar radiation incident on the same concentrator, using the same efficiency is $6 \times 0.65 \times 4.2 = \$16/m^2/year$. Sunlight delivered as light is therefore potentially worth 7 to 14 times more than sunlight converted to and delivered as heat.

The question then is, How much is the sunlighting system going to cost? The key element in the system is the heliostat which serves to "stop the sun." Heliostats employing extremely precise gears and massive rigid structures have long been manufactured for astronomers at a cost on the order of $10,000/m^2, clearly a prohibitive figure in solar applications. In these instruments, a tracking accuracy of a fraction of an arc-second was achieved, which of course is a superfluous luxury in solar applications. In designing heliostats for projects such as the "solar tower," a new philosophy is being followed. Mirrors will be mounted on relatively crude mounts which will be controlled by a small computer, and this computer, probably aided by reflected beam position sensors, will compensate for the impreciseness of the mount. Fortunately, small computers are also inexpensive; with many mirrors sharing one computer, a computer cost of about $20/m^2 of heliostat can be anticipated in sunlighting a commercial building.

Estimates of projected heliostat costs in mass production fall in the range $50–200/m^2. Since a ballpark figure of $100/m^2 has been quoted by solar advocates and critics alike (3,9), it will be adopted here for the mirror, M_1, in Figure 4. A cost of $100/m^2 would imply a lightweight device (about 25 kg/m^2 if the current cost of $4/kg for automobiles is used), which would make the transparent cubicle in Figure 4 a welcome addition, insuring a stable reflected beam on windy days and providing protection. For a 1 m^2 heliostat, a 4 m^2 cubicle could be built, as a sort of little "greenhouse," at an estimated cost of $100 in mass production (using a figure of $25/m^2 for greenhouse building materials). Other optical elements in Figure 4, plus installation and building modifications (holes, lens holders, etc.) might add another roughly $200/m^2 of heliostat. The total system cost would add up to about $420/m^2.

In this system, silicon solar cells would be used to convert the separated solar infrared to electricity. The higher quantum efficiency of silicon in the 7300–11,000 Å band would most likely make possible overall conversion efficiencies of 20–30% in that band for cells optimized at these wavelengths. Adopting the lower figure and assuming that 35% of the incident solar radiation falls in the 7300–11,000 Å band (the exact percentage will depend on local atmospheric conditions), then 7% of the energy delivered by the concentrator would be converted to AC electricity. This is somewhat less than in the hybrid system we described for Figure 3, but the two-dimensional focusing allows reduction of solar cell cost. Choosing 200 for the geometrical concentration ratio, 50 cm^2 of silicon cells would be needed—amounting to $7.5 at the prevailing price of silicon cells. Inverter cost (prorated to output) would be $10/m^2.

In analogy with the hybrid system discussed earlier, the useful delivered light would pay for the concentrator and distribution optics, and the electrical output would pay for the solar cells. Assuming a 15% fixed charge rate, the levelized cost of solar light at the East Coast site is

$$\frac{0.15 \times 420}{6 \times 0.65 \times 0.5} = \$32/GJ_i$$

and the cost of the solar electricity

$$\frac{0.15 \times (7.5 + 10)}{6 \times 0.65 \times 0.07} = \$9.6/GJ_e$$

$$= 3.4 ¢/kWh$$

A load factor (the percentage of sunshine hours that are usage hours) of unity was assumed in this calculation. This would apply in buildings that are in constant use, such as airports, restaurants, and some stores. An average load factor typical of places of work such as offices might be 70%, which increases the cost of sunlighting to $46/GJ_i$. This amount is still cheaper than fluorescent lighting ($65/GJ_i$ when paying 4¢/kWh for electricity) and much cheaper than incandescent lighting ($130/GJ_i$). Solar electricity from this type of system, which is already competitive with present daytime electricity, would become more so following anticipated reductions in solar cell costs. Of course, here it is a by-product of solar lighting and represents only a small output.

Another by-product of the sunlighting system is the heat that appears along with the end product of lighting. (In large well-lit buildings, a sizable fraction of the heating is currently derived from the lights.) An additional, less expected advantage is that the solar system can help keep buildings cool during the warm season. Visible sunlight reflected by a "cold" mirror has a luminous efficacy of about 200 lumens/Watt, compared to 50 lumens/Watt for fluorescent lights, including ballasts. Thus a solar lighting system would reduce the cooling load of a building, thereby improving indirectly the economics of solar lighting and electrical generation.

Potential of solar electricity

The simplified analysis presented here indicates that it is not unreasonable to think that electricity could be generated economically in hybrid solar systems. Of course, only a small percentage of the total demand for electricity could be satisfied at first. Heating all the hot water consumed in residential and commercial buildings in the U.S. by electric thermal hybrid systems, for example, would produce, as a by-product, solar electricity in an amount equal to only about 4% of the total U.S. electrical power output. A realistic appraisal of conversion possibilities in existing buildings might reduce the amount to perhaps 1%. Though this sounds like an insignificant quantity, on an absolute scale it represents an expenditure of approximately one billion dollars for imported oil used to generate the same amount of electricity in the course of a year.

Similarly, the sunlighting hybrid system at first would only make a very small contribution to national electrical production. Even though over half the electricity consumed in commercial buildings is used for lighting, that represents only 15% of the total U.S. electrical consumption (*19*). Since each day the probability of sunshine is typically 55–65% on the East Coast and 75–85% in the Southwest, the sun could probably be made to shine inside commercial buildings all across the country during roughly one-half of the normal working hours. If all these buildings were to be retrofitted with solar-lighting hybrid systems, about 7% of the electrical consumption could be saved. The by-product electricity from this wholesale shift to sunlighting would, at first, amount to only 0.2% of the present total electrical output. Of course, again, practical considerations of building and installing the system would reduce the solar lighting market in existing buildings, but demonstration of economic generation of *some* solar electricity in this restricted but sizable market would open the door for more.

There are at least two practical ways in which the market for solar hybrid systems could be expanded. First, new buildings could be designed in such a way as to improve the economics of solar hybrid systems. Just to give one example, buildings might be topped by a sort of greenhouse or solarium, made of low-absorption glass, which would provide a natural haven for sun-tracking concentrators, thereby reducing the cost when the solar systems were installed at a later time. Second, as experience and cumulative production increase in the hybrid systems industry, costs can be expected to decrease (move down the learning curve), while at the same time the systems will gradually expand their functions.

Electric thermal systems could expand into the large space heating market, where economics are tougher because of the small load factor. As people came to like having the sun in their shops and offices, solar lighting systems could be designed with an excess capacity sufficient to provide good lighting even at times when the sun is dimmed by haze or thin clouds (the experiment at Sandia Laboratories showed that the sun could sometimes be seen in the office even when the whole sky was blanketed by hazy clouds). On bright sunny days the excess visible sunlight could be diverted into a second set of solar cells (besides those for infrared) optimized for the visible. Overall conversion efficiencies in the 20–30% range could most likely be achieved that way, and the system might grow to the point where sunlighting itself would become a by-product of solar electric generation.

In this paper, we have been discussing solar electric generators that would operate in a fuel- and water-saver mode—that is, on sunny days, fossil fuels and water which would otherwise have been consumed in gas turbines or in hydroelectric plants are saved for cloudy days. Eventually, if solar electricity generated on sunny days came to exceed the intermediate load demand, excess daytime output could be stored, much as excess nighttime electricity from base-load plants is now stored in hydroelectric pumped storage stations (*21*). Only then—obviously far in the future— would solar electricity begin to displace base-load nuclear or coal-fired power plants.

On one acre (4,047 m²) of roof area there falls every year an amount of solar radiation worth approximately one million dollars in terms of its potential for lighting, heating, and electrifying a building. By comparison, typical U.S. uranium mines have 2–6 million dollars worth of uranium oxide per acre of ore-bearing land (*20*) (at the current $100/kg price for U_3O_8), and, obviously, it is a rare acre that has economically recoverable uranium. It is a long and arduous way from uranium-bearing ore to economic electricity, heat, and light in a building, yet, after an enormous effort, the U.S. has almost achieved that end. It is a short distance from sunlight on the roof to light, heat, and electricity under the roof by way of solar systems, but, of course, much research remains to be done. There are, however, reasons to believe that a vigorous development *and* research program, carried out in many nations, could lead to economically and otherwise attractive solar electric systems.

References

1. F. Daniels. 1964. *Direct Use of the Sun's Energy.* Yale Univ. Press.
2. A. B. Meinel and M. P. Meinel. 1976. *Applied Solar Energy: An Introduction.* Addison-Wesley.
3. *Astronautics and Aeronautics* 13, Nov. 1975. This issue contains a series of review articles on solar energy.
4. *Photovoltaic Devices,* Jan. 1977. A special issue of the IEEE Transactions on Electron Devices.
5. *Proceedings* of the 12th IEEE Photovoltaic Specialists Conference, Baton Rouge, LA, Nov. 1976.
6. A. F. Hildebrandt and L. L. Vant-Hull. 1974. Tower-top focused solar energy collectors. *Mech. Eng.* 96:23–27.
7. P. F. Pittman. 1976. *Conceptual Design and Systems Analysis of Photovoltaic Power Systems.* Westinghouse Electric Corp. Research Laboratories Report.
8. H. Bethe. 1975. The necessity of fission power. *Sci. Am.* 234:21–31.
9. W. G. Pollard. 1976. The long-range prospects for solar energy. *Am. Sci.* 64:424–49 and (correspondence) 608–9.
10. *Solar Energy Intelligence Report* 2:203–5, Dec. 1976.
11. D. G. Schueler, J. G. Fossum, E. L. Burgess, and F. L. Vook. 1975. *Eleventh Photovoltaic Specialists Conference Record,* pp. 327–30. IEEE Catalog No. 75CH0948-OED.
12. E. C. Boes, I. J. Hall, R. R. Praire, R. P. Stromberg, and H. E. Anderson. 1976. *Distribution of Direct and Total Solar Radiation Availabilities for the USA.* Sandia Labs. Report No. SAND76-0411.
13. J. N. Ott. 1976. *Health and Light.* Simon and Schuster.
14. R. J. Wurtman. 1975. The effects of light on the human body. *Sci. Am.* 233:68–77.
15. L. Thorington, L. Parascandola, and L. Cunningham. 1971. Visual and biologic aspects of an artificial sunlight illuminant. *J. IES* 67:33–41.
16. M. A. Duguay and R. M. Edgar. 1977. Lighting with sunlight using sun-tracking concentrators. *Appl. Optics* 16:1444–46.
17. W. Basich. Pers. comm.
18. D. G. Carson. Pers. comm.
19. L. Schipper and A. J. Lichtenberg. 1976. Efficient energy use and well-being: The Swedish example. *Science* 194:1001–13.
20. D. F. Irving. Pers. comm.
21. M. G. Morgan. 1975. *Energy and Man: Technical and Social Aspects of Energy,* pp. 154–61. IEEE Press.

William G. Pollard

The Long-Range Prospects for Solar Energy

Solar electricity can make only a limited contribution to the nation's large-scale energy needs

There is, in the United States and many other countries today, a rapidly developing conviction among the public at large that energy from the sun can and will ultimately replace fossil fuels and nuclear electricity in meeting mankind's energy needs. This conviction expresses itself in almost every public discussion of energy problems and at the Congressional level in increasingly vehement demands that federal agencies sponsor major research and development programs on solar energy. The public at large has an implicit faith in the capability of science and technology to achieve any desired end if only adequate funding is provided. This faith was amply justified by the dramatic successes of radar and atomic energy during World War II and the space program after the war. These scientific achievements are not, however, appropriate models for solar-energy research because none of them involves considerations of cost or competitive merit with respect to alternative ways of achieving the same desired end.

The problem in using the sun to meet mankind's energy needs is not primarily scientific or technological. Solar-energy systems can, without question, be developed and made to work from an engineering standpoint.

Dr. Pollard, a theoretical physicist, was Executive Director of Oak Ridge Associated Universities (formerly Oak Ridge Institute of Nuclear Studies) from its incorporation in 1946 until August 1974. Since then, and until his retirement at the end of April 1976, he was a staff member of the Institute for Energy Analysis, where, in addition to solar energy problems, he did studies on nuclear energy centers, or "parks," and on a non-nuclear future for the United States. Address: 191 Outer Drive, Oak Ridge, TN 37830.

Rather, the central issue is whether once achieved they would be commercially feasible as a practical component of the nation's energy system in competition with alternative means for providing the same energy. It is from such an economic and commercial standpoint that I shall attempt in this article to evaluate the long-range potential of various ways of using solar energy. In effect, the technology is assumed to be successfully established, and I shall be concerned with visualizing the resulting system in operation by industry in the normal business of selling electricity, fuel, or equipment. The first section of the article discusses solar radiation used directly, and the second section analyzes indirect uses.

Direct solar energy

Systems designed for the direct use of solar radiation can be broadly grouped into those for solar heating and those for solar electricity. The primary concern in this article is with the latter, but a brief summary of the former application will be useful.

Solar hot-water heaters were used extensively earlier in this century and were displaced only by the availability of cheap natural gas. As natural gas supplies become uncertain and the prospect for large price increases accompanying deregulation becomes more certain, they are coming back. They are commercially available, and although the initial investment is several times that of a conventional electric or gas water heater, the fuel savings (even at present prices) easily make up for it in a few years. There seems little doubt that as the price of electricity and fuel increases, solar hot-water heaters will be quite widely installed. A similar application is in

the heating of swimming pools, where only slightly warmed water is needed and thus very inexpensive plastic solar panels can be used.

Next in order of feasibility is the solar heating of buildings. Even in buildings equipped with fuel heating systems, architectural design can gain sufficient heat from the sun to reduce fuel consumption materially. For greater utilization of solar heat, there are systems commercially available that use flat, absorbing solar panels to heat water which is stored in an insulated storage tank and used as the source of heat for a standard hot-water or circulating-air heating system. For sunless days an auxiliary heater must be provided to replace the solar panels for heating the water in the storage tank. In areas with a large number of solar-heated buildings it would probably be necessary to use an oil- or gas-fired auxiliary heater rather than an electric one, since the latter could result in an unacceptably large additional demand on the electric utility.

Solar heating systems are presently being used in a number of residences and some office buildings and schools. Very few of them save enough fuel to amortize the large initial investment and to pay for maintenance. There are problems of cracked glass, corrosion and clogging of water tubes, leaks, and other difficulties. A number of companies, however, market such systems, and all have active development programs to improve efficiency and durability and reduce cost. From the very extensive literature on the subject, three recent review articles are cited for additional information (1).

Several factors govern the extent to

which solar heating of buildings will be utilized in the long range. There will always be a number of people for whom the idea of tapping solar energy is very appealing and who will therefore go to considerable lengths to have it. It will be most attractive in areas of the country with a maximum amount of sun or excessive fuel costs, and least attractive in areas with frequent prolonged periods of sunless days or frequent snows requiring snow removal from the solar panels and in cold northern climates in which requirements for supplementary heating greatly exceed the solar heating. The initial capital investment will prove a barrier to many; others will be reluctant to take on the possibility of excessive maintenance burdens of cleaning and repairing large areas of solar collectors or clogged or leaking water channels. Some will be concerned about the possibility of future building construction blocking direct exposure to the sun or the incompatibility of solar panels and trees.

Turning now to the prospects for solar electricity, two major approaches are presently under vigorous research and development: solar-thermal and photoelectric systems. Solar-thermal systems use mirrors or lenses to concentrate the solar radiation on a receiver or receivers where its energy is absorbed as heat in a working fluid such as steam. The hot working fluid then drives a turbine or some other form of heat engine to generate the electricity. Photoelectric systems use silicon, germanium, cadmium sulfide, or some other photosensitive material in thin sheets to convert some fraction of the solar spectrum directly into electrical energy.

Both these systems, and any other that might be developed in the future, are constrained by certain limitations imposed by the nature of solar energy that cannot be circumvented by any amount of research and development. There are four categories of such constraints: the efficiency of the system in converting solar radiation into electricity, the low intensity of solar radiation coupled with the resulting land-area requirements, and the intermittent supply of solar energy. Any conceivable land-based direct solar-electric system can be evaluated under these four headings. A reference system of more or less standard size and cost is useful for such evaluations, and for this purpose a coal-fired or nuclear electric plant with an installed capacity of 1,000 megawatts generating 5 billion kilowatt-hours per year (corresponding to a capacity factor of 57%) will be used. An arbitrary figure of $1,000 per kilowatt of electrical energy, in 1975 dollars, or $1 billion, will be used for the cost of constructing the plant. In specific applications the actual values could, of course, deviate considerably from these, but they are rough guides to assist the reader in visualizing the scope and cost of substitute solar-electric plants.

Conversion efficiency. For the purpose of this analysis, the simplest measure of the conversion efficiency of a solar-electric system is the amount of heat, measured in British thermal units, received from the sun from all directions on the solar collectors of the system for each kilowatt-hour of electricity generated. This measure is the annual average gross heat rate, H, of the system in Btu/kWhr. In solar-thermal systems the instantaneous heat rate is subject to daily and seasonal variations with solar elevation and intensity that are not nearly as pronounced in photoelectric systems. In both cases, fluctuations are averaged over a full year. The smaller the value of H, the more efficient the system is in converting solar radiation into electricity.

One of the more promising solar-thermal systems under active development consists of a tall tower on which is mounted a solar-radiation receiver with a field of heliostats spread out from its base, each of which reflects an image of the sun into the receiver. For such a system the minimum achievable value of H is 15,000. This value could be realized only on a site at which over 80% of solar radiation throughout the day is direct from the sun, over 85% of that falling on the heliostat mirrors over long periods is reflected into the tower-receiver, and the loss in the receiver in converting received heat into high-pressure steam is less than 5%. In that case the product of these factors allows two-thirds of the 15,000 Btu received on the solar collectors, or 10,000 Btu, to be retained in high-pressure steam. The steam pressure and temperature contemplated for such systems are sufficient for the generation of 1 kWhr of electrical energy from this 10,000 Btu of thermal energy. On other sites with a higher percentage of diffuse radiation that cannot be reflected by mirrors—or with systems of lower continuous reflectivity, greater receiver heat loss, or lower efficiency of steam-electric conversion—the heat rate could easily be double this minimum value, or 30,000 Btu/kWhr. We will assume that, in the long range, R&D on systems of this type will achieve a gross annual average heat rate of 30,000 or less but no smaller than 15,000.

Photoelectric cells convert a fraction of the solar spectrum directly into electricity. At a heat rate of 15,000 this fraction would be 23%, which is probably more than can be achieved by even the most optimistic research breakthrough. At a heat rate of 30,000 the fraction is 11%, which is realized and somewhat exceeded in present photoelectric cells. Thus an annual average minimum heat rate of 15,000 and a maximum of 30,000 can be taken to bracket both systems—and probably any other solar-electric system that might be developed in the future.

Low intensity of solar energy. The total solar radiation received per year on a unit area of the earth's surface varies considerably over the United States. We designate this annual insolation per unit area by J and measure it in Btu/ft^2/yr. In the deserts of the American Southwest there are sites for which J is 750,000, but this is a maximum value for the country. (At a site 40 miles north of Phoenix in the Sonora Desert annual solar insolation was measured for over 20 years; the average for 1955–74 was 696,000 Btu/ft^2/yr; the minimum value of 628,000 occurred in 1966, and the maximum of 734,000 in 1956, 2.) In the lower Midwest and the Southeast a typical value of J is 525,000. Other areas with a value of J less than 500,000 have so many sunless days that no type of solar-electric system would probably be practical.

A solar-electric plant with a heat rate H built on a site with annual insolation J will generate J/H kWhr of electricity annually for each square foot of solar collector. Thus a plant of maximum efficiency on a Southwestern desert site might deliver 750,000/15,000, or 50 kWhr/ft^2 of electricity annually, while the same plant on a Midwestern or Southeastern site would deliver only 35 kWhr/ft^2 annually. A plant with a heat rate of 30,000 would deliver only half as much electricity, or 25 and 18 kWhr/ft^2, respectively.

In order to match the output of the reference plant (5 billion kWhr/yr) with a system heat rate of 15,000, 100 million ft^2, or 2,300 acres, of solar collectors would be required on the most favorable site, and 140 million ft^2, or 3,300 acres, on the less favorable site. The least efficient systems considered in this analysis would require double these areas. Assuming that the capital cost of the reference system is $1 billion, and that it is desired that the solar-electric plant be no more costly, then the cost of the most efficient plant should not exceed $10/ft^2 of solar collector on the desert site or $7/ft^2 on other sites. These cost limits become $5 and $3.50/ft^2, respectively, when the system heat rate is 30,000.

For the tower-heliostat plant about half the cost is for heliostats and the other half for the tower, receiver, steam lines, turbine, generator, cooling towers, and land. Thus the heliostats should not cost more than $5/ft^2 of mirror area even on the best site. For the photoelectric-cell system the facilities other than solar collectors (consisting of land, mountings, electrical-collection network, and DC-AC converters) are considerably less costly, so that the solar cells could cost perhaps $9/ft^2 at the best site.

The heliostats might consist of a 25-ft^2 mirror on a suitable foundation and mounting and be equipped with an accurate two-axis tracking mechanism responsive to a central control continuously monitoring the position of the sun. Protection of the mirrors against damage by storms, hail, and windblown sand would be necessary to ensure a long, trouble-free life expectancy over which high mirror reflectivity would be maintained. For the reference system, on the sunniest site with premium efficiency, 4 million heliostats would be required. The cost limitation of $5/ft^2 of collector area means that they must be manufactured, delivered, and erected on the site (on an adequate foundation) for no more than $125 each, a requirement that calls for great inventiveness but does not constitute research in the usual sense of the term. In the case of the photoelectric system the requirement is for 100 million

ft^2 of solar cells manufactured and installed on the site for $9/ft^2 or less. This requirement does call for the highest order of research in solid state physics combined with cost-saving mass production techniques, since present solar cells cost several hundred dollars per square foot at best. If the best heat rate achievable turned out to be 30,000 Btu/kWhr for either system, and it was built in the Midwest or Southeast, 11 million 25-ft^2 heliostats would be required at a maximum cost of under $2/ft^2, or $50 each, or 280 million ft^2 of solar cells at a maximum cost of $3.50/ft^2.

Land requirements. The low intensity of solar energy means that very large areas of land are needed for the construction of a solar-electric system. To provide maintenance access to heliostats or solar-cell panels and to avoid shading of one set by an adjacent set, a land area of over twice the area of solar collectors is required. At a heat rate of 15,000 this means 7 to 10 square miles of land for the reference system, the smaller area being for the Southwestern desert sites. At the 30,000 heat rate, twice the area would be required.

Considering that a 1,000-MW plant is a relatively small component in most electric utility systems, these large land areas constitute a major barrier to future large-scale deployment of solar-electric plants. In the long range we are considering here— when larger populations lead to increased competition for land for agriculture, forestry, recreation, wilderness preservation, and community development—the procurement of such large blocks of land for electric-power generation is likely to be a severe problem in parts of the country other than some desert sites. There is also the environmental and aesthetic problem of covering hundreds of square miles of American deserts or the lower Midwest and South with mirrors or photocells.

Intermittency of solar energy. Any direct solar-electric system can, of course, generate electricity only when there is sun. No generation can take place between sunset and dawn on any day, and even in the Southwestern desert there can be 4- to 5-day periods of bad weather, during which the system would be inoperative. The result is that even on the sunniest sites electricity can be generated only

40% of the time, and on most other sites 30% of the time, or less, annually.

One way to deal with the intermittency of solar energy is to use solar electricity to supply an industry that would operate only when the sun was out. Some industrial processes requiring large amounts of electricity could be designed to operate intermittently. At night and on sunless days the plant would simply close down, and the employees, like farmers, would occupy themselves in other pursuits. Such applications are, however, clearly very limited. The major portion of the demand for electricity must be met in more continuous and dependable ways.

A second alternative is to operate the solar-electric plant in conjunction with other electric-generation facilities in such a way that the output of the combined system can match the demand for electricity at any time. A particularly attractive system of this type is being planned for the Salt River Project of Phoenix, Arizona (3), where a 100-MW(e) tower-heliostat system is to be built adjacent to the Horse Mesa Dam. The solar plant will have eight towers receiving concentrated solar heat from eight fields of heliostats covering the steep slope of the mesa. The calculated heat rate for the system is 16,100 Btu/kWhr. The Horse Mesa hydroelectric facility would be operated for electrical generation only when the solar-electric plant was not operating. The purpose of the solar plant is to conserve water for the hydroelectric plant. In this case the land area is available, and the practicality of the project depends on holding the cost of the heliostats to $5/ft^2, keeping the gross heat rate down to the target 16,000, and assuring the security of the investment in the system against weather damage for 20 years or more. If these conditions can be met, combined solar-electric and hydroelectric systems might prove attractive elsewhere in conjunction with impoundment dams of sufficient excess storage capacity, although the maintenance of an adequate water supply for downstream run-of-the-river dams might be an added problem.

Another alternative is to build a coal-fired steam generator to feed the steam lines of the generating plant when steam is not being supplied by the adjacent tower receivers. To make the combined system economic, the amortized capital cost of the heliostats and tower receivers would have to be less than that of the coal-fired steam generator plus the cost of the coal consumed. In view of the cost constraints on the heliostats, this may be unachievable. The same cost criterion would apply to any solar facility integrated into a utility network primarily using fuels for electric generation.

The only other way to deal with the intermittency of the solar supply is to store large amounts of the energy produced either as heat or as electrical or mechanical energy. A variety of both types of energy-storage facilities are currently being studied, and a few examples will serve to clarify the problem. A high-temperature heat-storage system developed by the Phillips Company uses fused fluoride salts as the storage medium. For a tower-heliostat system the salts could be fused directly in the solar receiver or indirectly with an intermediate heat-transfer medium such as a molten metal. The fused salts would later give up their stored energy to generate high-temperature steam. Another storage method that could be used with either solar-thermal or photoelectric systems, in areas where the terrain is suitable, would use 60 to 70% of the electrical output to pump water from a depressed reservoir into an elevated reservoir. From this reservoir, electrical generation could take place as required to meet demand as it does in any hydroelectric power plant.

In both storage modes the efficiency is presently about 75%, so that if 70% of the direct solar energy goes temporarily into storage, the area of solar collectors must be increased about 20% to yield the same annual output of electricity. The capital cost of either type of storage facility, with a capacity for storing eight days' output from the system, is about $800 million for the reference system, assuming $6/kWhr stored. This means that the capital cost of the reference system with this amount of storage and the additional collectors is double that of the system without storage.

Since photoelectric cells generate low-voltage electricity, this type of system is particularly suitable for storage batteries or the electrolysis of water to produce hydrogen. The ef-

ficiency of battery storage is about the same as that of pumped-hydro storage, but the cost is about five times as great, although present research can be expected to lead to higher efficiencies and lower costs per kilowatt-hour stored. In the other method, part of the photocell output is used to electrolyze water, the hydrogen produced is liquefied and stored, and fuel cells are then used to generate electricity on demand. However, when the energy losses in electrolysis and liquefaction are combined with the 50% efficiency of fuel cells, the electricity recovered from the fuel cells is only 30% of that originally supplied by the photocells. Thus if 70% of the photocell generation went through this storage mode, the area of photocells would have to be doubled to maintain the same annual output of electricity. The capital cost of electrolytic cells, liquefaction equipment, and fuel cells would be at least as great as that of storage batteries for the same storage capacity. Even allowing for improved efficiencies of batteries, electrolytic cells, and fuel cells, and for lower costs for the same capacity as a result of present and future research, the capital cost of a solar system with either method of storage would be from three to five times that of the same system without storage.

It should be evident from this analysis that the direct use of the sun is simply not a very suitable way to provide electricity in utility systems designed to supply it on demand. Any one of the factors discussed might well support this conclusion by itself, but all of them taken together are decisive. It seems quite unlikely that it will ever be possible to manufacture and install adequate 25-ft² heliostats for as little as $50–125 or to provide adequate maintenance of mirror reflectivity and tracking accuracy for millions of them spread over several square miles without excessive costs for operation and maintenance. It seems equally unlikely that photocells with an efficiency of 23% can be produced and installed for as little as $9/ft².

If such economic achievements are in fact impossible, then the initial investment in a reference solar-electric plant would be considerably more than the $1 billion assumed in the foregoing analysis. Moreover, such an investment would have to be made in

delicate equipment spread over 10 to 20 mi², which, during a 30-year life expectancy, would be subject to massive losses through storm damage. For the utility investing in such a system, sufficient energy storage to make it equivalent to a 1,000-MW coal-fired or nuclear reference plant would at least double—probably quadruple—the total capital investment required. Taking all these factors into consideration, it seems clear that electric utilities in the business of generating large blocks of central-station electricity will almost certainly decide not to invest in a solar-power plant, no matter how well engineered, as long as less costly and hazardous alternatives are available. To the utility, the investment would appear excessive for the expected return, and there would be little assurance that it would survive a normal 20- to 30-year operation period without major loss.

Indirect solar energy

For thousands of years man has been using water and wind power for mechanical work. Both are, of course, derived from the sun, one through lifting vaporized water to high altitudes and the other through the atmospheric turbulence that accompanies the maintenance of thermal equilibrium between the earth and the sun. Water power has long been a major source of electricity at modest cost, and hydroelectric resources will certainly be developed further as the world population and its need for electricity increases in the decades ahead. Wind power was extensively used for pumping water in the Netherlands until cheaper, less demanding, and more convenient ways to do so became widely available. In rural America in the 1920s and 1930s, windmill electrification systems were common but were abandoned when central-station power became available through the Rural Electrification Program. Now, impending energy crises have reawakened interest in wind power as a source of electricity.

The NASA-Lewis Laboratory, at Cleveland, Ohio, has built a prototype test unit that uses a two-bladed propeller, each blade of which is 62½ ft long. Mounted on a high tower and turning at 40 revolutions per minute, it generates 100 kW(e) when the wind speed is above 18 miles per hour. It is structurally strong enough to withstand occasional gale-force winds. The initial test unit is quite expensive, but it is hoped that this cost can be substantially reduced as production models are built in quantity. In areas of frequent wind, such as Oklahoma and West Texas, wind speeds above 18 mph may occur up to 30% of the time. An actual unit tested in Vermont generated electricity 14% of the time, and one in Denmark 23% of the time. If electrical generation occurred 20% of the time on an annual basis, the output of the NASA-Lewis generator would be 175,000 kWhr/yr. It would require nearly 30,000 of such installations to equal the electrical output of the reference plant used in the solar-electric analysis.

The intermittent nature of wind leads to many of the same problems of the solar-electric systems. The addition of large-capacity energy-storage facilities to an already large capital investment in wind machines and generators would militate against their commercial adoption. Among the various ways in which wind-generated electricity might be incorporated into existing electric utility systems, three seem the most promising. In conjunction with hydroelectric dams, large wind machines might be used to pump water into existing impoundment reservoirs whenever wind is available, with corresponding increases in the hydroelectric generating capacity. In the second mode, wind electric generators could be built along utility transmission lines with line control of the frequency and phase of the electricity generated so that the output could be fed into the line at selected points by matching transformers. The third method would use wind machines to compress air in the vicinity of gas turbines used when utility power needs peak. The compressed air would be stored and drawn upon for injection with fuel to the turbine when needed, thus saving the energy otherwise taken from the turbine to drive a compressor on the same shaft.

Electricity generated by the wind in any of these ways would simply replace what the electric utility system would have supplied without it by burning more fuel. The economics of such an application therefore depends on whether the annual costs of capital recovery, operation, and maintenance of the wind system are

less than the annual cost of the fuel it saves. At present fuel costs and expected capital costs of wind systems, this would hardly be the case. But in the long range, the situation might be reversed, and we can anticipate some use of wind electricity in utility systems, though it seems doubtful that it would ever constitute more than a minor fraction of the total electrical output of the country.

None of the considerations advanced so far in this analysis is intended to apply directly to rural areas in the United States and developing countries or to remote places in the world. Small, self-contained total energy systems using local wastes to supplement solar heat, combined with wind or photoelectric generation and limited storage and no transmission costs, seem destined to play an increasingly important role in the total world energy system. There are attractive possibilities in such applications, and their economic and practical evaluation rests on quite different criteria from those for central-station power.

Two other solar electric systems that have been proposed should be described briefly. The one under most active development at present—the Ocean Thermal Energy Conversion (OTEC) system—uses the difference in temperature between warm surface water and cold deep water in the tropical oceans. Several contractors of the Energy Research and Development Administration are testing engineering design and components for such systems. A plant would be hung from a large moored platform floating on an appropriate site in the tropical ocean. The working fluid would be pressurized ammonia gas driving a turbine generator. The exit gas would be condensed in a large condenser cooled by water drawn by marine propellers through a concrete pipe some 40 ft in diameter, from a depth of 1,000 ft or greater. The cool liquid ammonia would then go to an evaporator heated by surface water fed through a similar but shorter pipe. Electric power generated in this way would either be transmitted to land by submarine cables or used to produce electrolytic hydrogen from seawater, which would then be shipped to land in tankers. About 30% of the power generated would be needed to operate the system, mainly for cold- and hot-water circulation. If hydro-

gen is generated, another 40% would be used in electrolysis and liquefaction.

A design by Lockheed of a 160-MW(e) OTEC calls for 235,000 tons of concrete and 26,000 tons of steel for the platform, 8,700 tons of cold-rolled steel and 1,800 tons of titanium for the power system, 11,000 tons of steel for the mooring line, and 32,000 tons of concrete for the cold-water pipe. The capital investment in such a system for each net electrical kilowatt would be very large. Moreover, the problems of repair and maintenance over a 20- to 30-year lifetime seem formidable. The titanium heat-exchanger elements, for example, have a surface area of over 150 acres, and any fouling of them by growth of biological organisms would make the system inoperative. It seems very doubtfull that the massive private investment required to construct such a system could ever be justified commercially by electric utilities.

A nonsolar application of the same system might, however, develop marginally out of the R&D program on OTEC. The waste heat from large nuclear power plants, gas diffusion enrichment plants, or other plants yielding large amounts of waste heat might be used in a land-based, ammonia-cycle electric-generation system using similar 40–50°F temperature differentials. Heat exchangers would be of aluminum rather than titanium, and a total system of the same electrical output would cost only about one-fifth that of the OTEC system and considerably less in annual costs for operation and maintenance.

The other solar electric system that has been proposed involves mounting mirrors or photocells on a satellite equipped with an electric-generation system and a microwave radiation converter which would beam the microwave energy continuously to a receiver on earth, where it would then be converted to 60-cycle power and fed into a utility grid. In stationary earth orbit, solar energy would be converted continuously at a constant rate, except for an hour and ten minutes each day around midnight when the satellite passed through the earth's shadow. The annual collection of solar energy would be much higher than that for any land-based system, since J would be 3,600,000 Btu/ft^2/yr.

However, because of the microwave conversion and transmission to earth, the heat rate would be considerably larger. Even 30,000 Btu received on the solar collectors in space for each kilowatt-hour delivered into the utility grid on earth seems optimistic. With this value of H, the annual electrical yield would be 120 kWhr/ft² of solar collector.

Using our reference system for comparison, 42 million ft², or ½ mi², of solar collectors would be required. However, if serious consideration were given to assembling such a satellite in space, it would doubtless be for a much greater installed capacity. The equivalent of 50,000 MW of nuclear capacity at a 60% capacity factor would require 75 mi² of solar collectors on the satellite. Even with a fully developed space shuttle, the assembly of that magnitude of solar collectors

(together with associated facilities for energy collection and generation and equipment for microwave conversion and beaming) in a satellite 22,000 miles above the earth's surface represents a monumental task. As long as electricity can be obtained in simpler ways, it is a task not likely to be undertaken.

It is unfortunate that so many people continue to entertain high hopes for satisfying all our needs for electricity through direct or indirect means of generating it from the sun. In remote locations, where cost is not a factor, a small amount may be produced with wind or solar cells and battery storage, and the potential for small self-contained total energy systems for rural homes and farms is significant. But for any appreciable contribution to future national requirements for central-station electricity, neither

direct nor indirect solar energy (other than hydroelectric) is really suitable. There is practically no chance of realizing such a contribution regardless of how vigorously it is promoted and funded by the Congress in response to public aspirations.

References

1. J. A. Duffie and W. A. Beckman, Jan. 1976, Solar heating and cooling, *Science* 191: 143–49; E. Faltermayer, Feb. 1976, Solar energy is here, but it's not yet utopia, *Fortune*, p. 102; and W. A. Shurcliff, Feb. 1976, Active type solar heating systems for houses: A technology in ferment, *Bull. Atomic Scientists*, p. 30.

2. The Desert Sunshine Exposure Tests, Inc., P. O. Box 185, Black Canyon Stage, Phoenix, AZ.

3. F. A. Blake and J. D. Walton. 1975. Update on the solar power system and component research program. *Solar Energy* 17:213–19.

William G. Pollard

The Long-Range Prospects for Solar-Derived Fuels

What are the possibilities of replacing oil and natural gas with fuels made from animal and agricultural wastes, forest biomass, and sawmill and cotton-gin residues?

The long-range potential of the sun as a source of both direct and indirect electric power was evaluated in my article in the July 1976 *American Scientist* (1). Here I am going to consider the corresponding potential of fuels derived from the sun through photosynthesis. My objective is to judge the potential for the commercial production of such fuels as an integral component of the world energy system in the long-range future when reserves of fossil fuels, especially petroleum and natural gas, have been depleted. The basic science underlying the production of such fuels was discussed by Calvin in the May 1976 *American Scientist* (2).

In one sense all fossil fuels, including coal, oil shale, and tar sands, are properly classified as derived from the sun. However, they were produced from the solar energy received by the earth many million years ago, and in the sense used here, solar-derived fuels are those produced annually by photosynthesis in currently growing plant material.

If it should prove possible to replace fossil fuels to a large degree with solar-derived fuels, there would be two important advantages. One is that the burning of solar-derived fuels, in contrast to burning fossil fuels, does not add to the accumulation of carbon dioxide in the atmosphere. All of the earth's biomass, whether converted into solar-derived fuels and burned or left to decay naturally, is soon converted back to CO_2, but when fossil fuels are burned

they release to the atmosphere CO_2 that was removed from it many millions of years before. Thus solar-derived fuels share with uranium and thorium the property of yielding energy without adding to atmospheric CO_2. Insofar as atmospheric build-up of CO_2 may, in time, lead to large planetary changes in climate, this is obviously a vitally important property of solar-derived fuel.

The other advantage is that the potential production of solar-derived fuels is an order of magnitude greater in the humid tropics lying between latitudes $\pm 20°$ than it is in temperate zones at much higher latitudes. This potential has been considered in detail by Makhijani and Poole (3). Since the countries lying in this tropical belt are now the most impoverished and suffer from the highest rates of population increase, the development of a large and lucrative export trade for this area could represent a major contribution to world economic welfare. If the countries of tropical Latin America, Africa, and Southeast Asia could, by the end of this century, begin supplying the rest of the world with a major portion of its demand for essential liquid fuels (and so acquire ample trade balances for the import of food and other goods), the result would be a contribution to world stability of the very highest importance.

Such an outcome would appear to run contrary to the current energy policy of the United States underlying Project Independence. There is, however, very little prospect that the United States will be able to insulate itself from the critical problems facing the third world countries in the tropics in the remainder of this century and after. As a consequence, very

different political and economic motivations must govern our policy for these countries from those responsible for our present desire to reduce dependence on the Middle East for oil. The prospect of future dependence of tropical countries on America for food and of America, Europe, and Japan on them for liquid fuel is not entirely unattractive either politically or economically. Such a "Project Interdependence" could well be made one objective of long-range energy research and development.

The world dependence on gasoline, jet fuel, and fuel oil is massive, growing, and essentially irreversible short of real catastrophe. Private transportation by automobile, transport of food and goods by truck, air passenger and freight traffic, and agricultural production now depend entirely upon these fuels. Yet sometime late in this century or early in the next the finite reserves in the earth's crust must inevitably shrink and finally dwindle to insignificance. This is the most predictable aspect of the world energy picture at the present time. The energy needs for electricity, industrial process heat, and heating buildings do not loom with quite the urgency as those for liquid fuels for transportation.

It is of primary importance, therefore, to inquire what can be done when gasoline and other petroleum-derived liquid fuels are no longer available. One answer, already resorted to in the past in wartime, is the alcohols—methanol and ethanol—but at present, even with current high world prices for crude oil, they are too expensive to compete with gasoline. In the relatively near future, however, when these prices have again quadrupled, as they will, to $50/barrel, or

This is the second of two articles that Dr. Pollard, recently retired Executive Director of Oak Ridge Associated Universities, has written for American Scientist.

$8/million Btu, in 1975 dollars, the alcohols will become quite competitive. At $8/million Btu for the final fuel, the price of gasoline would be $1.04/gal, and that for the same energy content of methanol 45¢/gal and ethanol 61¢/gal. Assured prices at this level would make the manufacture of these alcohols, by currently available methods, an attractive commercial option.

The price of oil and natural gas will be governed in the same way as that of other highly valued but scarce commodities as world reserves decline. The price of alcohols, on the other hand, will be governed in the same way as that of food or timber in terms of an acceptable return to the farmers growing, harvesting, and delivering the required feedstocks. In general, when resource-limited commodities become scarce, substitutes are found whose cost is only production-limited. Just when in the long-range future this crossover between oil and alcohol will occur is difficult to predict. But if the price of oil had already reached $50/barrel, pure alcohol by presently available methods of production would certainly be attractive as a motor fuel.

The transition from a petroleum-based economy—with its vast network of supply, transportation, refining, and public distribution—to one based on alcohols would be extremely difficult to manage. The initial phase in the United States could well be the production of methanol from municipal solid wastes or coal, with the objective of using it as an additive to gasoline, called *neat*, of up to 15%. However, such a fuel has not as yet had extensive fleet tests and may well prove impractical. An alternative for the short-range that seems assured at present is the use of pure methanol in place of #2 fuel oil in utility peak power generators. These possibilities are discussed in detail in a report by Thomas and his co-workers (4). Alternatively, if shale oil or synthetic crude oil from coal proves to be commercially practical, these could well become the primary substitute for petroleum in the United States for some time, in which case the transition to alcohol might occur much earlier in Europe and Japan.

This article is not concerned with elucidation of the complex problems surrounding the transitional phase of the next half century but rather with visualizing the feasibility of a world wholly dependent on solar-derived fuels. This implies the question of the feasibility of pure alcohol as the sole fuel for transportation and solid and gaseous fuels produced from biomass for electricity and heat. There seems to be little question that both methanol and ethanol would make entirely satisfactory liquid fuels. Questions concerning the practicality of mixed gasoline-alcohol fuels previously noted do not apply to either pure alcohol or pure gasoline. With only minor modifications in engine design, both alcohols can be used effectively as high-octane fuels in internal combustion engines. In automobiles, trucks, and jet aircraft, larger fuel tanks would be required (double for methanol, 50% larger for ethanol) because of the lower heat values.

Solid fuel

The basic method for the production of both solid fuel and methanol from biomass is pyrolysis: wood, leaves, grass, or similar materials, including the organic component of municipal solid wastes, are heated in a closed container or converter by partial burning of a portion of the feedstock with air or oxygen introduced in a controlled manner at the base of the converter. The products are a low-Btu gas, volatile vapors, and a solid carboniferous char. The gases are mostly hydrogen and carbon monoxide; the condensable vapors are steam and a variety of volatile organics; and the char is a solid like charcoal. In low-temperature pyrolysis the air intake is reduced and its rate adjusted in proportion to the feedstock rate to minimize the production of gas and maximize that of char.

A major program for the development of low-temperature pyrolysis has been carried out in the Engineering Experiment Station of the Georgia Institute of Technology over the last eight years (5). The system has been tested extensively with a variety of organic feed materials in a 25-ton-per-day pilot plant. A commercial plant of 50 tons per day has been operating at a sawmill in Cordele, Georgia, for two years, producing charcoal briquettes from the sawmill wastes. As a result of this work the technical design and operating parameters for commercial plants are now well established.

The product of the process is a free-flowing, dry black powder similar to a finely ground coal, which has a heat value of 11,000–13,000 Btu/lb, the same as that of Eastern bituminous coal. It has essentially zero sulfur content and much lower ash than coal. The powder can be used in the same way as coal to form a slurry with fuel oil for use in oil-fired boilers, or it can be mixed with high-sulfur coals to form a reduced-sulfur fuel for use in coal-fired steam generators. At present coal prices, this fuel is economic only when the feedstock wastes are provided without charge. When fossil-fuel prices go much higher, it would become economically attractive even with ample payments to the supplier of the biomass feedstock.

The raw feedstock, generally having 50% moisture content, is fed through a hammer mill and thence through a dryer heated with low-Btu gas from the converter. From the dryer it is fed continuously into the converter at a rate that governs the flow rate of air to the combustion portion of the converter. The volatile organics are condensed at a temperature greater than the off-gas dew point to avoid moisture condensation, and the oil is mixed with the ground char to form the free-flowing powder, or char-oil. Power to operate the system is provided by a derated gasoline engine running on the low-Btu pyrolysis gas. Start-up of the engine and dryer is powered by an auxiliary propane supply. Typically, 450 lbs of char-oil are obtained from each ton of undried raw waste. A mobile pyrolysis unit of a 200-ton-per-day capacity (at 50% moisture) is being designed for mounting on two standard 55-ft trailers that can be moved from one collection site to another.

Increasingly stringent sulfur-emission standards are likely to hasten the acceptance of char-oil in the solid-fuel market. The initial production will most probably be at the sites of major continuous waste production, where a means of disposing of the wastes would be an economic benefit to their producer—for example, sawmills, cotton gins, sugar mills, and plants generating such wastes as peanut hulls, rice hulls, and bagasse. Perhaps as much as 200 million dry-weight tons of wood waste has accumulated over the years from sawmill operations in the Southeast, Maine, Michigan, and Oregon, including whole canyons filled with it in northern California. The processing of these wastes could stimulate the growth of a sizable industry for the manufacture of the mobile pyrolysis units as well as a growing market for their product. Once both have become established, demand for no-sulfur, very low-ash, high-Btu char-oil fuel is likely to increase rapidly as fossil-fuel prices continue to rise and environmental protection measures become more stringent.

The potential for producing char-oil fuel in the United States is undoubtedly great but also largely indeterminate, primarily because there has been no motivation to use biomass as an energy resource. Agricultural wastes are simply plowed back into the ground, and forestry wastes are left in place after logging and pulpwood collection or accumulated in huge sawdust piles thousands of acre-feet in volume. There are vast stretches of uncultivated forests where the ratio of culls to marketable trees is very high and logging operations not practical. In Maine 5 million acres infested with the spruce budworm are entirely suitable for energy harvesting but not for lumber or pulp. Considering all these sources, U.S. forests probably produce annually some half-billion dry-weight tons of woody material suitable for char-oil production. Canadian forests could add greatly to this annual production.

Our perspective on the magnitude of the energy resource in solar-derived fuels is distorted by the abundance of cheap fossil fuels we have enjoyed since the start of the Industrial Revolution two centuries ago. In this period, plant material has been grown exclusively for food, fiber, lumber, and paper; everything unsuitable for these end products we call weeds, culls, or wastes. When oil and gas are dwindling in supply and the cost of energy approaches $8/million Btu, the situation will be radically different: farmers will derive nearly as much income from the sale of their annual product to an energy-conversion industry as they do from the sale of food or fiber; forest management will yield an annual income from energy feedstocks comparable to that

from logging and pulpwood production.

Agricultural research will then be devoted to maximizing both food *and* biomass yields. Hybridizing to reduce the height of corn will be reversed to return to the corn that grows "as high as an elephant's eye." Harvesting machinery will be developed to minimize the cost and effort of harvesting food and biomass simultaneously. Wetlands not suitable for producing food or trees will be used for growing tall grasses, reeds, or other plants with a high yield in dry-weight tons per acre. A joint program of the University of Georgia and the Southeastern Forest Experiment Station has been devoted to growing sycamore, sweetgum, or yellow poplar trees in rows that are harvested after 1–4 years of growth with equipment similar to corn silage cutters (6). After cutting, the trees resprout from stumps, and growth continues. Land in the Piedmont and Georgia Coastal Plain unsuitable for other purposes yields nearly 6 tons of sycamore per acre annually on a 4-year growth cycle.

In the tropics, with their much higher annual production of biomass per acre, the production of char-oil fuel for local use in electric generation or industrial process heat is a very attractive possibility at the present time. These countries can ill afford to import oil from the Middle East, and an aid program to supply them with the relatively simple equipment for char-oil production and specially designed machinery for harvesting and delivery could pay off handsomely for a minimal outlay of aid dollars.

Liquid fuel

When a pyrolysis unit is operated at a high temperature with increased air supply, most of the char and oil is converted into the gaseous component, which can then be used as the synthesis gas for the production of methanol. For this purpose, two moles of hydrogen are combined in the presence of a catalyst with one mole of carbon monoxide at a pressure of some 100 atmospheres and a temperature around 700°F to produce one mole of methanol (CH_3OH). The process has been used for many years in the United States to produce nonfuel industrial methanol. In 1972, 26 million barrels were produced, and

the present market price for tank car lots is 40¢/gal.

In a recent book devoted to the world food problem, Eckholm gives an excellent analysis of the status and prospects of agriculture in the humid tropics (7). Every effort to achieve sustained tropical production of food in the past has been disappointing at best and catastrophic at worst. The most promising approach is an agrisilviculture that combines food and forestry in a closed cycle of food crops on cleared land followed by a planting of fast-growing trees before the last food crop is harvested. Fuel methanol would be the major economic crop, with food and timber production for auxiliary income. This joint cropping system could be supplemented with biomass grown exclusively for methanol production, for which a major tropical pest such as the giant elephant grass might be a prime candidate. Another possibility would be selected species of reeds and similar plants that grow with exceptionally high yields in tropical bogs and marshes unsuitable for food production.

The other alcohol that could serve as a motor fuel is ethyl alcohol, or ethanol. Industrial ethanol (190 proof) is produced in the United States from ethylene, but for fuel use in the long-range future a source other than petroleum would have to be found. Such a source has been developed recently by the U.S. Army Natick Laboratory using a fungus that causes jungle rot in wood and other cellulosic material in Vietnam (8). An enzyme produced by selected strains of this organism converts properly prepared cellulose into glucose with very high yields, and it is estimated that 5 million barrels of ethanol could be produced by the enzymatic hydrolysis of 3 million tons of cellulose from municipal or agricultural wastes.

Calvin (2) discusses the production of ethanol in the tropics by fermentation and distillation and cites the example of sugarcane in Brazil, where it is being used for ethanol production under a government subsidized program that plans to add 10% ethanol to all gasoline by 1980. Sugarcane, which is one of the most efficient natural photosynthesizers, requires ample water and is best grown in river valleys where extensive irrigation is feasible. Another tropical plant suitable for ethanol production—and one much easier to grow—is cassava, which is called *manioc* in Africa and *yuca* in South America (9) but is best known in the United States by its product, *tapioca*.

The cassava plant, a woody perennial that can grow to a height of 10 ft, produces several below-ground tubers growing radially from the root stem to a length of a foot or more and weighing up to 30 lbs each. It is extensively used as a food by natives in the Amazon basin, the Congo, Madagascar, Indonesia, the Philippines, and elsewhere, mainly because of the ease in growing it. Plots in the forest are slashed and burned, leaving stumps and large trees; stems of cassava plants from the previous harvest are simply thrust into the ground through the coating of ashes, and new plants grow from these cuttings in rather poor soil unsuitable for other, more nourishing food plants. Cassava has the further advantage of great drought resistance. In a prolonged period of drought it sheds its leaves so that transpiration is greatly reduced; when rains come again it puts out new leaves and continues growing.

Because cassava roots are high in starch and low in protein and other nutrients, other foods are better for raising nutritional levels in tropical native populations. If cassava could be grown as an energy crop of high monetary value, the income would amply cover the means of replacing it with higher-quality food. The production of fuel ethanol would constitute just such an application. For this purpose, the tubers can be sliced and sun-dried for 2–3 days to produce a durable starch product called *gaplek* in Indonesia. Molds that develop during drying, plus fiber and other constituents of the whole root, make gaplek unsuitable for human use, but it would be entirely satisfactory for fermentation to industrial alcohol.

In agri-silviculture management of tropical ecosystems there would be the possibility of combining methanol and ethanol production. In clearing a forest, the good logs would be sold and the remaining branches, culls, and residual biomass delivered to a nearby methanol plant. Cassava grown in the clearings and later interplanted with fast-growing trees would be delivered as dried gaplek to an adjacent ethanol plant. In order to

be economic, both plants should have capacities of the order of 15,000 barrels per day, which would require an assured supply for each of around 2 million tons of feed material per year. At a production rate of 2 tons per acre—low for the tropics—of both biomass and gaplek, the alcohol-production plants would require a service area of a million acres and a delivery radius of 20 miles. At a more likely tropical production rate of 16 tons per acre of each product, the service area is reduced to 125,000 acres and the delivery radius to 8 miles. It is not possible at this stage to make reliable estimates of yields from this type of management of tropical ecosystems.

The alcohol-production plants themselves are an established technology. Arrangements for supplying such plants with the required feedstocks will, however, be complex and should be put into operation only after the completion of research on safe modes and rates of removal of biomass from tropical areas and on selected species and culture of cassava for maximum production in partially cleared areas. An R&D program with these objectives is agricultural rather than technological. It is urgently needed if the topics are to become the future source of liquid fuels.

The extent to which alcohols would be produced domestically for motor fuel in the United States will depend on the price of imported fuel from the tropics. It is doubtful whether domestic ethanol production from the enzymatic hydrolysis of cellulose to glucose could ever compete with imported ethanol from sugarcane or cassava. On the other hand, methanol production from coal would probably be markedly cheaper than imported methanol. As long as coal is abundant, the production of 500 million tons per year for conversion to methanol or other liquid motor fuel is not an unreasonable expectation. This would constitute a significant fraction of U.S. requirements in the long range, assuming greatly improved automotive efficiencies and somewhat lower per-capita use of automobiles.

Gaseous fuel

Equipment for deriving the gaseous fuel methane from agricultural wastes by anaerobic digestion (10) has been

available and used for some time by municipalities for sewage sludge digestion. Essentially the same equipment could be used near animal feedlots for the digestion of urine and manures. The Chemical Engineering Department of the University of Missouri at Rolla has developed and demonstrated a process to use a slurry of ground agricultural crop wastes in similar facilities. The anaerobic digestion produces a mixture of methane, carbon dioxide, and small amounts of other gases. After scrubbing the CO_2, the product is a pipeline quality gas.

Domestic production of natural gas has been declining for the last several years. Price deregulation will stimulate the opening of new fields, the eventual exploitation of new sources such as extraction of methane from coal beds before the coal is mined, and the tapping of large reserves in Devonian shales. When the price of natural gas reaches $5/1,000 ft^3$ and higher, it will become economically attractive to build anaerobic-digestion plants, with a capacity of 100–500 tons/day, in the vicinity of animal feedlots and poultry farms and, later, at selected sites throughout the Midwest and elsewhere, at which producers contract with farmers within a 6- to 15-mile radius to supply their feedstock. The methane produced could be pumped into an existing natural gas pipeline. At present, the entire operation is technologically feasible and awaits only a favorable product price to begin commercial deployment without government assistance or intervention.

There are two major advantages in anaerobic digestion over char-oil production. Animal wastes have too high a moisture content to be suitable as feedstock for char-oil, but for the same reason are ideal for anaerobic digestion. Agricultural wastes, on the other hand, can be digested without loss of the original plant nutrients, and the digested sludge is rich in humus. Thus farmers delivering feedstock to the plant could return with a corresponding amount of dried sludge for spreading on their land. If the same wastes were used for char-oil production, some portion of them, depending on soil character, would have to be retained for plowing back to preserve the quality of the soil. For these reasons, methane rather than char-oil will probably prove to be the preferred solar-derived fuel from animal and agricultural wastes, whereas char-oil will predominate from forest and marginal-land biomass and from sawmill and cotton-gin wastes. The demonstration equipment at the University of Missouri at Rolla, when scaled up to a large plant, has shown that such a plant could produce 50 million ft^3 of methane a day from an annual delivery of 1.5 million tons of crop waste.

In addition to animal and crop wastes, there is a large potential in growing plants such as water hyacinths in wetlands and marshes specifically for the purpose of anaerobic digestion to methane. Another possibility might be to grow forests of kelp in the ocean to be harvested in large quantities for methane production.

The prospects

The long-range prospect for the use of solar energy through photosynthesis to produce solid, liquid, and gaseous fuel seems reasonably well assured. The technology for such fuels is already well developed. Their extensive use awaits a favorable price level for delivered biomass and a major shift in agricultural and silvacultural practice through which energy crops become comparable in value to traditional crops. The transition to this new mode will take place gradually and at varying rates in different parts of the world. In countries like the United States with abundant reserves of coal and oil shale it is likely to be delayed, while an early onset seems most likely in the humid tropics where the production of solar-derived solid fuel and alcohol can eliminate all dependence on imported oil. When tropical agri-silvaculture has become widely deployed, the countries in this region can begin to benefit from a growing, lucrative, and assured market for alcohol to replace gasoline.

The prospects are also promising in the case of the direct use of solar energy for hot water and for heating buildings. As the cost of oil, natural gas, and electricity escalates and scarcities develop, such solar-heating systems seem certain to be increasingly adopted. Only in the case of the direct or indirect use of the sun for the generation of central-station electricity in utility systems does the long-range prospect for solar energy

seem economically unfeasible, mainly because of the high costs for equipment and maintenance and the insecurity of the investment.

References

1. W. G. Pollard. 1976. The long-range prospects for solar energy. *Am. Sci.* 64:424–29.
2. M. Calvin. 1976. Photosynthesis as a resource for energy and materials. *Am. Sci.* 64:260–78.
3. A. Makhijani and A. Poole. 1975. *Energy and Agriculture in the Third World*. Energy Policy Project of the Ford Foundation. Cambridge, MA: Ballinger.
4. C. O. Thomas et al. 1976. *Methanol from Coal: Fuel and Other Applications*. Oak Ridge Associated Universities Publication 126, IEA 75-2. Springfield, VA: National Technical Information Service, U.S. Department of Commerce.
5. J. W. Tatom et al. 1975. *Clean Fuels from Agricultural and Forestry Wastes: The Mobile Pyrolysis Concept*. New York: American Society of Mechanical Engineers.
6. K. Steinbeck and C. L. Brown. 1976. *Yield and Utilization of Hardwood Fiber Grown on Short Rotations*. Applied Polymer Symposium, No. 28. Wiley. Also, AIChE Symposium Series 70:62–66 (1976) and *J. Forestry* 74 (1976).
7. E. P. Eckholm. 1976. *Losing Ground: Environmental Stress and World Food Prospects*. Norton.
8. L. A. Spano, J. Medeiros, and M. Mandels. 1976. Enzymatic hydrolysis of cellulosic waste to glucose. *J. Washington Acad. Sci.* 66:279–94.
9. W. O. Jones. 1959. *Manoic in Africa*. Stanford Univ. Press. Also, Processing of cassava and cassava products in rural industries. FAO Agricultural Paper No. 54. Washington, DC: FAO (1956).
10. A. Poole. 1975. In *The Energy Conservation Papers*, ed. R. H. Williams. Energy Policy Project of the Ford Foundation. Cambridge, MA: Ballinger.

"What's the big deal about solar heat?"

Melvin Calvin

Photosynthesis as a Resource for Energy and Materials

The natural photosynthetic quantum-capturing mechanism of some plants may provide a design for a synthetic system that will serve as a renewable resource for material and fuel

Photosynthesis is the biological process by which green plants use the energy of sunlight—the quanta—to convert carbon dioxide and water to the higher-energy carbon-hydrogen bonds of carbohydrates and other organic compounds. Green plants (and a few nongreen ones as well) can capture the quanta and convert them to food, material, and fuel. The principal product of photosynthesis is of course food, but material, in the form of cellulose (either wood or fiber), is also important. Very little fuel, however, is produced directly by photosynthesis, and a major effort is now under way to try to use both the natural process and, eventually, a synthetic process to provide fuel on an annually renewable basis.

Until the mid-fifties carbohydrate from sugarcane and other plants was converted by fermentation to alcohol, and this process may again become economic in light of improved fermentation technology and rising costs of other fuels. Even the direct photosynthetic production of hydrocarbon by existing plants such as the

Melvin Calvin is well known for his scientific achievements in fields ranging from metal-organic chemistry to the chemical origin of life and for his contributions to the understanding of photosynthesis in green plants and, more recently, of chemical oncogenesis. In 1961 he was awarded the Nobel Prize for Chemistry, in 1964 he received the Davy Medal of the Royal Society, and in 1975 the Virtanen Medal in Finland. This paper was originally presented, in somewhat different form, at the Second Philip Morris Science Symposium and will be published in The Recent Chemistry of Natural Products, Including Tobacco: Proceedings of the Second Philip Morris Science Symposium, *ed. Nicholas J. Fina, New York: Philip Morris, Inc., 1976. This work was supported in part by the U.S. Energy Research and Development Agency. Address: Laboratory of Chemical Biodynamics, University of California, Berkeley, CA 94720.*

rubber tree, or by newly bred ones, seems possible in view of the large number of species (some 2,000) that produce hydrocarbons and the new techniques of plant-cell cloning that have already proved successful in breeding new varieties of sugarcane. It may also be possible to produce materials, such as plastics and fibers, from the products of direct fermentation of relatively labile carbohydrates, in seaweed, for example. And someday, in the more distant future, synthetic systems constructed on the basis of our growing knowledge of the photosynthetic processes may produce fuel, fertilizer, and power.

Several renewable sources of energy offer alternatives to the fossil fuels, but photosynthesis seems to hold great promise (Calvin 1974a,b, 1975; Sarkanen 1976). This process originally gave rise to the starting material for the geologic formation of coal, oil, and natural gas. Today, of course, the rate of consumption of the fossil fuels is thousands of times greater than the rate at which the initial photosynthetic product is being converted into hydrocarbon. The exact percentage of current photosynthetic product that is being deposited in the sedimentary rocks of the earth is not known, but it must be a very small fraction of the total photosynthetic rate. One way to get around the slow geologic processes necessary to produce energy sources is to look at current photosynthetic rates and current impingements of solar energy and see what can be done now, rather than waiting for geologic processes to accumulate the material.

Solar energy collection

The first question we must answer is where on the earth's surface sunshine (insolation) is found today. There are

three areas of high solar intensity: the Sahara Desert, the southwestern United States (mainly Arizona), and the Kalahari Desert of South Africa. The mean annual insolation in these high-intensity areas is approximately 250 watts per square meter. If all that was needed was the impingement of sunshine, the problem would be simple, but there is another essential requirement for photosynthesis.

In contrast to areas of high insolation, high points for the rate of photosynthesis and carbon fixation are found along the Amazon Basin, equatorial Africa, and in equatorial Southeast Asia. In these regions the fixation rate is approximately 1 kilogram of carbon per square meter per year. The reason for the disparity between the areas of high carbon fixation and those of highest insolation is that the regions of high carbon fixation contain adequate soil and water as well as sunshine. At least two of the areas with the highest solar impingement are deserts. Both requirements—solar impingement and soil and water—coalesce in the Amazon Basin, Southeast Asia (Malaysia and Indonesia), and equatorial Africa, and these areas, not unexpectedly, have the highest productivity in terms of kilograms of carbon (in the form of carbohydrates) per square meter of land area per year. Since they have an insolation of something less than 200 w/m^2, the maximum overall efficiency in fixation rate is less than 1%.

The next question to which we must address ourselves is the *quality*—what we see as *color* with the naked eye—of the solar energy that impinges on the surface of the earth. The energy that reaches the surface of the earth has lost some wavelengths in the invisible infrared re-

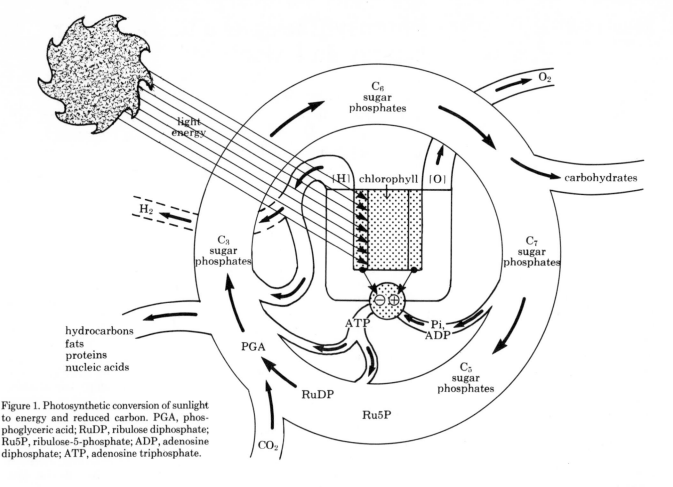

Figure 1. Photosynthetic conversion of sunlight to energy and reduced carbon. PGA, phosphoglyceric acid; RuDP, ribulose diphosphate; Ru5P, ribulose-5-phosphate; ADP, adenosine diphosphate; ATP, adenosine triphosphate.

gion, which are absorbed by water vapor and carbon dioxide as they pass through the atmosphere. The light between about 4,000 and 7,000 Ångstroms is the only light that is useful in photosynthesis, and it is distributed more or less evenly over the earth's surface. The energy of a chemical bond is approximately 50–90 kcal/mole, and the energy for exciting an electron in a molecule in the visible region is of the order of 40–70 kcal/mole, or more. Thus, with visible quanta it would be possible to excite electrons and allow them to do specific chemistry if they could be kept from returning to the ground state. If, however, you degrade the quanta—i.e. convert the quanta between 4,000 Å (70 kcal) and 10,000 Å (28 kcal) to much smaller ones (for example, 2 kcals, corresponding to 1,000°K)—then it is not possible to perform specific chemistry with them; all that is possible is to excite the vibrational or rotational modes in the molecule and to perform chemistry that depends on random thermal events.

Most discussions of solar energy describe processes for collecting it as heat and using it in various ways: to

heat or cool buildings; to operate thermal engines, which could run as high as perhaps 600–700°C; to evaporate water for hydroelectric power; to operate engines using the small temperature differentials in the ocean (25°C at the surface down to 10°C at 300 m); or for wind power. These uses of solar radiation degrade all the high-quality quanta in order to use the heat. I am concerned with a much less discussed aspect of solar energy —collecting the quanta as quanta, the way they come to the surface of the earth, and using them in their initial, high-quality form. The sun, after all, has a surface temperature of nearly 5,000°C, and, in effect, those quanta in the range of 4,000 to 7,000 Å could be considered as running a thermal engine with an upper temperature of about 5,000°C.

Quanta can be collected in at least three different ways—photoelectrically, photosynthetically, and photochemically. Photoelectric quantum collection is used today on a relatively important scale. An example is the photovoltaic device, usually solid state, in which direct quantum conversion is used to produce a voltage and a current. It is the other two

methods—photosynthetic quantum collection and photochemical quantum conversion—that I want to examine in closer detail in this paper.

Photosynthetic quantum collection: Green plants

Let us look at the photosynthetic process, shown diagrammatically in Figure 1. The light energy of sunshine is absorbed by chlorophyll and results in charge separation. The plus charge becomes oxygen, and the minus charge goes into the active hydrogen atoms that eventually reduce the carbon dioxide to sugars. If there is no carbon dioxide present, and at low O_2 pressures, it is possible that these active reductants would appear as molecular hydrogen. Carbohydrates are the principal product of photosynthesis; however, the plant can, of course, make other products from the cycle, even hydrocarbons.

Carbohydrate has many forms; one of the most common, wood, was for millennia burned as a fuel until it was replaced by coal and then oil and natural gas. But there are better ways than burning for using the carbohydrate productivity of the plant for

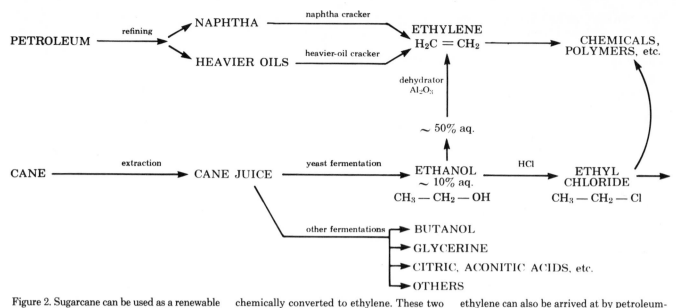

Figure 2. Sugarcane can be used as a renewable resource for chemicals and other materials. The cane juice is fermented to ethanol, which is chemically converted to ethylene. These two products are the raw materials for the chemicals most widely used in industry. Ethanol and ethylene can also be arrived at by petroleum-product cracking.

fuel. Polymerized sugars of one kind or another are the dominant form in which photosynthetic products are stored. Sugars such as sucrose are carbohydrates made up of only two hexose molecules, and they can be fermented directly to produce various types of chemical products—including alcohols. A simple equation gives the stoichiometry of the fermentation conversion of carbohydrates to alcohol:

$$C_6H_{12}O_6 \rightarrow 2C_2H_5OH + 2CO_2$$

The fermentation is accomplished with yeast or bacteria, and the weight of the material is reduced by a factor of two (180 gm to 92 gm), with practically no loss of energy (673 kcal to 655 kcal) in the process. The form of the material has been changed from an unwieldy solid to an easily used liquid—alcohol. In this fermentation reaction, it takes about 1.3 kilograms of sugar to produce 1 liter of alcohol (equivalent to 5,300 kcal), and the cost of conversion is 5 cents (1974) per liter, to be added to the cost of the raw materials.

Figure 2 shows a way of using sugarcane as a renewable resource for chemicals and other materials (Patarau 1969). Through a fermentation process it is possible to obtain ethanol, which, in turn, through suitable chemical processes, is converted to ethylene. These are the raw materials for most of the chemicals used industrially on a large scale—polyester, nylon, polyurethane foam, glycol, polyglycol. The boundary, or common material (ethanol or ethylene), can be arrived at through sugar fermentation or petroleum-product cracking.

Before World War II, yeast fermentation of a variety of plant products was the main source of industrial alcohol in the United States. Currently, most industrial alcohol is made from ethylene, as a by-product of petroleum refining, and not from renewable fermentables. Because the price of ethylene made from petroleum was only about 6¢/kg from 1950 to 1970, it did not pay to make alcohol from anything else. With the price of ethylene from petroleum today standing at approximately 28¢/kg and projected to rise to 40¢/kg in the near future, it again seems feasible to make ethylene from alcohol, at least in parts of the world where the raw materials are available in large quantities at a relatively low cost.

Let us look at three specific examples of plants that could be used for photosynthetic quantum collection—sugarcane (*Saccharum officinarum*), kelp (*Macrocystis spp.*), and the rubber tree (*Hevea brasilensis*) (see Table 1). Sugarcane is the most efficient natural quantum converter we know today that both collects and stores the energy in a relatively useful form, with a conversion efficiency to sucrose of about 0.25%, which is very good for a field crop. Actually, if you do not insist on making sucrose from the cane, and if you take the soluble hexoses, for example, which are directly fermentable, the yield will rise, thus raising the efficiency by a factor of about two. To make sucrose, it is necessary to use energy to hook two hexose molecules together, with a net reduction in total yield. If the final product, however, is alcohol instead of sugar, and the cellulose is also used, the apparent yield will increase, with a solar energy efficiency conversion of over 1%. I do not know of any other

Table 1. Annual productivity of rubber, cane, and kelp (yields, in metric tonnes, are for Hawaii)

	Present productivity	% of incident sunlight captured	Potential productivity
Rubber	2.2 t rubber/ha	0.2	4.5–9 t/ha
Cane	25 t sugar/ha 10 t ethanol/ha 6 t ethylene/ha	1.2	30 t fermentables/ha
Kelp	9 t dry weight carbohydrate/ha	~2.0	90 t/ha

field crop that comes anywhere near that overall efficiency. Unfortunately, the end product is a mixture of carbohydrates, and it is necessary to convert that to something more useful, such as alcohol.

As I pointed out above, photosynthetic carbon fixation is extremely high in the Amazon Basin, and even in the rest of Brazil it is higher than almost anywhere else in the world. Brazil, obviously, is an area ripe for exploration of the production of fermentables. And it turns out that Brazil is the largest sugarcane-growing area in the world: in 1974 it produced 9 million metric tonnes of raw sugar. To make this much sucrose it is necessary to make about 2.2 million tonnes of fermentables in the form of molasses as a by-product. The molasses can then be used as a raw material to produce alcohol. The Brazilians used no ethylene to make industrial alcohol in 1974; using only molasses, they produced about 740 million liters of alcohol. About 500 million liters of it were used for industrial products of various kinds, and the remaining 250 million liters extended the total gasoline supply by an average of 2% (see Wigg 1974 for a review of alcohol as a gasoline extender). The goal is to raise the average alcohol content of gasoline to 10% by 1980. In fact, in November 1975 the government set the price of alcohol to be equivalent to that of sucrose in order to stimulate production, and they expect to produce 4 billion liters by 1980.

Cane yields in Brazil are quite high in the area along the San Francisco River, which runs from the southern part of the country northward, turning into the Atlantic Ocean just south of the Amazon River. The San Francisco is an enormous river, and much of the hydroelectric power of Brazil is located on it. Sugarcane is grown in a large irrigated region along the river, where it is possible to obtain a yield of 150 tonnes of cane per hectare, containing at least 20% fermentables. This is equivalent to about 15 tonnes of ethanol per hectare. Thus the area required for very large-scale production is not great; for example, production of 10% of the annual gasoline requirements of a country as large as Brazil will take only about 150,000 hectares. In Brazil there are several million suitable hectares undeveloped at the present time.

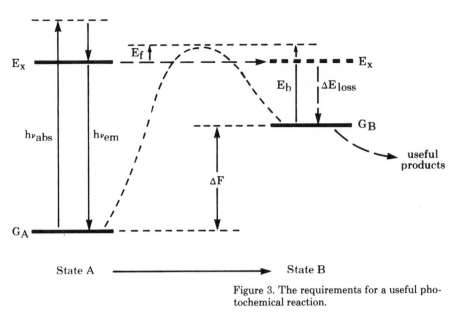

Figure 3. The requirements for a useful photochemical reaction.

Since sugarcane can be grown from cuttings, a good deal of ordinary plant breeding has used clonal as well as sexual reproduction to develop strains with high sucrose yields. The genetic variability is obtained for the most part through sexual breeding. However, the improvement of yields in cane as well as in other crops is not attained without added costs, both in economic terms and in terms of energy input. One of the largest costs of cane production in Brazil is for fertilizer—about 17% of the total, a very large fraction of that being the cost of ammonia. Unfortunately, there does not appear to be a tabulation of energy inputs into the production of a crop of sugarcane. However, the energy balance for corn production was examined from 1945 to 1970 in the United States (Pimentel et al. 1973, 1975), and since corn is a close relative of cane, similar results might be expected. According to this study, nitrogen fertilizer accounted for about 6% of the energy input in 1945, and this rose to over 33% in 1970 and is probably still higher today.

It thus becomes apparent that in the development of any natural biological system for the collection of solar energy for fuel or materials, the nitrogen requirement had best be met in some way other than fresh synthetic fixed-nitrogen application, since any synthetic process would require the availability of molecular hydrogen or equivalent reducing power. The obvious thing is to recycle the nitrogen, insofar as it is possible to do so. A second alternative would be to use biological nitrogen fixation, like that by root nodule bacteria in legumes, if that is possible for cane. Only recently it has been found that there may be microorganisms associated with grasses like corn and cane that might be usefully developed for this purpose (Newton and Wyman, in press). Specifically, a nitrogen-fixing *Spirillum* species has been found associated with certain maize genotypes growing in Brazil (Von Bülow and Döbereiner 1975; Döbereiner and Day, in press).

A third possibility, and one in the more distant future, is the introduction of the nitrogen-fixing enzyme system directly into the genotype, or at least phenotype, of some highly nitrogen-dependent plants such as cane and corn (Hardy and Havelka 1975). Transfer of the ability to synthesize the nitrogen-fixing enzyme system from one organism to another, with a much higher transfer rate than is possible by interchromosomal transfer, has been accomplished by Postgate (1975, in press), who used extrachromosomal information-carrying plasmids in organisms. Postgate and his co-workers have transferred a nitrogen-fixation gene containing plasmid from *Klebsiella pneumoniae* to *E. coli*, which lack the nitrogen-fixing genes. Whether this nitrogen-fixing information could eventually be integrated into the recipient's chromosomes and thus be stabilized in the recipient is not yet known. In fact, even the location of some of the nitrogen-fixation genes in the root-nodule bacteria is still in question. Still further off lies the possibility of using these plasmids to

Figure 4. This detailed view of the chloroplast shows lamellae (lower portion of picture) and two types of particles (large, in a band across the center, and small, at top left). These structures may play a part in the hydrogen and oxygen generating complexes of the green plant. X 102,000.

introduce the nitrogen-fixing gene from certain bacteria into higher plants, such as corn and cane.

Unlike the production of cane in Brazil, the use of kelp as a biological solar energy converter is still under consideration, and kelp production is an extrapolation. We know, for example, that present production of kelp along the California coast is about 9 tonnes dry weight/ha, whereas the potential for kelp is believed to be about 90 t dry weight/ha. If that is indeed the case, kelp would be a good source for fermentables. However, the question as to whether that much kelp can be grown, harvested, and processed economically and commercially still remains.

Rubber, which grew originally in Brazil and has been transplanted very successfully to Malaysia, produces hydrocarbon, rather than carbohydrate. Here, then, is a tree that is a source of hydrocarbon. Although it has the same type of chemical composition as petroleum, *Hevea* hydrocarbon has a somewhat different atomic arrangement and a different molecular weight distribution from hydrocarbon in petroleum. If we could control the molecular weight of the *Hevea* hydrocarbon—for example, keep it down to about 10,000 instead of 100,000 and 1,000,000 (Subramaniam 1972)—and harvest the product, we might have a "fuel" tree. I suspect this actually could be done. One of the things we would have to do is to find out what the control mechanism is for the molecular weight of the hydrocarbon product in the rubber tree.

Such a discovery may not be impossible. After all, in an attempt to make natural rubber as economical as the synthetic substitutes developed during World War II, rubber growers in Malaysia found methods to improve the yield of rubber, raising it from approximately 225 kg/ha to 2.2 t/ha in twenty years (1945–65). It is believed possible to achieve yields as high as 5.5 t/ha, and some very unusual *Hevea* trees may go as high as 9 t/ha. The accomplishments so far have been due to a combination of agronomy, breeding, and grafting. These high producers are really triple trees, with two grafts: onto unusually hardy root stock is grafted a large trunk which has many latex tubes, and onto that is grafted a crown with a large number of leaves to collect the sunshine.

Of course, *Hevea* cloning from plant cuttings has long been practiced, and for many characteristics these cuttings breed relatively homogeneously. However, genetic introduction of properties such as nitrogen-fixation and other useful traits into *Hevea*, or other latex-producing plants, from noninterbreeding sources must await the development of the techniques mentioned in connection with the introduction of nitrogen-fixation into sugarcane.

Knowing what we do about the way the green plant manipulates carbon (see Fig. 1), it should be possible to devise additional chemical methods to control the direction in which the synthetic sequences occur. Eventually we may be able to alter the storage products of the plants—for example, to produce more hydrocarbons and fewer carbohydrates.

Photochemical quantum conversion: A synthetic system

The third type of quantum collection—photochemical quantum conversion—is the only one that is not in use today for either energy storage or material production. Schematic requirements for a useful photochemical reaction are shown in Figure 3 (Calvin 1974d). If we begin with a material in "state A" and we want to store some energy in that material in a "state B" via a quantum conversion process, certain conditions must be fulfilled to do this efficiently. The amount of energy stored is represented by the free energy storage (ΔF)—the difference between the energy state of the materials in states A and B. The quantum that is absorbed is usually to some higher energy, either some virtual state or some other excited state, which immediately degrades to the lowest excited state (E_x). The system now must transform from the excited state of A to the ground state of B with an energy loss (ΔE_{loss}). To some extent, the larger this loss, the higher is the quantum efficiency (not the energy

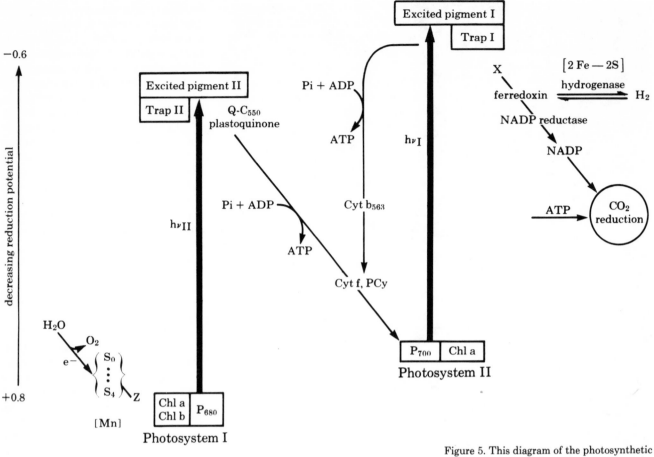

Figure 5. This diagram of the photosynthetic electron transfer system shows details of the process that takes place in the chloroplast.

efficiency). Most of the photochemistry we have been doing in the laboratory has involved enormous ΔE losses, in fact so great that for the most part the ground state of B may actually be below the ground state of A. The net result may be a downhill reaction in which the quantum is providing the activation energy and no energy at all is stored.

The real trick in photochemistry is to get state B up as high as possible and still have a high quantum efficiency for its production. The requirements are to find systems for which the activation energy for the forward or storage reaction (E_f) is as small as possible, while the activation energy for the reverse reaction (E_b) from an otherwise relatively stable storage state (G_B) is as high as possible, and the energy efficiency (Δ F/hν_{abs}) is kept as high as possible. The last two requirements are, of course, antithetical.

Photochemistry is primarily useful for direct fuel production (Lichtin 1974; Archer 1975), though it can also

be used to produce electric current through photogalvanic cells and to produce heat in various ways. When we speak of fuel we usually think in terms of a material that, upon combination with oxygen, can be used to produce useful work, whether it be electrical, mechanical, or in some other form. This means that the material must be in a reduced form so that it can be oxidized and, by extension, that photochemistry must, in general, produce a reduced material. One useful reducing material is hydrogen, which can be produced not only directly photochemically but also photobiologically; both methods are, of course, based on solar energy. The quantum conversion processes do not pass through a thermal step. There are quantum steps, either for the direct production of hydrogen or for the production of something that, in turn, produces hydrogen in subsequent chemical reactions that do not require light.

The other type of reduced material commonly used as fuel is a reduced carbon compound of some kind; the

most reduced one is methane—carbon attached only to hydrogen. The only method we now have to produce reduced carbon photochemically is the same one used by the green plant—reducing oxidized carbon in the form of carbon dioxide. Unfortunately, most green plants do not produce totally reduced carbon (methane), although hydrocarbon production by green plants, such as the rubber tree, has been known for some time. Most of the product of green plant photosynthesis is carbohydrate, or only partly reduced carbon; the carbon dioxide is not reduced all the way to hydrocarbon. The first products of photosynthesis are active hydrogen and oxygen, and then a partially reduced carbon compound, which is the carbohydrate. Both hydrogen and oxygen can be used as fuel, and the carbohydrate, which is the principal reduction product of carbon by photosynthesis, can be further converted into more reduced products, such as methane. Although methane production is generally not a photochemical step, hydrogen and carbohydrate production can be very close to the direct quantum conversion processes.

In the green plant, we know that quantum conversion takes place in the chlorophyll-bearing organelles of the plant, the chloroplasts, where all the hydrogen and oxygen generating complexes are located. Figure 4 shows a chloroplast in detail: the lamellae are coated with protein particles of different sizes. It is reasonable to suppose, and a current hypothesis suggests, that the low-oxidation catalysts that generate hydrogen and the high-oxidation catalysts that generate oxygen are on opposite sides of these lamellae and may be contained in two different types of particles, both visible in the figure. Whether these do indeed represent the oxidizing and reducing sides of the photosynthetic system remains to be established. They may represent two aspects of the same particle in which the two catalysts are separately mounted.

Some details of the quantum conversion process that takes place in the chloroplasts are shown in Figure 5, in which the emphasis of Figure 1 is reversed. The major carbon reduction cycle is placed in a circle to the right, and the box in the center of Figure 1 is expanded to be the entire focus of interest. It is evident that two successive quanta are required to move electrons from the water to the highly reduced acceptor (Trap I), even more reduced than molecular hydrogen itself. The beginning of the electron flow from the water is induced by the absorption of light by pigment P680, leading to an excited state (Photosystem II). This excited electron then flows back down a potential gradient through plastoquinone and cytochrome f. In the course of this movement, some high-energy phosphate is generated from ADP and orthophosphate, the electron returning eventually to the hole left by an excitation in pigment P700 (Photosystem I). Excitation of this pigment brings the electron to a very high reduction level, which can give rise to molecular hydrogen or, eventually, reduced carbon dioxide.

The structure of the pigment and electron transport system is not yet known in detail, but the fact that there are two quite different enzymatic processes at the two ends of the system seems established. At the oxygen end of the process (on the left

side of the figure), a manganese complex (probably polynuclear) is involved in the generation of the molecular oxygen. Details of the structure of the polynuclear manganese compound have yet to be determined, but we do know that it is rather labile in terms of its complexing properties, since almost every effort to isolate it as a complex has so far resulted in the isolation of uncomplexed manganese ion. A few cases have been reported in which a rather labile protein complex of manganese has been claimed as the oxygen-generating catalyst, but its structure has yet to be determined (Cheniae and Martin 1973; Chen and Wang 1974; Blankenship and Sauer 1974). A suggested process for the evolution of photosynthetic oxygen involves a binuclear manganese complex in which two Mn^{+2} ions move up to two Mn^{+4} ions, as a result of 4 e^- transferring quanta, and return, with the evolution of oxygen (Benemann et al. 1973).

While photosynthetic oxygen evolution is still not clearly understood, a good deal more is known about the other end of the photosynthetic system—Photosystem I and the production of hydrogen (the right side of Fig. 5). In this process in some plants, if oxygen is excluded or minimized and carbon dioxide withheld, the potential reducing agent may appear as molecular hydrogen itself: an enzyme in the plant (hydrogenase) gives rise to the molecular hydrogen (Benemann et al. 1973). The electron passes through an iron-sulfide protein, known as ferredoxin, on the way either to molecular hydrogen or carbon dioxide. The basic structure of this group of complexes is known.

Following excitation by Photosystem II, the excited electron passes through the connecting cascade to the hole left behind in Photosystem I (pigment P700) to return the pigment to its initial condition. The electron excited by Photosystem I is transferred eventually to ferredoxin and hydrogenase on its way to molecular hydrogen. The structure of the group of iron-sulfide proteins that contain ferredoxin, hydrogenase, and perhaps even nitrogenase is rapidly being determined (Holm 1975; Hall et al. 1974; Orme-Johnson 1973; Eady and Postgate 1974).

The crystalline structure of bacterial

ferredoxin (as opposed to plant ferredoxin) has been elucidated, and the active center of this material appears to be a 4-iron/4-sulfide distorted cube (Adman et al. 1973). A model compound representing this distorted cube has been synthesized, using benzylmercaptide to replace the cysteine groups of the ferredoxin protein, and its crystal structure has also been determined (Herskovitz et al. 1972). Although the structure of green plant ferredoxin has not yet been so completely determined, it appears to be a 2-iron/2-sulfide protein, for which models have been synthesized. In fact, the hydrogenase itself seems to be a 2-iron/2-sulfide protein of a similar type.

While a model substance of the 2-iron/2-sulfide protein has indeed been made, using two orthoxylylene dimercaptide molecules, the two halves of the molecule are not tied together, and when the model substance is reduced, the molecule is unstable. For this reason we are presently attempting to produce a model substance for the hydrogenase in which all *four* mercaptide groups are part of a single molecule. This is most readily achieved in terms of a dodecapeptide, modeled on bacterial ferredoxin, in which each of the four cysteine groups is separated from its neighbors by two intermediate peptides (Que et al. 1974).

We believe that if and when such a (one hopes cyclic) 2-iron/2-sulfide/ 4-mercaptide compound is constructed it will be capable of adding two electrons without falling apart and may give rise to the evolution of hydrogen when two electrons are added to it in the presence of protons. Recently, the use of iron-sulfide complexes with a reducing agent has appeared to be successful in the catalysis of the carboxylation of thiocarboxylic esters to form α-keto acids that were subsequently stabilized by pyridoxamine amination (Nakajima et al. 1975). If our knowledge of the manganese compound could be developed to the same extent as our knowledge of the iron-sulfide compounds, it is conceivable that such a compound could be placed in a system so that the two successive quanta involved in the transfer of an electron from water to hydrogen would go through two separate stages, one through the manganese and the other through the iron.

Photochemical membrane

The question remains whether the electrons are physically able to move from one side to the other of the molecule (or membrane) while, at the same time, the two sides are prevented in some way from reacting with each other. Prevention of back reaction may be achieved by mounting the electron donors and the electron acceptor molecules on opposite sides of a membrane (or molecule) and requiring that only the excited electron can actually pass through the membrane (perhaps by a tunneling mechanism), while the ground state electron on the acceptor side cannot, thus providing a barrier to the back reaction (Kuhn 1972; Kuhn and Möbius 1971; Miller 1975). An alternative structure would be to surround each individual molecule with a non-conducting hydrocarbon-rich barrier through which the excited electron must pass and through which the ground state acceptor electron cannot return (Alkaitis et al. 1975; Ford and Calvin, unpubl.). Whether either or both of these ideas play a role remains to be seen; however, the fact that mobile charges are generated in such a lamellar system seems to be established.

Using the notion of charges mobile through a membrane, one can construct a synthetic system in which hydrogen might be evolved on one side of the membrane and oxygen on the other. Such a concept, shown in part in Figure 6, is still hypothetical (Calvin 1974b, 1975; Tien and Verma 1970). The idea is that the separated electron is taken off the membrane by means of an iron-sulfide catalyst that will generate hydrogen with protons from the medium. The hole left by the electron is filled by the manganese complex (as yet unknown), which, in turn, removes an electron ultimately from water to generate molecular oxygen and a proton. Similarly, a proton carrier, such as alkylnitrophenol, is built into the membrane to transmit the protons through the membrane. The basic principle of this system is the tunneling of the electron from the excited state, represented by S* in Figure 6, through some 20 Ångstroms to the conducting system, represented, in this case, by the carotenoid. At the other end of the carotenoid will be another photosensitized electron transfer to an acceptor

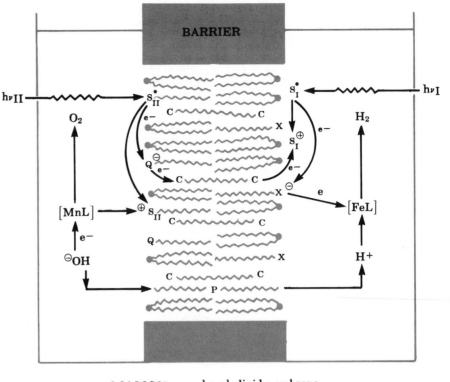

phospholipid membrane

S   sensitizer (electron donor)

Q, X   electron acceptor

P   proton carrier

C   carotenoid

Figure 6. The natural photosynthetic membrane served as the model for this hypothetical photochemical cell.

molecule, and the electron will then pass out through it to the iron-sulfide catalyst. The manganese and iron might very well be mounted on the membrane itself and not be free to move in the medium.

Since, in general, the energy of at least two visible quanta would be required to generate hydrogen and oxygen from water, two such membrane boundaries with somewhat different sensitizers could be placed in a series arrangement and connected by a molecular electron transport system. The directionality of the movement of electrons would be determined by the relative heights and thicknesses of the insulating barrier that must be traversed (Kuhn 1972).

Finally, it is conceivable that each of these catalysts might be individually surrounded by the electron and proton permeability barriers, so that the back reaction is prevented. It is also possible to remove the excess electrons on one side of the barrier and to neutralize the hole on the other side by means of a redox system, which, in turn, through a pair of electrodes

could deliver useful current and voltage in an external circuit (Valenty 1974).

As we learn the detailed structures of each of the crucial components in the quantum conversion process, it should be possible to reconstruct the photochemical system for generating molecular hydrogen from water in a relatively stable synthetic system without the need of an agricultural environment. The hydrogen so generated could be used directly as fuel, or as a component in the construction of hydrocarbon fuels from hydrogen-poor sources such as coal and shale. It might even be possible to construct a carbon dioxide–reduction system as well and thus be able to generate reduced carbon in a synthetic system.

This kind of totally synthetic system which emulates and simulates in some ways the chloroplast membrane activity of the green plant may very well be, in the future, another way to capture the energy of the sun and produce a storable fuel directly from it. Such a device would not compete with either the land or water regions of the earth that are required to fulfill the food needs of the growing population of the world and the increasing expectations of that population.

References

Adman, E. T., L. C. Sieker, and L. H. Jensen. 1973. The structure of a bacterial ferredoxin. *J. Biol. Chem.* 248:3987.

Alkaitis, S. A., M. Grätzel, and A. Henglein. 1975. Laser photo-ionization of phenothiazine in micellar solution. II. Mechanism and light induced redox reactions with quinones. *Ber.* 79:541.

Archer, M. D. 1975. Electrochemical aspects of solar energy conversion. *J. Appl. Electrochem.* 5:17.

Benemann, J. R., J. A. Berenson, N. O. Kaplan, M. D. Kamen. 1973. Hydrogen evolution by a chloroplast-ferredoxin-hydrogenase system. *Proc. Nat. Acad. Sci.* 70:2317.

Blankenship, R. E., and K. Sauer. 1974. Manganese in photosynthetic oxygen evolution. I. Electron paramagnetic resonance studies of the environment of manganese in triswashed chloroplasts. *Biochim. Biophys. Acta* 357:252.

Calvin, M. 1974a. Solar energy by photosynthesis. *Science* 184:375.

———. 1974b. Solar energy by photosynthesis: Are we able to raise enough cane to get it? *Kagaku to Seibutsu* 12:481.

———. 1974c. Solar radiation and life. In *Progress in Photobiology* (Proc. 6th Int. Photobiol. Congress, Bochum, Germany, 1972), ed. G. O. Schenck, p. 1. Hamburg: Deut. Gesell. Lichtforsch.

———. 1974d. In *The Current State of Knowledge of Photochemical Formation of Fuel* (Report of NSF Workshop, Osgood Hill, Boston, Sept. 1974), ed. N. Lichtin, p. 139.

———. 1975. Photosynthesis as a resource for energy and materials. *Kemia e Kemi* 2:46.

Chen, K.-Y., and J. H. Wang. 1974. Effect of manganese extraction on oxygen generation and EPR signal II in spinach chloroplasts. *Bioinorganic Chem.* 3:339.

Cheniae, G. M., and I. F. Martin. 1973. Absence of oxygen-evolving capacity in dark-grown *Chlorella:* The photoactivation of oxygen-evolving centers. *Photochem. Photobiol.* 17:441.

Döbereiner, J., and J. M. Day. In press. Associative symbioses in tropical grasses. Characterization of microorganisms and dinitrogen fixing sites. In *Dinitrogen Fixation,* ed. W. E. Newton and D. J. Wyman. Washington State Univ. Press.

Eady, R. R., and J. R. Postgate. 1974. Nitrogenase. *Nature* 249:805.

Ford, W. G., and M. Calvin. Unpublished results from Laboratory of Chemical Biodynamics, Univ. of Calif., Berkeley.

Hall, D. O., R. Cammack, and K. K. Rao. 1974. The iron-sulphur proteins: Evolution of a ubiquitous protein from model systems to higher organisms. *Origins of Life* 5:363.

Hardy, R. W. F., and U. D. Havelka. 1975. Nitrogen fixation research: A key to world food? *Science* 188:633.

Herskovitz, T., B. A. Averaill, R. H. Holm, J. A. Ibers, W. D. Phillips, and J. F. Weiher. 1972. Structure and properties of a synthetic analog of bacterial iron-sulfur proteins. *Proc. Nat. Acad. Sci.* 69:2437.

Holm, R. H. 1975. Iron-sulphur clusters in natural and synthetic systems. *Endeavour* 34:38.

Kuhn, H. 1972. Electron tunneling effects in monolayer assemblies. *Chem. Phys. Lipids* 8:401.

Kuhn, H., and D. Möbius. 1971. Systems of monomolecular layers: Assembling and physico-chemical behavior. *Angew. Chem. Internat. Ed.* 10:620.

Lichtin, N. N., ed. *The Current State of Knowledge of Photochemical Formation of Fuel* (Report of NSF Workshop, Osgood Hill, Boston, Sept. 1974).

Miller, J. R. 1975. Intermolecular electron transfer by quantum mechanical tunneling. *Science* 189:221.

Nakajima, T., Y. Yabushita, and I. Tabushi. 1975. Amino acid synthesis through biogenetic type CO_2 fixation. *Nature* 256:60.

Newton, W. E., and D. J. Wyman, eds. In press. *Dinitrogen Fixation.* Washington State Univ. Press.

Orme-Johnson, W. H. 1973. Iron-sulfur proteins: Structure and function. *Ann. Rev. Biochem.* 42:159.

Patarau, J. M. 1969. *By-Products of the Sugar Cane Industry.* Elsevier.

Pimentel, D., W. Dritschilo, J. Krummel, and J. Kutzman. 1975. Energy and land constraints in food protein production. *Science* 190:754.

Pimentel, D., L. E. Hurd, A. C. Bellotti, M. J. Foster, I. N. Oka, O. D. Sholes, and R. J. Whitman. 1973. Food production and the energy crisis. *Science* 189:443.

Postgate, J. R. 1975. In *Genetic Manipulation with Plant Materials,* ed. L. Ledoux. Plenum.

Postgate, J. R. In press. In *Dinitrogen Fixation,* ed. W. E. Newton and D. J. Wyman, Washington Univ. Press.

Que, L., Jr., J. R. Anglin, M. A. Bobrik, A. Davison, and R. H. Holm. 1974. Synthetic analogs of the active sites of iron-sulfur proteins. IX. Formation and some electronic and reactivity properties of Fe_4S_4 glycyl-L-cysteinylglycyl oligopepetide complexes obtained by ligand substitution reactions. *J. Am. Chem. Soc.* 96:6042.

Sarkanen, K. H. 1976. Renewable resources for production of fuels and chemicals. *Science* 191:773.

Subramaniam, A. 1972. Gel permeation chromatography of natural rubber. *Rubber Chem. and Tech.* 45:346.

Tien, H. T., and S. P. Verma. 1970. Electronic processes in bilayer lipid membranes. *Nature* 227:1232.

Valenty, S. J. 1974. The use of membranes to prevent recombination in photochemical reactions. In *The Current State of Knowledge of Photochemical Formation of Fuel* (Report of NSF Workshop, Osgood Hill, Boston, Sept. 1974), ed. N. Lichtin.

Von Bülow, J. F. W., and J. Döbereiner. 1975. Potential for nitrogen fixation in maize genotypes in Brazil. *Proc. Nat. Acad. Sci.* 72:2389.

Wigg, E. E. 1974. Methanol as a gasoline extender: A critique. *Science* 186:785.

G. H. Heichel

Agricultural Production and Energy Resources

Current farming practices depend on large expenditures of fossil fuels. How efficiently is this energy used, and will we be able to improve the return on investment in the future?

The primary energy source for modern agriculture, as for agriculture in ages past, remains the sun. Sunshine still provides the stimulus for the basic biochemical processes that reduce carbon dioxide in the air to the building blocks of food: the sugars, starches, and proteins in crop plants. Today, however, man has learned that he can increase the productivity of his food crops by modifying the plant's environment for maximum capture of sunlight. The technology developed for this purpose depends on fossil fuels, and thus modern agriculture may be thought of as a partnership of solar energy and fuel energy.

There are two main ways in which fuel energy is consumed in the production of food. Fuel is used off the farm to manufacture goods used for farming: natural gas for nitrogen fertilizers, coal for steel production, and petroleum for herbicides, fungicides, plastics, and the manufacture of machinery. And fuel is used on the farm as manufactured products

G. H. Heichel is a Plant Physiologist with the Plant Research Unit of the U.S. Dept. of Agriculture's Science and Education Administration and is Adjunct Professor of Agronomy at the University of Minnesota. Formerly he was with the Connecticut Agricultural Experiment Station, New Haven, as a research scientist after receiving a Ph.D. from Cornell University. His research interests include the physiological and environmental modulation of crop productivity and the energetics of agricultural ecosystems. Address: Plant Science Research Unit, Dept. of Agronomy and Plant Genetics, Univ. of Minnesota, 1509 Gortner Ave., St. Paul, MN 55108.

Table 1. Energy consumed by agriculture in the U.S., in kilocalories and as a percentage of the total U.S. energy budget (18,900 × 10¹² Kcal/yr)

	10^{12} Kcal	%
Farm production		
Crops		
Petroleum	189	1.0
Electricity	3	0.02
Livestock		
Petroleum	63	0.3
Electricity	9	0.05
Purchased inputs		
Fertilizer	140	0.7
Petroleum	49	
Feeds and additives	25	
Animal and marine oils	9	0.5
Farm machinery	8	
Pesticides	3	
Total	498	2.6

SOURCE: Economic Research Service 1974.

are applied to assist the growth of crops: by tractors during planting, cultivation, pest control, and harvesting, and by frost protection equipment and irrigation pumps.

The crops that supply our food need these fuel expenditures for a variety of reasons. Some have special nutrient demands, some have specific tillage requirements, and others need protection from the competition of weeds and the depredations of insects. The total energy requirement for crop production has been described as "cultural energy" (Heichel 1973) or as an "energy subsidy" (Slesser 1973). It is the aggregate of the fuel energy expenditures in Figure 1.

With fossil fuels becoming ever scarcer and more expensive, it is perhaps time to consider whether

we are getting a good return on our fuel energy investment and how we might improve both productivity and efficiency. In this article I will examine the use of fuel energy in U.S. agriculture, including the energy requirements of major crop production systems. To place the energy needs of specific crops in perspective, I will compare their yields of food energy and of protein with the amount of fuel energy required to produce them, and then describe some technological alternatives which could reduce the amount of fossil fuels consumed in crop production and increase the usefulness of plants as food and fuel sources. I will conclude with some speculations on agriculture's future energy needs.

Energy efficiency of cropping systems

Production of raw agricultural commodities ranks third in energy consumption after the steel and petroleum refining industries (Heichel 1974). Excluding the energy used by farmers for domestic needs, agriculture accounts for about 2.6% of the U.S. energy budget (Table 1), or about 10% as much energy as is burned in cars, trucks, trains, and airplanes, and 14% as much energy as is used for heating buildings. About equal amounts of fuel energy are used on and off the farm. Among the purchased inputs, the manufacture of fertilizers uses the most energy—nearly 28%—and the production of pesticides the least—only 0.6%. Consumption of petroleum products on the farm accounts for 51% of the energy used by agriculture.

We now know rather precisely how

much cultural energy it takes to produce many of the principal U.S. food crops, and these "energy accounts" range from 1 to 30 × 10³ megacalories (Mcal)/acre/year among the 24 important field, fruit, and vegetable crops shown in Figure 2. Field crops like oats, corn, soybeans, and wheat consume least cultural energy, and they account for most of the acres in crop production. Interestingly, perennial deciduous fruits and sugar crops, largely destined for human consumption, are only moderately energy-intensive and rank between soybeans and corn, at 3 to 6 × 10³ Mcal/acre/year. Citrus fruits, which are slightly above the median, use more energy than field crops and deciduous fruits because of specialized requirements for irrigation and frost and pest control. Peanuts, rice, and potatoes, important sources of protein and carbohydrates in the human diet, appear about midway on the scale, at 11 to 14 × 10³ Mcal/acre/year.

The most energy-intensive of the 24 crops in Figure 2 are the annual vegetables and fruits for human diets. These crops need more fertilizer and pesticide than do field crops destined for animal consumption, and they grow best under frequent irrigation. In addition, there are exacting market requirements for the uniformity and quality of these perishable foods. In fact, the energy used by the sophisticated processing and marketing technology geared to consumer preferences can far exceed that needed to produce the crop (Hirst 1973). It is worth noting that it takes as much energy to build a 6-passenger car (Berry and Fels 1972) as to grow an acre of cauliflower for a year. In comparison, about 5 acres of corn and 20 of wheat can be grown on the same amount of energy.

Like their cultural energy requirements, the yields of food from different cropping systems vary greatly. To ascertain the return on investment, we can relate the biological value of the crop to the technological expenditure in production. One criterion of biological value is the yield of food energy (measured in calories) per unit of cultural energy spent to produce it. Since balanced nutrition requires protein as well as energy, another criterion of biological value is the yield of protein per unit of energy spent in production.

Figure 1. The diagram represents the flow of cultural energy required both on and off the farm to produce raw food and feed commodities.

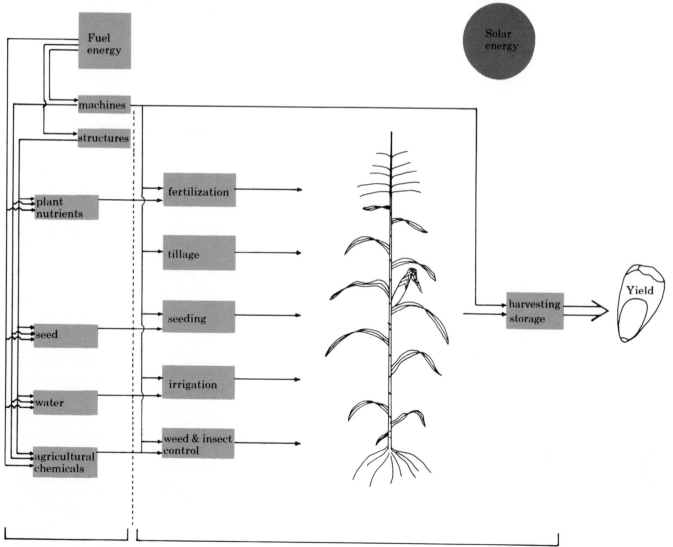

Purchased inputs   Farm production

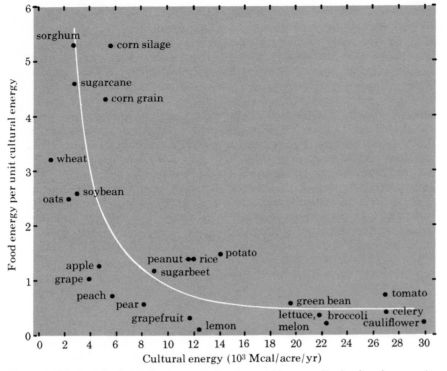

Figure 2. The graph shows the amount of cultural energy used to produce 24 grain, forage, fruit, and vegetable crops, and the amount of food energy (in calories) that each plant yields per unit of cultural energy invested. The ratio is a measure of food energy efficiency. (From Heichel 1973, 1974; Cervinka et al. 1974.)

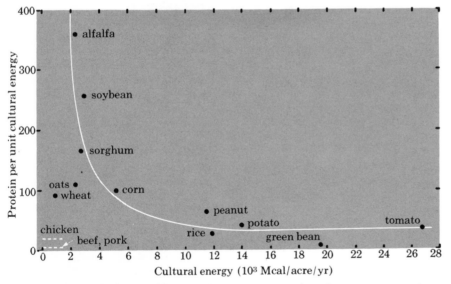

Figure 3. The graph shows yield of protein per Mcal of cultural energy for 11 forage, grain, and vegetable crops. The dashed lines, indicating ordinate values for 3 animal protein sources, show that many crops return more protein per unit of energy than do animals.

Caloric gain (Rappaport 1971), the ratio of food energy harvested to cultural energy consumed, is a measure of the energy efficiency of farming. Among the 24 crops listed in Figure 2, sorghum, sugarcane, and corn offer the greatest caloric gain: they return 4 to 5 calories of food per unit of cultural energy. Although grown in systems that demand similar investments of fuel energy, wheat, oats, and soybeans yield 2 to 3 calories of food per unit of cultural energy—less food energy return than sorghum, sugarcane, and corn. Among the crops using less than 5×10^3 Mcal/acre/year, there are several examples of very efficient food energy production. In contrast, grapes and apples barely return the energy they consume in production.

Among the 14 cropping systems that require more than 5×10^3 Mcal/acre/year, few return in calories of food the amount of energy used to produce them. Some intermediate crops, including sugarbeets, peanuts, rice, and potatoes, yield 1.0 to 1.5 calories of food energy per unit of cultural energy, while others, for example peaches, pears, and grapefruit, use 1.5 to 4 calories of cultural energy to produce one calorie of food. The 7 most energy-intensive crops in Figure 2 require 19 to 30×10^3 Mcal/acre/year, and they return only 0.25 to 0.75 calories of food energy per unit of cultural energy. From these examples it is apparent that highly energy-intensive cropping systems are not very efficient sources of food energy for animals and humans.

Food energy can be produced very efficiently with corn or sorghum, but Americans prefer to eat meat; thus hogs and cattle are gathered into feedlots and fed grain to produce this more desirable human energy source. By introducing an intermediate consumer such as a cow or hog into the food chain, the energy efficiency of corn and sorghum is reduced by an order of magnitude—a result which may be surprising but is nevertheless a straightforward outcome of animal metabolism. Like the same plants consumed directly by people, the corn and soybeans fed to animals are produced with a net energy gain. But in hog and broiler metabolism, for example, only 10% of the energy in the feed is converted into edible meat (National Academy of Sciences 1975). As a result, the caloric gain of beef and pork production ranges from 0.2 to 0.6, which is comparable to that of the most energy-intensive vegetable and fruit crops consumed directly by humans. Again, consumer preferences often result in less food energy from crops at the farm gate than is spent in fuel energy to produce them.

There are similar marked differences in protein production per unit of cultural energy for the 6 cropping systems of Figure 3 using 5×10^3 Mcal/acre/year or less. The legumes alfalfa and soybeans produce 250 to 350 grams of protein, or the daily requirement of 5 to 7 people, per Mcal of cultural energy. The grains wheat, oats, corn, and sorghum produce 100 to 150 grams of protein per Mcal, sufficient food for 2 or 3 people, except that they are deficient in amino acids like lysine, tryptophan, and methionine that are essential for human metabolism. Legumes yield about 60 times more protein per Mcal than beef or pork and 15 times more protein than chicken.

There is a relatively narrow range of efficiency of protein production per Mcal in the energy-intensive crops requiring 11 to 29 × 10³ Mcal/acre/year. Among the intensive crops, only the peanut, with 60 g protein/Mcal cultural energy, is substantially more efficient than beef, pork, or chicken. We find that energy-intensive cropping systems are as inefficient as producers of protein as they are as producers of food energy. And again, consumer preference and tradition prevent the substitution of energy-efficient systems of protein production for others less frugal of energy. We do not consume alfalfa and soybeans as protein sources but use them instead to upgrade the amino-acid-deficient grains used to feed the animals we eat.

Efficiency of feed conversion in animals

The production of animals from plants involves the conversion of calories of feed into calories of meat and other animal products. The live weight gain of animals per unit of feed consumed is known as the feed efficiency. We will express this factor in terms of energy in order to measure the efficiency of animals in converting calories of plants into calories of beef, broilers, or hogs. Some trends are exemplified in Figure 4. Although the conversion of various feed units to corn equivalents has limitations (Byerly 1966, 1967), these estimates reflect the performance of present livestock production systems as reported by the USDA in 1973. The efficiency of conversion varies about twofold among animal species, and the conversion efficiency of monogastric animals like swine and broilers, about 0.12 calories of food per calorie of feed, is more than twice that of ruminants like cattle.

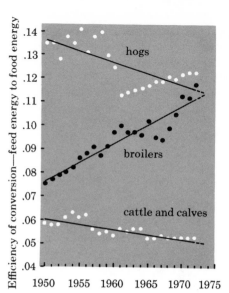

Figure 4. Some trends in the metabolic efficiency of converting calories of animal feeds into calories of human food are shown for cattle and calves, broilers, and hogs. During the 22-year period shown, the feed efficiency of hogs and cattle has declined, while that of broilers has improved. Cattle and calves includes beef cattle, dairy heifers, and dairy calves. (From Heichel and Frink 1975.)

The economic and social consequences of reducing the amount of grain consumed by animals to release more grain for human use have been analyzed and debated (Lockeretz 1975; Van Vleck 1975). Limiting animal consumption of corn, sorghum, barley, and oats would be unlikely to make significant amounts of food available to hungry people. Although wheat and rice are staple human foods, the physical and nutritional character-

Table 2. Contribution of fuel costs to the price of raw agricultural products

	Retail price ($)	Fuel cost (%)
Raw dairy products		
Purchased inputs	0.744	1.4
Farm production	0.256	18.4
	1.000	5.7
Raw meat products		
Purchased inputs	0.801	1.4
Farm production	0.199	16.7
	1.000	4.4
Raw grain products		
Purchased inputs	0.635	1.4
Farm production	0.365	10.0
	1.000	4.6

SOURCE: Economic Research Service 1974.

istics of most feed grains limit their wide acceptance except as fermentation and starch products. On the other hand, reducing the amount of grains consumed by animals might result in the replacement of grain crops on farms by food crops like rice and vegetables that may excel animal products in caloric gain. Whether this would happen, however, is difficult to predict.

The important point is that animal agriculture competes successfully on an energy efficiency basis with many food crops, including broccoli, melons, tomatoes, and cauliflower, that are consumed directly by people. This competitiveness is largely due to the use of energy-efficient grains like corn, sorghum, and soybeans as feed. Because the feed grains are an order of magnitude more energy-efficient than the horticultural crops consumed directly, they compensate for the energy losses in animal metabolism. If hybrid corn were as prodigal of energy as cauliflower is, beef and pork would be prohibitively expensive for any consumer.

Prospects for improving energy efficiency

The inflationary rise of food and fuel prices has led people to hope that the dependence of agriculture on fuel energy might somehow be abated. To learn how improvements in energy efficiency might affect the price of food, we must ask how the price of farm products changes with fuel prices.

The contribution of fuel costs to the price of raw agricultural products varies by type of commodity and method of production (Table 2). Fuel costs are about 1.4% of the retail price of purchased inputs to dairy, meat, and grain production. Fuel costs on the farm are 18.4 and 16.7% of dairy and meat production costs, respectively, but only 10% of the costs of grain production. Since fuel accounts for 4 to 6% of the cost of the raw agricultural products, frugal use of increasingly expensive fuels might slow the rise in food prices.

One way of achieving more frugal use of fuel is to modify production practices. For example, in corn production, chosen because of its cen-

Table 3. Generalized energy budget for producing a 100-bushel/acre corn crop by conventional fertilization and tillage practices

	Gals fuel/acre	Mcal/acre	%
Purchased inputs			
Fertilizer			45
155 lb N, 55 lb P, 45 lb K		1,295	
Limestone, 1,000 lbs/3 yrs		20	
Pesticides			2
Insecticide (1 lb), herbicide (3 lb)		44	
Machinery, miscellaneous		731	25
Farm production			
Crop establishment			4.5
Broadcast fertilizer		21.8	
Broadcast lime		11.6	
Disk	0.56		
Plow	1.73	97.6	
Plant	0.56		
Weed control	0.20		
	3.05	131.0	
Harvesting			24
Combine-sheller	2.0		
Transport			
Dry	20.0		
Shell			
	22.0	704.0	
Total cultural energy		2,925	100

Table 4. Generalized energy budget for producing a 100-bushel/acre corn crop using conventional fertilization practices and minimum tillage

	Gals fuel/acre	Mcal/acre	%
Purchased inputs			
Fertilizer			48
155 lb N, 55 lb P, 45 lb K		1,295	
Limestone, 1,000 lbs/3 yrs		20	
Pesticides			3
Insecticide (2 lb), herbicide (4.5 lb)		72	
Machinery, miscellaneous		609	22
Farm production			
Crop establishment			2
Broadcast fertilizer		21.8	
Broadcast lime		11.6	
Plant (+ starter fert.)	0.60	19.2	
	0.60	52.6	
Harvesting			25
Combine-sheller	2.0		
Transport			
Dry	20.0		
Shell			
	22.0	704.0	
Total cultural energy		2,753	100

tral role in milk, meat, and grain production, the manufacture of fertilizers, machinery, and pesticides accounts for about 72% of the cultural energy requirement. Crop establishment and harvesting consumes the remaining 28% (Table 3). The large quantities of fuel used for tillage and for fertilizers have made this an area of particular concern, but technological alternatives do exist which might possibly reduce the fuel expenditure for these items.

In one strategy, minimum tillage, the disking, plowing, planting, and weed-control operations of conventional tillage are combined into one operation by a specialized implement which plants and fertilizes in untilled soil (Table 4). Widely used at present in Kentucky and Ohio, minimum tillage is potentially useful in many other states with suitable growing conditions. About 2.5 gal/acre less fuel is required to establish the crop, and fewer implements are needed for farming. The energy savings are less than they might be because more insecticides are needed to cope with insect problems arising from the crop residues left on the land in minimum tillage (USDA 1975). Contact herbicides are also needed to control weeds at planting on untilled soil. About 6% less total energy, the equivalent of about 5.4 gallons of fuel per acre, might be needed for all the purchased inputs and farming operations of minimum rather than conventional tillage operations. Minimum tillage promises to increase the energy efficiency of crop production if yields can be maintained.

Renewed interest is also being directed toward substituting animal manures for commercial fertilizers to conserve much of the energy required for the synthesis of ammonia from natural gas. Of course, only about 50% of domestic animal manures, about 10^9 tons annually, are produced in confinement and thereby considered accessible (Stephens and Heichel, in press). If an application rate of 15 tons/acre is needed to supply the nitrogen needs of corn, about 70×10^6 acres, or nearly all the land now devoted to corn, might potentially be fertilized by animal manures.

If manures were substituted for commercial fertilizers, the energy requirements for manufactured products would be greatly reduced but those for application greatly increased (Table 5). Elimination of fertilizers would reduce energy use by 1,295 Mcal/acre, equal to about 40 gallons of gasoline. However, additional equipment needs and the loading and spreading of 15 tons/acre of manure within 3 miles of the feedlot would require, by a conservative estimate, the energy of 30 gallons of gasoline. Within this 3-mile radius of the feedlot, animal manure might compete with fertilizers, but beyond that distance the energy costs of loading and spreading manure would exceed those of conventional fertilizers. Moreover, adding labor costs to the energy costs of application limits the economic substitution of manures for fertilizers to distances of 1 mile or less. Since most manures are produced in feedlots many miles from major areas of cropland, it is clear that substitution would not significantly reduce the energy and economic costs of cropping except in situations like dairy farms, where animals are raised and crops are produced in very close proximity.

Other methods of increasing the energy efficiency of crop production involve changes not in technology but in land use and in the crop plants themselves. Traditionally, the sun's energy has been considered a "free good." However, the farmer must own or rent land to support the plants that capture sunlight, and the increasing price of farmland directly reflects the increasing value of photosynthetically active sunlight.

The solar energy incident on U.S. cropland varies from a high of 260 watts/m^2/yr in most of New Mexico, Arizona, and parts of California, to a low of 150 watts/m^2/yr in the dairy regions of upstate New York, Vermont, and Oregon. The area receiving 260/watts/m^2/yr is obviously the best choice for maximizing the photosynthetic productivity of crops (Calvin 1974), if water were plentiful instead of scarce. In this arid region of bright sunlight, the competition of manufacturing industries with agriculture makes water nearly prohibitively expensive. For example, an acre-foot of water in Arizona generates about 1,000 times more personal income when used for manufacturing than for agriculture (Young and Martin 1967). And, of course, the irrigation of arid lands to increase productivity consumes a good deal of fuel—and could even lead to decreases in the overall energy efficiency of farming (Heichel and Frink 1975). Therefore, it is logical to protect from development the productive farmland of semiarid and humid regions that is suitable for rainfed agriculture and to encourage agriculture in rainfed areas like New England, where farming is declining.

Table 5. Generalized energy budget for producing a 100-bushel/acre corn crop using conventional tillage practices and animal manures in lieu of conventional fertilization

	Gals fuel/acre	Mcal/acre	%
Purchased inputs			
Fertilizer			1
Manure, @ 15 tons/acre		0	
Limestone, 1,000 lbs/3 yrs		20	
Pesticides			2
Insecticide (1 lb), herbicide (3 lb)		44	
Machinery, miscellaneous		783	30
Farm production			
Crop establishment			40
Broadcast lime		11.6	
Load, spread manure (≤3 mi)	28.50		
Disk	0.56		
Plow	1.73	1,010.0	
Plant	0.56		
Weed control	0.20		
	31.55	1,021.6	
Harvesting			27
Combine-sheller	2.0		
Transport, Dry, Shell	20.0		
	22.0	704.0	
Total cultural energy (≤3 mi)		2,573	100

A counterproposal is to use the income from manufacturing to purchase farm products grown more cheaply elsewhere. However, we are uneasy that we may exhaust the "elsewhere"—the land that remains more valuable for agriculture than for alternative uses. Factories may reach many stories above ground with little effect on the efficiency of production, but the geometry of agriculture requires crops to be distributed over the land in a layer a few leaves thick for maximum utilization of sunlight.

With some food plants grown by modern farming methods incurring an energy deficit in production and real-estate developers competing with farmers for land, some advocates have proposed (e.g. International Research and Technology 1973; Johnson 1975) that food plants be grown in confinement rather than in open fields. A preliminary energy account for a modern, sunlit greenhouse continuously producing lettuce reveals that a cultural energy expenditure 70 times that for lettuce grown on open farmland would be needed. Another accounting for a large growth chamber, this one an artificially illuminated, climate-controlled lettuce-

growing facility, reveals that a cultural energy input nearly 300 times that for farm-grown lettuce would be required. Analysis suggests that the artificial light for growing lettuce might cost 33 cents/head at current costs for electricity of about 4 cents/kilowatt-hour. The caloric gain of greenhouse and growth-chamber lettuce is 0.01, less than one twenty-fifth the efficiency of lettuce produced on farms (Fig. 2). Clearly, growing food plants on farms with free and plentiful sunlight is a more efficient way to use scarce resources than growing them in confinement under either natural or artificial sunlight.

Another strategy for increasing agricultural energy efficiency is to modify the carbon and nitrogen metabolism of plants. Food plants are often classified as having inefficient or efficient photosynthetic CO_2 metabolism: efficient photosynthetic species like corn and sorghum produce more dry matter per unit of absorbed sunlight than inefficient species like oats, wheat, and soybeans. Clearly, therefore, using efficient photosynthetic species like corn or sorghum as feed crops instead of inefficient oats or soybeans would conserve energy resources

(Heichel 1973). Because photosynthetically inefficient soybeans are a desirable protein source for man and farm animals, genetic changes in the metabolism of the plant are being sought to increase the capture of sunlight and enhance photosynthetic efficiency without requiring additional cultural energy.

Reducing the rate of photorespiration, a light-stimulated loss of CO_2 that is believed to account for a growth depression of 15 to 50% in soybeans, is one possibility (Chollet and Ogren 1975; Zelitch 1975). Photosynthetically efficient crops like corn lack significant photorespiration, but in these species a screening program for strains having more rapid photosynthesis promises to increase yield (Heichel and Musgrave 1969). Differences in dark respiration, the energy-producing loss of CO_2 shown during night and day by all plant tissues, among strains of a species having similar photosynthetic efficiencies can also contribute to yield differences (Heichel 1971). Thus, the effort to identify wasteful components of dark respiration and to reduce their effects in photosynthetically efficient and inefficient species may stimulate yields without incurring increased demand for cultural energy (Zelitch 1974).

Leguminous crops like soybeans, because they are capable of biological nitrogen fixation, consume little cultural energy in the form of fertilizer. The development of corn and other grains with this characteristic might substantially reduce their requirements for fertilizer nitrogen (Evans 1975). This advance would lessen the fertilizer industry's demand for diminishing natural gas supplies, release scarce fuels for other needs, and possibly increase the protein content and nutritional quality of cereals.

Production efficiency could also be increased by putting to use portions of the crop that are now rejected as waste. About 60% of the primary production of U.S. agricultural crops is unsuitable for human food or animal feed and is used only for soil improvement and erosion control. These agricultural residues could be used as a fuel substitute for petroleum if the energetics and economics of residue retrieval were sufficiently favorable (Gifford and

Table 6. Annual dry matter production of crop and animal residues in the U.S.

Agricultural residue	Dry matter (10^6 metric tons)
Crop residues	
Cereal straw (wheat, rice, oats, rye, barley)	141
Corn cobs, fruit pits, nut shells	55
Corn and sorghum stover	90
Soybean hulls and stover	50
Sugarcane bagasse	5
Other crop residues	89
Total crop residues	430
Animal manures (50–80% in confinement)	300
Total agricultural residues	730

SOURCE: Stephens and Heichel, in press.

Millington 1974; Stephens and Heichel, in press). Approximately 430 × 10^6 metric tons of crop residues were produced in 1972 (Table 6), most of it by the major cereal and oilseed crops. These residues are for the most part distributed over millions of acres of cropland. Little sugarcane residue remains on the land since most of the bagasse is used to fuel boilers in sugar mills.

Another source of agricultural residues is animal manures (Table 6). About 300 × 10^6 metric tons of dry animal manures, or 2 billion metric tons of fresh manure with 85% moisture content, are produced annually. As noted earlier, only about 50% of animal manures are produced in confinement and are easily collected.

Distributed over cropland or accumulating in feedlots, plant and animal residues represent a potential energy resource for which the price of production has already been paid. Using hydrogenation with carbon monoxide and steam (Anderson 1972), a net yield of 1.3 barrels of oil can be produced from 1 metric ton of dry organic waste. The total potential, therefore, if all agricultural residues could be recovered, would be about 10^9 barrels of oil annually. Of course, inefficiencies in the collection of residues would greatly reduce the amount of energy available, but recovery of only 15% of the energy potential would yield 150 × 10^6 barrels of oil annually. This

amount of fuel would be sufficient to meet current on-farm petroleum needs, 189 × 10^{12} kilocalories.

Another promising technology is the enzymatic hydrolysis of cellulose-rich residues to produce glucose that, after fermentation, yields an alcohol substitute for gasoline (Reese et al. 1972). The anaerobic fermentation of animal wastes as a primary treatment for pollution control might yield a large amount of energy in the form of methane gas (Jewell 1974): the daily waste from 8 to 12 cattle would produce methane gas having the heat content of a gallon of gasoline.

There may be costs of utilizing agricultural residues that are less conspicuous than the benefits. For example, increased use of residues for fuel might speed the removal of plant nutrients from the soil. Normally a portion of the fertilizer applied to a crop accumulates in the unused parts of plants—in cornstalks and cereal straw, for example—and is returned to the soil as they decay. Collection of nutrient-rich residues could deprive the next year's crop of some "free" fertilizer, but as long as commercial fertilizer is routinely applied, the removal of residues is unlikely to lead to a decline in the productivity of cropland (Stephens and Heichel, in press).

Developments in biotechnology to convert energy- and protein-efficient crops like corn, sorghum, soybeans, and alfalfa into nutritious food analogs without the intervention of animals in the food chain might also help to lessen the competition of animals with people for scarce feed grains and slow the rise in food prices. For example, the extraction, coagulation, and drying of plant juices from alfalfa forage yields a concentrate that is 40 to 50% protein and suitable for human food after refining. Textured vegetable protein made from soybeans is already being used as a meat extender and substitute (Pour-el 1974), and single-cell protein harvested from bacteria that degrade cellulosic plant wastes is another promising food source. If calories of energy only are considered, corn might be grown barren of grain but with stalks nearly as rich in sugar as sugarcane. This potential reservoir of food energy in a crop more widely adapted than either tropical sugarcane or temperate sugarbeet remains unexploited for want of adequate technology. Dried corn stover, like corn grain, barley, oats, and sorghum, can now be changed to sugar by enzymatic or chemical hydrolysis.

The new technologies are exciting, but their energetics are poorly understood. If they are to conserve fuel, these systems must be capable of converting plants into nutritious food with an energetic efficiency superior to that of animal metabolism.

Future energy needs for agriculture

Having reviewed present patterns of energy consumption in agriculture, we turn to projecting agriculture's future energy needs. Estimating population growth is chancy, but I will follow the practice of projecting various futures (U.S. Department of Commerce 1972). By the year 2000, depending on assumptions about birth rates, the U.S. population is expected to number from 271 to 322 million. A population of 300 million will be used in my examples.

Prediction of per capita food consumption is equally difficult. A recent projection for the National Water Commission (Heady et al. 1972) suggests that U.S. beef consumption will increase from its present level of 117 pounds per person per year to 158 pounds by the year 2000. (This prediction is based upon a model of consumer demand for beef and other meats; it is approximately a linear projection of past patterns, though not strictly. It is perhaps worth noting, for comparison, that beef consumption in 1950 was about 62 lb/person.) Although reaction in the marketplace to high beef prices suggests that there is an upper limit to what consumers can or will pay, these predictions are nevertheless sufficient for an illustration, and they will allow us to project the per capita and national energy requirements for food production for the year 2000.

The energetic equivalent of 38 gallons of oil is needed to grow on farms the 667 pounds/year of animal products in an individual's present diet (Table 7). Although the

Table 7. Present food consumption patterns in the U.S. and estimated energy requirements for food production, 1972

	Food consumption (lbs/person/yr)	Energy requirement (gals oil/person/yr)
Animal product		
Beef	117	19.1
Dairy	356	4.7
Pork	67	4.8
Fats and oils	15	3.5
Poultry	52	1.6
Eggs	39	1.3
Veal and lamb	6	0.8
Fish	15	2.1
Subtotal	667	37.9
Plant product		
Flour and cereal	140	2.2
Sugar	122	1.4
Fats and oils	42	3.7
Fruits	132	1.9
Potatoes	105	0.6
Beans, peas, nuts	16	1.8
Green, yellow vegetables	215	2.3
Miscellaneous	15	0.2
Subtotal	787	14.1
Total	1,454	52

SOURCE: Heichel and Frink 1975.

Table 8. Projected food consumption patterns in the U.S. and estimated energy requirements for food production in the year 2000

	Food consumption (lbs/person/yr)	Energy requirement (gals oil/person/yr)
Animal product		
Beef	158	25.8
Dairy	242	3.2
Pork	68	4.9
Fats and oils	15	3.5
Poultry	54	1.7
Eggs	27	0.9
Veal and lamb	6	0.8
Fish	15	2.1
Subtotal	585	42.9
Plant products total	787	14.1
Total	1,372	57

SOURCE: Heichel and Frink 1975.

consumption of dairy products and eggs is expected to decline by 2000 (when a greater proportion of the population will be the cholesterol-conscious age groups), increased beef, pork, and poultry consumption will require additional energy equivalent to 5 gallons of oil. Thus, energy equal to nearly 43 gallons of

oil will be needed to grow the animal products in the diet projected for the year 2000 (Table 8).

Energy equal to 14 gallons of oil is needed to grow on farms the 787 pounds/year of plant products in the 1972 diet. Since direct consumption of plant products is expected to remain practically unchanged (Heady et al. 1972), 14 gallons of oil should also suffice for the year 2000. Thus, changes in dietary patterns might be be expected to increase the energy needs of food production by 10%, equal to 34 × 10⁶ barrels of oil, within 25 years.

Changes in dietary patterns are but one variable in projecting agriculture's energy requirements. If the population reaches 300 × 10⁶ people, the increase in national feed and cultural energy requirements to grow the animal products in the projected diet will be 64%, or the equivalent of 120 × 10⁶ barrels/year of oil, above the 1972 level. Although per capita consumption of plant products is not expected to change substantially by 2000, a population of 300 × 10⁶ might require a 44% increase in food from this source, leading to an additional 30 × 10⁶ barrels/year of oil used in production.

The conservative projection that production agriculture might need the equivalent of an additional 150 × 10⁶ barrels of oil annually to supply the food needs and dietary preferences of an expanding population does not take into account the energy needed to produce food exports and nonfood agricultural products. It assumes that the efficiency with which animals convert feed to food and the caloric gain of cropping systems will be maintained at current levels or improved. How realistic is this assumption? The subject of converting feed energy to food energy has been debated extensively, particularly for beef production (Council for Agricultural Science and Technology 1974; Van Vleck 1975). It is instructive to consider how the cultural energy requirements of agriculture in the United States might change if the feed efficiencies of hogs, broilers, and cattle continue to change according to the trends illustrated in Figure 4 and the predicted dietary and population trends materialize. If the apparent decline in feed efficiency of cattle and hogs and improvement in efficiency of broiler production in the next 25 years follow the trends established during the last 22 years, the resulting feed energy needed to produce the animal-protein-augmented diet of Table 8 will require an additional 29×10^6 barrels of oil by the year 2000.

Similarly, it is useful to ask whether the efficiency of energy utilization in major crop production systems is declining as technology is adopted (Pimentel et al. 1973; National Academy of Sciences 1975). The increased yields of the year 2000 are often projected with irrigation and perhaps other energy-intensive technologies not now widely used in crop production (Heady et al. 1972). Fully developing the irrigation potential of American agriculture might increase its cultural energy requirements by 30% (Batty et al. 1974). For example, the response of corn yields to increasing energy inputs suggests that yield increments are being obtained with progressively larger energy investments. While extensive irrigation might increase corn yields by more than 20%, energy investment would exceed yield return, and the caloric gain might decline by 50% (Heichel and Frink 1975). Evidence about the responses of other cropping systems is scant, and unfortunately we are unable to predict from experience what may lie ahead. However, if future yield increases of 20% are obtained at the expense of a 50% reduction in energy efficiency, production agriculture will need the equivalent of 417×10^6 barrels of oil yearly to deliver the projected diet of Table 8, or 180% more cultural energy than the 150×10^6 barrels projected without a decline in energy efficiency.

The possibility that growing raw agricultural commodities might require 60 to 180% more fuel energy in the next quarter-century emphasizes the need to develop energy-frugal and environmentally sound food production strategies for the future. Significant differences in the efficiency with which energy is used exist among our present cropping systems, and widespread adoption of the more efficient systems could save fuel resources. Additional ways of producing more food with less energy are urgently needed.

Clearly, we must learn to reduce the amount of energy used in food production by substituting plant for animal food, by altering plant metabolism to enhance yields, by developing new energy-efficient schemes for the nurturing of plants and animals, and by exploiting underutilized plant by-products as an energy resource. Ensuring a plentiful and relatively inexpensive supply of food for the next quarter-century is a persuasive stimulus.

References

Anderson, L. L. 1972. *Energy potential from organic wastes: A review of the quantities and sources.* U.S. Dept. of the Interior, Bureau of Mines Info. Circ. 8549.

Batty, J. C., S. N. Hamad, and J. Keller. 1974. *Energy inputs to irrigation.* U.S. Agency for International Development 211-d ser. no. 8. Agricultural and Irrigation Engineering Dept., Utah State Univ., Logan.

Berry, R. S., and M. F. Fels. 1972. *The production and consumption of automobiles.* Rpt. Illinois Inst. Environ. Qual., Dept. of Chemistry, Univ. of Chicago.

Byerly, T. C. 1966. The role of livestock in food production. *J. Animal Sci.* 25:552–66.

———. 1967. Efficiency of feed conversion. *Science* 157:890–95.

Calvin, M. 1974. Solar energy by photosynthesis. *Science* 184:375–81.

Cervinka, V., W. J. Chancellor, R. J. Coffelt, R. G. Curley, and J. B. Dobie. 1974. *Energy requirements for agriculture in California.* Calif. Dept. Food and Agr., and Univ. Calif., Davis.

Chollet, R., and W. L. Ogren. 1975. Regulation of photorespiration in C_3 and C_4 species. *Bot. Rev.* 48:137–79.

Council for Agricultural Science and Technology. 1974. *Efficiency in animal feeding with particular reference to non-nutritive feed additives.* Ames, Iowa.

Economic Research Service, USDA. 1974. *The U.S. food and fiber sector: Energy use and outlook.* Prepared for the Committee on Agriculture and Forestry. U.S. Senate, Washington, D.C.

Evans, H. J., ed. 1975. *Proc. workshop on enhancing biological nitrogen fixation.* Nat. Sci. Found., Washington, D.C.

Gifford, R. M., and R. J. Millington. 1974. Energetics of agriculture and food production with special emphasis on the Australian situation. In *Energy and How We Live,* Proc. Aust.-UNESCO Man and the Biosphere Symposium, Flinders Univ., Adelaide.

Heady, E. O., H. C. Madsen, K. J. Nicol, and S. H. Hargrove. 1972. *Agricultural and water policies and the environment: An analysis of national alternatives in natural resource use, food supply capacity, and environmental quality.* CARD Rpt. 40T, Center for Agr. and Rural Develop., Iowa State University, Ames.

Heichel, G. H. 1971. Confirming measurements of respiration and photosynthesis with dry matter accumulation. *Photosynthetica* 5(2):93–98.

———. 1973. *Comparative efficiency of energy use in crop production.* Conn. Agr. Exp. Sta., New Haven, Bul. 739.

———. 1974. Energy needs and food yields. *Tech. Rev.* 76:18–25.

———, and C. R. Frink. 1975. Anticipating the energy needs of American agriculture. *J. Soil Water Cons.* 30:48–53.

———, and R. B. Musgrave. 1969. Varietal differences in net photosynthesis of *Zea mays* L. *Crop Sci.* 9:483–86.

Hirst, E. 1973. *Energy use for food in the United States.* Rpt. no. 57, Oak Ridge National Laboratory, Tenn.

International Research and Technology Corporation. 1973. *Production of vegetables in controlled environments.* Rpt. IRT-318-R, Arlington, Va.

Jewell, W. J. 1974. *Energy from agricultural wastes—methane generation.* Cornell Univ. Agr. Engr. Ext. Bull. 397.

Johnson, L. E. 1975. *Increasing agricultural production with hydroponics and solar energy in urban areas of New England.* Rpt. CP-39, Center for Environ. and Man., Hartford, Conn.

Lockeretz, W. 1975. *Agricultural resources consumed in beef production.* Report no. CBNS-AE-3, Center for the Biology of Natural Systems, Washington University, St. Louis, Mo.

National Academy of Sciences. 1975. *Agricultural production efficiency.* National Research Council, Washington, D.C.

Pimentel, D., L. E. Hurd, A. C. Bellotti, M. J. Forster, I. N. Oka, O. D. Scholes, and R. J. Whitman. 1973. Food production and the energy crisis. *Science* 182:443–49.

Pour-el, A. 1974. Soy proteins in human nutrition. *Proc. 1st World Congress of Medicine and Environmental Biology,* Paris.

Rappaport, R. A. 1971. The flow of energy in an agricultural society. *Sci. Amer.* 225:117–22, 127–32.

Reese, E. T., M. Mandels, and A. H. Weiss. 1972. Cellulose as a novel energy source. In T. Ghose, A. Fiechter, and N. Blakebrough, eds. *Advances in Biochemical Engineering, 2.* N.Y.: Springer-Verlag.

Slesser, M. 1973. Energy subsidy as a criterion in food policy planning. *J. Sci. Food Agr.* 24:1193–1207.

Stephens, G. R., and G. H. Heichel. In press. Agricultural and forest products as sources of cellulose. *Biotech.& Bioeng. Symp.,* no. 5.

U.S. Department of Agriculture. 1975. *Minimum tillage: A preliminary technology assessment.* Office of Planning and Evaluation, Washington, D.C.

U.S. Department of Commerce. 1972. *Statistical Abstract of the United States, 1972.* Washington, D.C.

Van Vleck, G. 1975. For human nutrition: Grain vs. meat. *Professional Nutritionist* 7:1, 6–10.

Young, R. A., and W. E. Martin. 1967. The economics of Arizona's water problem. *Ariz. Rev.* 16(3):9–18.

Zelitch, I. 1974. Improving the energy conversion in agriculture. In D. E. McCloud, ed. *A New Look at Energy Sources.* Spec. pub. 22, Am. Soc. Agron. Madison, Wis.

———. 1975. Improving the efficiency of photosynthesis. *Science* 188:626–33.

"I find that an hour in the compost pile provides my minimum daily requirement of everything."

C. H. KASTER

Geothermal Systems and Power Development

A. J. Ellis

Natural hot water and steam fields are a valuable source of useful energy in several countries, and in many others they are being developed rapidly

For most people, geothermal fields probably call to mind visions of hot-spring areas like those of Yellowstone National Park, Iceland, or New Zealand, with geysers like Old Faithful or the Great Geysir, and fumaroles like Karapiti blowhole at Wairakei (see Fig. 1). But natural hot waters are now being used in many countries for space heating, bathing, and agricultural purposes, and the power stations in Italy, the United States, and New Zealand that produce electricity from geothermal steam have received considerable publicity. Plans to utilize geothermal energy are being made worldwide, and, although the technology is still at an early stage of development comparable to that of the petroleum industry at the turn of the century, rapid development may be expected as a result of the current energy crisis.

This article provides a brief outline of the nature of geothermal fields,

Dr. Ellis is Assistant Director-General of the New Zealand Department of Scientific and Industrial Research. A graduate of Otago University with a background in physical chemistry and geochemistry, he was responsible for organizing the geochemical work of the project team of engineers and scientists undertaking the development of the New Zealand geothermal fields. He was active in devising new field and laboratory techniques and in the attempt to gain an understanding of the chemical behavior of high-temperature water solutions. Dr. Ellis has also served as a scientific adviser to several overseas geothermal development projects, and has published widely in hydrothermal chemistry and geochemistry. The views expressed in this article are his own and do not necessarily reflect the official attitudes of New Zealand Government Departments. Address: Department of Scientific and Industrial Research, Private Bag, Petone, New Zealand.

geothermal power projects, and some development problems. The topic was covered thoroughly in a series of papers presented in a United Nations Symposium on the Development and Utilization of Geothermal Resources at Pisa, Italy, in late 1970 (*Geothermics* 1970) and again at a similar symposium held in San Francisco in May 1975.

Hot springs are found throughout the world, but systems with water temperatures above about 100°C within 1–2 kilometers of the surface are limited to regions with heat flows several times greater than the crustal average. Frequently these regions are in zones of current or recent crustal plate divergence, with associated basaltic volcanism (as along the Mid-Atlantic Ridge or the African rift valleys), or in subduction zones where plates meet and crustal material is forced down to depths where it melts to produce andesitic volcanism, such as is common around the margins of the Pacific and in the Caribbean. A belt of high-temperature hydrothermal systems stretching from Italy to Greece and Turkey, through the Caucasus, and to southwest China is associated with major tectonic activity and mountain building (Ellis, in press). Water temperatures may reach 120–150°C at depths of 3–4 km in coastal areas of the Gulf of Mexico basin (Jones 1970) and in western Siberia, in thick sedimentary beds filling a wide rift zone created by the separation of Siberian and European platforms (Tamrazyan 1970). Figure 2 shows the distribution of major hydrothermal systems throughout the world.

The geothermal fields now being investigated or used for power pro-

duction are found in a wide variety of geological environments and rock types. The western Pacific fields are predominantly within recent andesitic or rhyolitic volcanic areas, whereas the Larderello steam field in Italy and the Kizildere area in Turkey are in regions of limestone, dolomite, marble, and shales. The Geysers steam field in northern California is situated in fractured graywacke, and, in the Salton Sea area of southern California, water at high temperatures is contained in deltaic river sediments.

All geothermal fields have individual peculiarities, but they also have features in common. In some sedimentary basins, or in areas of current metamorphism, heated waters or brines of moderate temperature (up to 150–200°C) may be accessible for geothermal exploitation. For the most part, however, with exceptions like Larderello and Kizildere, the known high-temperature (200–350°C) geothermal fields are found in, or are associated with, nearby Quaternary or Recent volcanic rocks. Extensive faulting, tilting, and graben and caldera formation are common in the volcanic hydrothermal areas, creating conditions of high permeability. Isotopic analysis of hydrogen and oxygen (Craig 1963) has shown that the hot water or steam discharged at the surface is derived from local cold surface water that penetrated the rocks from above. The circulation of water through the systems is thought to be extremely slow (at least 10^4 years). Geothermal systems are long-lived geological features, several having indicated lifetimes of the order of 10^6 years (White 1974), but fluid output during this period may be discontin-

Figure 1. The fumarole, one of many in the Karapiti area to the south of the Wairakei Geothermal Field in New Zealand, is typical of the naturally occurring hot water–steam vents found in hot spring areas throughout the world. The steam emerges from the center of a minor hydrothermal explosion crater. (Photo by R. B. Glover.)

uous, due to the tendency of systems to seal themselves off through mineral deposition (Facca and Tonani 1967; Ellis 1970).

A model system

In a typical high-temperature geothermal system (Fig. 3), cold surface water penetrates to levels many kilometers deep through fractured or porous formations, becoming heated and chemically altered as it flows slowly through the rocks. The lower density of the heated water creates a convection cycle, and the upflowing hot water selectively follows routes of highest permeability. In porous horizons, the water can spread laterally for distances up to several kilometers. Impermeable formations such as igneous rocks or fine-grained sediments act as barriers between aquifers, but faults or igneous intrusions offer a passage through these strata. Various kinds of cap rocks, sometimes supplemented by the deposition of silica or calcium carbonate from the water, prevent rapid dissipation of both water and heat.

In areas where surface outflows are restricted, the geothermal system often contains a liquid throughout, with boiling water emerging at the surface under a "thermo-artesian" pressure. Where the surface outflows of water are of similar magnitude to the limited upflows permitted by faults in deep impermeable formations, a subsurface hot water–steam interface may form and persist. Steam leakage at high surface levels supplies fumaroles; water seepage at lower levels forms hot springs.

Wells in the field tap hot water, steam, or a mixture of the two, depending on their depth and site. They may force the hot-water level to recede deeper into the system, locally or over the whole area, creating or extending a steam phase. The pressures in the hot-water phase are equal to the overlying hydrostatic head of the water plus the pressure of the saturated steam phase above it (the steam pressure is temperature-dependent). The water output from wells in the field gradually decreases until the total output equals the input to the aquifer under the prevailing pressure gradients. A new inflow–outflow steady state is thereby reached, with new hot-water levels, temperatures, and saturated steam pressures.

In a system where subsurface permeabilities (see Fig. 3) are high compared to deep permeabilities, the liquid water levels may be depressed to great depths during pro-

Figure 2. Dots indicate the major high-temperature hydrothermal areas of the world. Most are located in regions of current or recent volcanism or tectonic activity.

duction. The overlying country rock may contain saturated steam at temperatures in the vicinity of 235°, the temperature of maximum enthalpy for saturated steam (James 1968a). Steam condenses at the cooler top of the permeable formations, and the condensate falls back to recycle within the top section of the system. The characteristics of this type of "vapor-dominated" system are described in greater detail by White, Muffler, and Truesdell (1971). A well tapping a vapor-dominated system first discharges saturated steam (or steam plus water), but gradually the steam discharged becomes superheated as pressures are lowered. (There is little change in temperature because most of the heat in the system is stored in the rock.) This behavior is typical of wells at Larderello and The Geysers.

In drilling into a hot-water system, temperatures increase with depth, at first in an erratic manner, as steam flows or near-surface steam-heated groundwaters are encountered. In general, temperatures then increase with depth in accord with the boiling point–depth relationship: each point in the column of water is at a temperature at which the saturated steam pressure (plus gas pressure) equals the confining hydrostatic pressure. In the central part of the field a base temperature is reached at some depth (about 500 meters for a 250° system; 1,000 meters for a 300° system), and, in a system of good permeability, there is little further change in temperature with increasing depth.

Studies of convective heat transfer in porous media (e.g. Elder 1965) suggest that the high heat flows found in many hot-water geothermal areas are maintained by a rapidly rising central column of hot fluid trending into lateral and convecting flows near the surface, which creates a mushroom-shaped distribution of isotherms. Measured temperatures in the Wairakei field at various depths support this flow model (Banwell 1961). Because of the mushroom-shaped profile, at some positions it is possible to drill through a hot-water zone into colder country rock at deeper levels. In the vapor-dominated section of a system, temperatures are rather homogeneous due to rapid recirculation and convection, but, again, near-surface temperatures may be erratic, because of the uneven distribution of shallow steam-heated water or accumulations of condensate.

Assessing field characteristics

The surface manifestations of underground hot water or steam are highly variable. Some geothermal fields, such as Wairakei or Otake–Hatchobaru, Japan, are sites of active hot springs and fumaroles, but other major systems, as at The Geysers or near the Salton Sea, have unimpressive surface steam or water flows. In some areas the surface rocks are extensively altered by the hot fluids, as at The Geysers, and in others there are major surface silica deposits, as in many of the New Zealand fields, but the extent of surface alteration has no simple relationship to the size of the underground system.

To assess the power potential of a geothermal field, we must measure the size of the hot-fluid system, its temperature, and the probable permeability of the rocks. Careful geological, geophysical, and geochemical work by experienced personnel should precede any sizable investment in drilling; the costs are minor compared to those for drilling operations. In making their recommendations, the scientists should have a sense of reality as well as scientific curiosity. Some drilling has been recommended and carried out in unlikely territory. The basic scientific investigations in a geothermal field are straightforward, but the conclusions seldom narrow down to the exact answer "Drill here."

A preliminary step in assessing a field is aerial photography, including infrared scanning, of the area under examination. Aerial photographs are useful in the preparation of a geological map representing the tentative structure and thermal anomalies of a region. Present and past areas of hydrothermal alteration are revealed and fault systems can be outlined. Gradations in vegetation types and growth offer a rough idea of near-surface temperature anomalies.

Next, a chemical study of the steam and water flows from springs, fumaroles, and local streams can provide much useful survey information at low cost (White 1970) (Fig. 4). Underground temperatures can be estimated by interpreting the analytical data in terms of calibrated chemical and isotopic reaction equilibria (Ellis 1970). For example, the concentration of silica and the relative proportions of ions such as sodium, potassium, calcium, and magnesium in spring waters are good geothermometers because they are controlled by various solubility and ion-exchange reactions (e.g. Fournier and Truesdell 1973). The concentrations of calcium and bicarbonate, together with the deep water temperature, enable us to measure the carbon dioxide concentrations in deep hot water. This is desirable because waters with high carbon dioxide concentrations tend to produce troublesome calcite scales in pipelines. In steam flows, the relative concentrations of carbon dioxide, hydrogen, methane, and water and the distribution of carbon or hydrogen isotopes among these molecular species enable us to estimate deep temperatures, assuming that the chemical equilibrium reaction is "frozen" during flow to the surface.

The ratios of relatively unreactive solutes such as chloride, bromide, boron, and cesium in water can be used to check the homogeneity of the underground system (whether a single body or several isolated aquifers), and the probable origins of

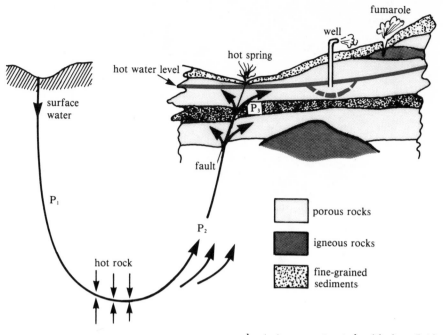

rock strata are saturated with hot fluid. Steam leaks out as fumaroles at high surface levels, and water leaks out in hot springs at low surface levels. Wells may initially tap both hot water and steam for power generation; discharge of fluid often forces the hot-water level (*colored line*) deeper into the system, at least in the vicinity of the well. P_1, P_2, and P_3 indicate zones of limiting recharge permeability, deep permeability, and subsurface permeability, respectively.

the water can be traced by analysis of its characteristic hydrogen and oxygen isotopic makeup. In arid areas the water source may be important in defining the long-term output of the system.

Geochemical analysis can suggest whether the surface activity results from a vapor-dominated or a hot-water system. In general, springs arising from hot-water systems have chloride concentrations in excess of 50 ppm, and the proportions of other involatile ions are reconcilable with equilibrium in a high-temperature liquid. High concentrations of volatiles such as carbon dioxide (or bicarbonate), boron, and ammonia in spring waters suggest steam heating. However, these factors are indicative only and must be considered in context with other survey results. For example, there is an association between known steam-producing areas and mercury mineralization in local country rock.

Measurements of thermal gradients and of electrical resistivities in the

field are probably the most profitable geophysical survey techniques (Banwell 1973). Thermal gradients should be measured over depths of at least several tens of meters in order to avoid surface anomalies due to water movement, vegetation, and ground contours. Many geothermal systems have been outlined successfully by this technique, and it is used extensively in steam-producing areas. In hot-water areas, convection and lateral near-surface spreading of the water may present a deceptive picture of the system's configuration, and in these areas thermal gradient measurements should be interpreted in conjunction with the results yielded by other techniques.

Electrical resistivity surveys have also been successful in determining the underground extent of geothermal areas. Usually dc techniques are used, with Schlumberger, Wenner, or dipole electrode arrays. One survey method is to traverse the area, measuring resistivity at a specific depth; another is to probe at a

Figure 4. Geochemical analysis of the water and steam discharged by hot springs can help to determine whether a region is suitable for development as a source of geothermal power. In the photo a geochemist samples the gases rising up through the boiling water of the Champagne Cauldron in Geyser Valley, Wairakei. (Photo by A. L. Tilbury.)

of scientific surveys alone. Although it may be of considerable underground extent and have high temperatures, the field may fail to produce useful quantities of hot fluid because of an unfavorable sequence of rock permeabilities or mineral deposition problems.

Characteristics of developed fields

The temperatures, surrounding rocks, and fluid types encountered in the developed geothermal areas are outlined in Table 1. Whereas the steam fields of Larderello, The Geysers, and Matsukawa have temperatures consistently in the range 230–250°, hot-water fields have temperatures covering a wide range, up to a maximum of 370° measured at Cerro Prieto, Mexico (Mercado 1970). The compositions of water discharged from wells in several fields are given in Table 2. The analyses represent concentrations in water at atmospheric pressure after 20–40% steam loss (depending on the original water temperature); pH values are 1–2 units higher than in underground water, because the acidic gases CO_2 and H_2S are lost into the steam (Ellis 1970).

At depths below the surface where there is no oxidation, and if boiling has not occurred, high-temperature geothermal waters are usually alkali-chloride solutions with a near-neutral or slightly alkaline pH. The salinity of the water varies widely from field to field, as do the relative concentrations of particular solutes. The most concentrated solutions found so far are in the Imperial Valley of California, where a near-saturated sodium–calcium–potassium–chloride brine reaches temperatures above 300° in the Salton Sea area. High-temperature brine was also found on the Reykjanes Peninsula of southwest Iceland in a geothermal system based on seawater. In areas of recent volcanic activity geothermal waters are frequently

single point to determine variations in resistivity with depth (Combs and Muffler 1973). The main variables controlling resistivity are water salinity, temperature, and rock porosity; low resistivity indicates an anomaly in one or more of these properties. The resistivity of hot saline water is very low, frequently of the order of a few ohmmeters, which may be compared with resistivities of many tens or hundreds of ohm-meters in the surrounding cold or impermeable country rock. Figure 5 is an example of a resistivity survey map, showing part of the Taupo Volcanic Zone of the North Island of New Zealand (Hatherton, Macdonald, and Thompson 1966). Broadlands and Wairakei appear to be fields of comparable size, but the natural heat outflow of Broadlands was less than one-seventh that of Wairakei (20,000 versus 150,000 kcal/sec).

Other features may also be measured, but for the most part they are less significant—for example,

the amounts of heat, mass, and ground noise that leak from the underground system. Or else—like gravity and magnetic anomalies—these characteristics are difficult to interpret at an early stage of prospecting. Test drilling provides the only real check on the production capabilities of the field, which should not be oversold on the basis

Figure 5. The survey map shows electrical resistivity contours, in ohm-meters, of part of the Taupo Volcanic Zone of the North Island of New Zealand. Resistivities were determined with a Wenner electrode spacing of 550 meters. Shaded areas have apparent resistivities less than 5 ohm-meters. Although the Broadlands and Wairakei fields are comparable in size on the map, the natural heat outflow of Wairakei was much greater. (Adapted from Hatherton, Macdonald, and Thompson 1966.)

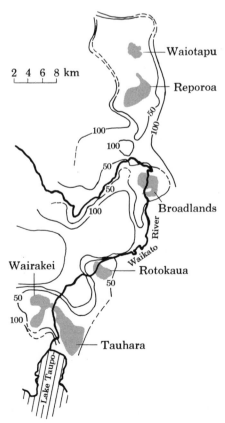

Table 1. Geothermal power installations in operation or at an advanced stage of development

	Geological situation	Average (max.) drillhole depth (m)	Average (max.) temperature (°C)	Discharge type (S = steam; W = water)	Total dissolved solids in water (g/kg)	Total generating capacity (MW) Installed	Planned addition
Chile							
El Tatio	Quaternary and Tertiary rhyolite, andesite; Mesozoic sediments	650 (900)	230 (260)	S + W	15	—	15
El Salvador							
Ahuachapan	Quaternary andesite	1,000 (1,400)	230 (250)	S + W	20	30	50
Iceland							
Namafjall	Quaternary basalt	1,000 (1,400)	250 (280)	S + W	1.0	2.5	—
Italy							
Larderello region	Triassic–Jurassic sediments	600 (1,600)	200 (260)	S	—	406	—
Mount Amiata	Triassic–Jurassic sediments; Quaternary volcanics	750 (1,500)	170 (190)	S(+W)	—	25	—
Japan							
Matsukawa, N. Honshu	Quaternary andesites; Miocene sandstones	1,000 (1,500)	220 (270)	S	—	20	—
Otake, Kyushu	Quaternary andesites	500 (1,500)	230 (250)	S + W	2.5	11	—
N. Hachimantai	Quaternary andesites, dacites	800 (1,700)	— (>200)	S + W	—	10	—
Hatchobaru, Kyushu	Quaternary andesites	1,000 —	250 (300)	S + W	5.5	—	50
Onikobe, Honshu	Quaternary andesites, dacites, granite	300 (1,350)	— (288)	S(shallow) + W (deep)	1.5	—	20
Mexico							
Cerro Prieto	Sandstone, shales, granite	800 (2,600)	300 (370)	S + W	17	75	75
New Zealand							
Wairakei	Quaternary rhyolite, andesite	800 (2,300)	230 (260)	S + W	4.5	192	—
Kawerau	Quaternary rhyolite, andesite	800 (1,100)	250 (285)	S + W	3.5	10	—
Broadlands	Quaternary rhyolite, andesite	1,100 (2,420)	255 (300)	S + W	4	—	100
Philippines							
Tiwi, S. Luzon	Quaternary andesites	920 (2,300)	—	S + W	—	—	10.5
Turkey							
Kizildere	Pliocene–Miocene sandstones limestones; Palaeozoic schists	700 (1,000)	190 (220)	S + W	5	—	10
U.S.A.							
The Geysers	Jurassic–Cretaceous graywackes and shales, basalt	1,500 (2,900)	250 (285)	S	—	600	300
U.S.S.R.							
Pauzhetsk	Quaternary andesite, dacite, rhyolite	— (800)	185 (200)	S + W	3	5	7

Table 2. Composition of waters from geothermal wells

	Depth (m)	Source temp. (°C)	pH (20°C)	Solutes separated from discharge at atmospheric pressure (ppm)										
				Li	Na	K	Rb	Cs	Mg	Ca	F	Cl	Br	SO_4*
Well 44 Wairakei, N. Z.	695	255	8.4	14.2	1,320	225	2.8	2.5	0.03	17	8.3	2,260	6.3	36
Well 7 El Tatio, Chile	878	255	7.3	47.5	5,000	840	8.6	17.9	0.09	203	2.5	9,100	—	29
Well E-205 Matsao, Taiwan	1,500	245	2.4	26.0	5,490	900	12.0	9.6	131	1,470	7.0	13,400	—	350
Well IID-1 Salton Sea, Cal.	1,600	~340	~5.5	320	54,000	23,800	100	20	100	40,000	—	184,000	700	10
Well 8 Mexicali, Mexico	1,310	355	—	31	11,200	3,420	—	—	11	720	—	21,600	26	0
Well 1A Kizildere, Turkey	415	200	9.0	4.5	1,280	135	0.0	0.33	0.11	3.0	23.7	117	1.2	770

SOURCE: Ellis, in press.
* Total of molecular and ionized acid/base species.

dilute alkali-chloride solutions (0.5–1.5% total dissolved solids) with temperatures in the range 200–300°.

The solutes in geothermal waters can be divided roughly into two categories: "soluble" elements (of which there are very few) and elements with concentrations controlled by mineral equilibria. Chloride, bromide, iodide, boron, cesium, lithium, and ammonia fall in the first category under high-temperature, deep-water conditions, whereas most of the other elements are present at concentrations controlled by temperature- and pressure-dependent chemical equilibria. For example, the concentration of silica in the water of known systems above about 150° is controlled precisely by the solubility of quartz (Mahon 1966), while the ratio of sodium to potassium is controlled by feldspar equilibria reactions (Fournier and Truesdell 1973).

Concentrations of sulfate and fluoride are limited by the low solubilities of the minerals anhydrite and fluorite, respectively, while Ca^{2+}, CO_2, and HCO_3^- concentrations are related to the solubility of calcite, with which geothermal systems are usually nearly saturated. Deep waters high in CO_2 are particularly prone to deposit calcite in discharge pipes, in some cases presenting a serious problem to development.

The common rock-forming elements aluminum, iron, magnesium, and manganese are held at low levels by alumino-silicate–water equilibria (e.g. in high-temperature dilute geothermal waters Mg^{2+} concentrations are controlled at the level of parts per billion (10^9) and in equilibrium with chlorite and montmorillonite). At constant temperature the maximum concentrations of many divalent metal ions appear to vary in proportion to the square of salinity (m_{Na}^2). High-temperature brines may contain appreciable (ppm) concentrations of iron, manganese, copper, zinc, and lead (e.g. at Salton Sea), while in dilute water systems (as in New Zealand) these metals form a very much lower proportion of total dissolved solutes (ppb) (Ellis 1969).

Geothermal waters often contain unusually high concentrations of lithium, rubidium, cesium, boron, arsenic, and ammonia, in proportions reflecting the composition and types of rocks in the locality. The rare alkalis are especially abundant in rhyolitic or andesitic areas, while waters of sedimentary rock systems often contain exceptionally large concentrations of boron and ammonia. The hydrothermal recrystallization of minerals results in the concentration of many highly soluble species into the water phase (Ellis and Mahon 1964), including elements such as lead, zinc, and copper if the waters are highly saline.

There is little evidence in the fields so far investigated that the water or the chemicals in the water originate from a deep magmatic source. Some chemicals are dissolved from the containing rocks, but supply waters, evaporite beds, and connate waters all contribute to the composition of discharged fluid.

Unusually acidic hot water is found in rock systems containing sulfur, as in the Tatun Volcanic Zone of Taiwan (see Table 2). Hydrolysis of sulfur by water at high temperatures produces a strongly acid, sulfate, sulfide solution (Ellis and Giggenbach 1972) which is too corrosive to be used for power generation. High-temperature brines are also highly corrosive, because their pH is lower than that of dilute water systems at the same temperature. The pH differences are due to rock mineral equilibria acting as a buffer, controlling ratios such as a_{Na}/a_H, a_K/a_H at a particular value.

The steam produced by wells at The Geysers and Larderello con-

As*	SiO$_2$*	B*	NH$_3$*	CO$_2$*	H$_2$S*
4.8	690	28.9	0.15	19	1.0
48	810	210	3.1	32	—
3.6	639	106	36	2	—
—	—	498	500	—	—
2.0	1,420	37.0	—	1,180	—
38	263	26.2	2.5	1,350	—

tains approximately 1% and 2% by volume, respectively, of total gases—predominantly CO_2—with minor amounts (1–2%) of H_2S, CH_4, H_2, N_2, and NH_3. In the hot-water areas the composition of the steam separated from the steam–water discharge at the surface contains gas of a similar composition, and up to the same total concentration (but frequently less and sometimes as low as 0.01%). In active volcanic areas H_2S may be present as a high proportion (10–15%) of total gases.

In hot-water fields in rocks of good permeability and at temperatures above about 200°, the original rocks are usually recrystallized into secondary equilibrium mineral assemblages. Typical assemblages found in quartz feldspathic rock systems are: at Broadlands from glassy rhyolitic rocks (260° water, $0.12m$ CO_2), the minerals quartz, K-feldspar, K-mica, albite, calcite, chlorite, and pyrite; at Salton Sea from deltaic sandstones (300° water, $0.01m$ CO_2), quartz, K-feldspar, K-mica, albite, chlorite, epidote, and pyrite; and in the Otake area from andesitic volcanics (250° water, $0.01m$ CO_2), the minerals quartz, albite, K-feldspar, montmorillonite-sericite, epidote, chlorite, pyrite, and anhydrite. In cooler, near-surface levels, calcium zeolites such as laumontite or mordenite may re-

place epidote or wairakite as the predominant calcium minerals, and at the depth of first boiling of the rising hot water there may be abundant deposits of calcite, silica, and K-feldspar (Browne 1970; Ellis, in press). High-temperature acidic waters produce alteration minerals such as kaolin, dickite, alunite, pyrophyllite, quartz, and montmorillonite.

Power generation

Some geothermal wells have a considerable energy output, and The Geysers, Larderello, Broadlands, and Mexicali fields all have wells with a steam discharge sufficient to generate at least 15 megawatts. In the last two fields the steam is accompanied by water having a similar energy output. A steam well at Travale, near Larderello, initially produced at the rate of 450 metric tons (tonnes) per hour (equivalent to about 45 MW).

In hot-water fields, steam is separated by cyclones from the mixed steam–water discharges at the wellhead and transmitted to the power station. The cover photograph of the Wairakei field shows several wellhead installations. In the foreground may be seen the steam–water separator, the twin-tower silencer, in which water expands to atmospheric pressure by steam flashing, and steam lines leading to the power station. Wellhead separation pressures range from approximately 50–200 pounds per square inch gauge (psig), and separation in two stages to supply steam to different turbine inlets makes the process both more economical and more flexible. Separated water, still at temperatures hotter than 100°, is usually rejected into a drainage system after passing through the silencer (Smith 1958). It now appears that, instead of separating water and steam at the wellhead, two-phase steam–water flows from wells could be piped to a large central separation plant without creating dangerous flow characteristics (James 1968b). This would enhance the possibility of using the residual heat of the separated water. For steam or hot-water transmission, conventional low-strength carbon steel pipes perform satisfactorily, as they are protected by the growth of an internal sulfide-oxide coating.

In the early stages of development in a field, generators using noncondensing turbines that vent their exhaust into the atmosphere are sometimes used temporarily to obtain immediate power production from individual wells. In permanent stations, condensing turbine generator sets, venting their exhaust to a vacuum, operate directly on geothermal steam at pressures up to 180 psig, but more usually in the range 50–100 psig. The largest generating unit, producing 106 MW, is in a recently established station at The Geysers. The direct use of steam containing H_2S and CO_2 has limited the choice of turbine blade materials to low-stress alloys, and, consequently, blade speeds are low (Marshall and Braithwaite 1973). No major erosion or corrosion problems have arisen in operating turbines. Turbine manufacturers are now taking an interest in designing special geothermal steam turbines, utilizing new alloys.

Direct-contact condensers use either local river water (as at Wairakei) or condensate recycled from cooling towers. With turbines taking in fresh steam continuously, geothermal stations are unique in not requiring the recycling of pure condensate as boiler "make-up" water. Due to the presence of noncondensable gases at 0.2–2% by volume in the steam, major gas-extraction plant is required, usually steam jet ejectors. In the condensing and gas-removal plant the oxidation of sulfide creates conditions of high acidity, and corrosion reaches a maximum.

At present there are only a few large power stations, and these are mainly in geothermal fields with temperatures in excess of about 220° (see Table 1). Although hot-water fields are encountered much more frequently than steam fields, the steam fields around Larderello and The Geysers currently produce more electricity than all of the hot-water fields together. The Larderello field comprises several production zones in an area covering approximately 250 km^2. Power development in its present form dates from post-1945 reconstruction, although some power was produced as far back as 1913. The Wairakei power station has been operating since 1958, with wells in two locations in an area of about 5 km². At The Geysers, power production commenced in 1960 with a 12.5 MW unit, but in recent years production capacity has increased dramatically, and now 100 MW per year is being added as new wells are drilled and new units installed. The field covers some 40 km², with wells and production units in several localities. The Japanese stations at Otake and Matsukawa both began to produce power in the 1960s, as did the stations at Pauzhetsk in Kamchatka, U.S.S.R., and at Namafjall, Iceland. Production at Cerro Prieto commenced in 1973, and the most recent station to come on-stream is at Hachimantai, which produces 10 MW.

Developments in several other areas are progressing rapidly and may lead to the establishment of power stations over the next few years. Especially promising high-temperature areas are found in Ethiopia at Dallol and in the Tendaho graben; in Greece on the island of Milos, 150 km southwest of Athens; on Guadeloupe; in Iceland at Krafla in the north and at Reykjanes, Krysuvik, and Hengill in the southwest; in Indonesia at Kawah Kamodjang, 40 km southeast of Bandung; in Italy at Monti Volsini, 80 km northwest of Rome; in Japan at Takinoue, North Honshu; in Kenya at Olkaria; in the Philippines at Tiwi, South Luzon; at Tongonan, North Leyte; and in the United States in the southern part of the Imperial Valley.

The production cost of geothermal power is variously quoted from about 0.3 to 0.8 U.S. cents per kWh (e.g. Armstead 1973), and installation costs range from $200–400 per kW. There is not a major difference between the cost of power from a small (20 MW) station and that from a large station (200 MW)—a factor with important implications for developing countries, where small needs may be met by inexpensive power.

The use of water at moderate temperatures (150–200°) is being investigated by pilot projects in several places, but at these temperatures conventional steam separation from two-phase flows is inefficient, and the water also commonly contains calcium carbonate concentrations at a level that causes rapid scaling in the pipes. Downhole pumps have been proposed to raise the water to the surface under pressure as a single-phase flow; at the surface, the heat is transferred to a secondary working fluid with a lower boiling point, such as isobutane or Freon. Projects focusing on the development of these binary-cycle systems are still in the experimental stage. For example, a 0.7-MW pilot power plant based on 81° water and Freon as a secondary fluid was opened at Paratunsk in southeastern Kamchatka in 1967 (Moskvicheva and Popov 1970).

Projects underway in New Mexico are aimed at utilizing the heat in hot dry-rock systems by artificial fracturing of granitic rocks at deep levels by high-pressure water and the creation of hydrothermal heat exchange systems (Smith 1973). Many practical problems have yet to be solved before this type of earth heat can be economically exploited, but the approach should offer new insight into the physical and chemical processes at work in geothermal systems.

Other uses

Geothermal power stations are inefficient in converting thermal energy to electric energy. The limiting thermodynamic efficiency is approximately 30% because of the low average steam temperatures, but the actual efficiencies of conversion are approximately one half this figure in steam-producing fields and about one quarter in water-producing fields. There is an obvious incentive to make more efficient use of the energy from high-temperature geothermal fields but, unfortunately, many of the fields are located in isolated areas with minor requirements for heating by waste hot water.

Two notably efficient projects using geothermal steam are at Kawerau, New Zealand, and Namafjall, Iceland. At Kawerau, a major paper and timber mill utilizes 200 tonnes/hr of steam produced from wells tapping water at temperatures up to 285°. Part of the steam is used in heat exchangers to produce clean process steam and part directly for timber drying and for the evaporation of black liquor. Surplus steam drives a noncondensing turbine to produce 10 MW of electricity for the industry (Smith 1970). At Namafjall, some 240 tonnes/hr of steam from four wells is used to dry diatomite from a nearby lake bed and to generate 2.5 MW of electricity, while the separated hot water supplies community space heating (Ragnars et al. 1970). The combination of electricity produced by steam and space heating by separated hot water is highly efficient, and plans are underway to utilize the fields of Reykjanes, Krysuvik, and Hengill to provide steam for power generation and industry and hot water for heating the populated areas around Reykjavik.

Space heating is usually not the primary objective of geothermal fields producing water at temperatures well in excess of 100°, but there are some exceptions. Reykjavik uses well water at up to 128°C to supply most of the city's heating needs. In the nearby Hengill field, water at up to 180–200° heats houses and greenhouses. In Rotorua, New Zealand, many wells produce water (up to about 200°) to heat houses and public buildings. A major hotel uses a geothermal well to produce air conditioning by means of a lithium-bromide absorption unit.

Geothermal wells can also be harnessed to serve other purposes. Production of heavy water with geothermal steam as an energy source has been proposed in New Zealand, Japan, and Iceland, but so far there are no practical developments. At Rotokaua, New Zealand, high-pressure steam derived from high-temperature geothermal water is used in a modified Frasch process to melt and raise to the surface sulfur contained in sediments at shallow depths.

The uses of geothermal water at lower temperatures are too diverse to review here. Lindal (1973) describes projects ranging from salt recovery and mining to animal husbandry, horticulture, and fisheries.

Problems of utilization

Vast quantities of fluid are expelled from geothermal production wells. For example, the rate of fluid production from the Larderello area is approximately 3×10^7 tonnes per

Figure 6. Ohaki pool, a large, deep, boiling spring of slightly alkaline chloride water, was the largest spring in the Ohaki-Broadlands area. Surrounding the pool is an extensive white silica sinter terrace with an intricate crenated edge. When wells were drilled and discharged in preparation for the production of electrical power, water levels in the pool dropped below sight. (Photo by R. B. Glover.)

year, and for Wairakei it is about 5 × 10⁷ tonnes per year. During the years of operation at Wairakei, a total of almost 1 km³ of water has been removed from the system. The extraction of this much fluid sometimes leads to local changes in the level of the field, depending partly on the strength of the near-surface rocks, and subsidence can produce local earth tremors. In the Wairakei area of low-strength ash and breccia formations, the ground level has generally fallen, exceeding 0.3 meters per year in the center of the field and tapering off to zero at the perimeter. Even so, the volume of the ground subsidence is only a small percentage of the total volume of water removed, and gravity surveys show that most of the water now being produced is being replaced by deep inflow (Hunt 1970).

In wells in rocks of low permeabili-

ty, hot water tends to be replaced by steam. The artificially stimulated production of large volumes of water from a hot-spring area is likely to cause a general lowering of subsurface water levels, and steam created to fill the voids in the hot aquifer sometimes finds its way to the surface through lines of weakness to form new fumaroles or cause hydrothermal blowouts. Several large explosion craters have been formed in this way on the periphery of the Wairakei field. The water output from hot springs is likely to decrease and eventually cease. After major well production, the hot springs of Wairakei Geyser Valley and Broadlands no longer discharge, and what were once tourist attractions are now gray holes in the ground (see Fig. 6). A good example of how deep aquifer pressures and water levels decline to a new equilibrium level was present-

ed by Bolton (1970) for ten years of operation of the Wairakei field. Most of the changes were due to varying water levels in the aquifer and consequent temperature–steam pressure adjustment to the new conditions.

In highly permeable fields, wells may produce water of almost constant composition for many years. At Wairakei, during fifteen years of operation the salinity as measured by chloride content did not change by more than 1–2%. In systems of lower permeability, however, although the general character of the water may not change markedly, concentrations are variable because of the effects of boiling and evaporation on rock surfaces (e.g. at Broadlands; see Mahon and Finlayson 1972).

In steam fields the concentration of

gas in the steam tends to decrease with time after the well is tapped, as would be expected if the steam were produced by deep-level boiling of water. At Larderello the steam temperatures have increased over the years of production, while pressures have generally stabilized after an initial period of decrease (Banwell 1974).

Deposition of silica or calcium carbonate sinters at the surface is characteristic of many hot-spring areas but, unfortunately, in some fields, minerals are also deposited along both natural and artificial flow channels, including fissures, pipes, and surface equipment. Calcium carbonate, the major scale-forming mineral in pipelines, is largely deposited when the water first boils, losing carbon dioxide and increasing in pH. The deposition of calcite in pipelines is particularly troublesome in fields with deep waters containing dissolved carbon dioxide at high concentrations (above $0.1m$). It has not caused trouble at Wairakei ($m_{CO_2} = 0.01$), but some deposition occurs in the Broadlands and Kawerau fields ($m_{CO_2} \approx 0.1$). Rapid rates of deposition have been recorded in other countries, and wells have become blocked in a matter of days or hours in severe situations. In some areas—Kizildere for example—calcite formation in pipes threatens to limit the development of the geothermal field.

Silica is deposited by high-temperature geothermal waters only when they become supersaturated with amorphous silica during cooling and concentration. The kinetics of silica polymerization and precipitation are highly complex (Rothbaum and Wilson, in press): the rate of polymerization depends on the degree of supersaturation, which in turn depends on the original water temperature and the amount of steam removed. Consequently, waters with original temperatures up to about 270–280° usually emerge into surface outflow drains before much silica is deposited. But in systems where water reaches very high temperatures—for example, Salton Sea and Cerro Prieto—some silica may be deposited in the outflow pipes. This problem was particularly troublesome at the Salton Sea wells (Skinner et al. 1967), where the cas-

ings became coated with a hard, black siliceous scale that contained high concentrations of iron, copper, and silver sulfides precipitated from the metal-rich brines.

Environmental impact

Geothermal power has been promoted in some quarters as "pollution-free." This is a fallacy: the problems are not major, but as with any other scheme to produce electric power, environmental problems must be considered (Axtmann 1975). A 100-MW power station in a dry-steam field requires fluid production from wells of about 10^7 tonnes per year, while in a hot-water field the required fluid production is greater by a factor ranging from about 3–10, depending on the water temperature. In the latter situation, rejection of separated water at temperatures of about 100° into local waterways could cause greater thermal pollution than would conventional power stations of equivalent capacity.

The magnitudes of annual chemical output from a 100-MW station for the representative hot-water fields of Wairakei and Cerro Prieto would be (in tonnes per year): alkali chlorides, 10^5; SO_4, NH_4, B, and F, each 10^2–10^3; SiO_2, 10^4; CO_2, 10^4–10^5; and H_2S, 10^3–10^4. In the steam fields of Larderello and The Geysers such a power station would produce comparable quantities of NH_4, B, CO_2, and H_2S. The major air pollutant from geothermal stations is hydrogen sulfide, but 30–50% of the total output of this gas may be dissolved in the cooling water.

In addition to simple thermal or salinity effects of geothermal effluents in local waterways, elements in the water such as boron, fluoride, and heavy metals may have their own harmful effects. In areas of intensive agriculture, like the Imperial Valley or Ahuachapan, El Salvador, disposal of effluents is a major development problem. At Ahuachapan, where local disposal of geothermal water is impossible because of its effect on nearby coffee plantations (coffee is a crop with low tolerance for boron), a long channel is being constructed to carry the waste water to the sea. And at El Tatio, Chile, the water contains 30–40

ppm arsenic, which could contaminate the limited local water supplies.

Some geothermal waters form metal-rich sulfide precipitates consisting mainly of antimony or arsenic sulfides but also containing high concentrations of metals such as mercury, thallium, gold, and silver. The precipitates may accumulate in local riverbeds, causing methyl mercury production, which can lead to unacceptably high mercury levels in fish (Weissberg and Zobel 1973).

An alternative to surface disposal is to reinject geothermal effluents into the field. Reinjection of condensate into the peripheral parts of The Geysers steam field is current practice and easily accomplished (Budd 1973) because of the relatively small volumes involved. Hot-water fields face a more serious water disposal problem, because every tonne of steam used requires the disposal of 3–10 tonnes of saline water. Reinjection of water through wells has been tried experimentally in several fields (e.g. Ahuachapan and Hachimantai), but the problem of scale formation in pipes and reservoir rocks may be a serious long-term limitation. The extraction of geothermal heat by the use of heat exchangers, either downhole or under pressure at the surface, should alleviate the problem of scale formation by avoiding the concentration of solutes and loss of carbon dioxide caused by steam flashing.

Chemical treatment of geothermal waters to remove scale-forming silica has been successfully attempted in a pilot project in New Zealand (Rothbaum and Anderton, in press). The result is a low density calcium silicate material suitable for use as an industrial filler or insulant, and an effluent which will not subsequently lay down silica scale. The treatment also removes arsenic—an important factor if water is to be released into local waterways. Desalination to produce a potable water and a concentrated brine of lower disposal volume is an alternative procedure in arid areas.

The next ten years

In the next decade geothermal fields will become an important source of electric power in many developing countries, particularly around the Pacific basin and in Latin America, but in developed countries they will rarely contribute more than a minor, though very economical, fraction of the total power production. In this period new geothermal power stations are likely to be established in field situations not very different from those being utilized at present. The number of producing fields will increase considerably, however, and new and improved technology, especially new turbine designs and deep-well pumps coupled with heat exchangers at the surface, will lead to much greater operational efficiency. Waste hot water from geothermal power stations will be used increasingly for community space heating, following the example of stations at Larderello, which are already supplying waste hot water for this purpose by pipeline for distances up to 105 km from the field (Haseler 1975). Water from geothermal fields having lower underground temperatures (80–150°C) is more likely to be used for space heating than for the production of electricity.

References

Armstead, H. C. H. 1973. Geothermal economics. In *Geothermal Energy*, H. C. H. Armstead, ed. Paris: UNESCO Earth Sciences Series, no. 12, pp. 161–74.

Axtmann, R. C. 1975. Environmental impact of a geothermal power plant. *Science* 187: 795–803.

Banwell, C. J. 1961. Geothermal drillholes—physical investigations. *Proc. U. N. Conf. New Sources of Energy*, Rome 1961. (Paper G.53, N.Y: U.N. 1963.)

Banwell, C. J. 1973. Geophysical methods in geothermal exploration. In *Geothermal Energy*, H. C. H. Armstead, ed. Paris: UNESCO Earth Sciences Series, no. 12, pp. 41–48.

Banwell, C. J. 1974. Life expectancy of geothermal fields. *Geothermal Energy* 2:12–13.

Bolton, R. S. 1970. The behaviour of the Wairakei geothermal field during exploitation. *Geothermics* (Special Issue 2), 2(pt. 2):1426–39.

Browne, P. R. L. 1970. Hydrothermal alteration as an aid in investigating geothermal fields. *Geothermics* (Special Issue 2), 2(pt. 1):564–70.

Budd, C. F. 1973. Steam production at The Geysers geothermal field. In *Geothermal Energy*, Paul Kruger and Carel Otte, eds. Stanford: Stanford Univ. Press. pp. 129–44.

Combs, J., and L. J. P. Muffler. 1973. Exploration for geothermal resources. In *Geothermal Energy*, Paul Kruger and Carel Otte, eds. Stanford: Stanford Univ. Press. pp. 95–128.

Craig, H. 1963. The isotopic geochemistry of water and carbon dioxide in geothermal areas. In *Nuclear Geology on Geothermal Areas*, E. Tongiorgi, ed. Pisa: Consiglio Nazionale della Richerche, Laboratoria di Geologia Nucleare. pp. 17–23.

Elder, J. W. 1965. Physical processes in geothermal areas. In *Terrestrial Heat Flow*, Geophysical Monograph no. 8, Amer. Geophys. Union. pp. 211–39.

Ellis, A. J. 1969. Present-day hydrothermal systems and mineral deposition. *Proc. Ninth Commonwealth Mining and Met. Congress* (Mining and Petrol. Sect.) London: Inst. Mining and Metall. pp. 1–30.

Ellis, A. J. 1970. Quantitative interpretation of chemical characteristics of hydrothermal systems. *Geothermics* (Special Issue 2), 2(pt. 1):516–28.

Ellis, A. J. In press. Explored geothermal systems. In *Geochemistry of Hydrothermal Ore Deposits* (2nd ed.), H. L. Barnes, ed. N.Y.: Wiley.

Ellis, A. J., and W. F. Giggenbach. 1972. Hydrogen sulphide ionization and sulphur hydrolysis in high-temperature solutions. *Geochim. Cosmochim. Acta* 35:247–60.

Ellis, A. J., and W. A. J. Mahon. 1964. Natural hydrothermal systems and experimental hot water/rock interactions. *Geochim. Cosmochim. Acta* 28:1323–57.

Facca, G., and F. Tonani. 1967. The self-sealing geothermal field. *Bull. Volcanol.* 30: 271–73.

Fournier, R. O., and A. H. Truesdell. 1973. An empirical Na-K-Ca geothermometer for natural waters. *Geochim. Cosmochim. Acta* 37:1255–75.

Geothermics. 1970. Special issues containing the proceedings of the U.N. Symposium on the Development and Utilization of Geothermal Resources, Pisa, 1970. Vols. 1 and 2 (pt. 1 and pt. 2). Pisa: Istituto Internazionale per le Richerche Geotermiche.

Haseler, A. E. 1975. *District Heating: An Annotated Bibliography*. London: Property Services Agency, Dept. of the Environment. p. 35.

Hatherton, T., W. J. P. Macdonald, G. E. K. Thompson. 1966. Geophysical methods in geothermal prospecting in New Zealand. *Bull. Volcanol.* 29:485–97.

Hunt, T. M. 1970. Net mass loss from the Wairakei geothermal field, New Zealand. *Geothermics* (Special Issue 2), 2(pt. 1): 487–91.

James, C. R. 1968a. Wairakei and Larderello: Geothermal power systems compared. *New Zealand J. Sci.* 11:706–19.

James, C. R. 1968b. Pipeline transmission of steam/water mixture for geothermal power. *N.Z. Engineering* 23: 55–61.

Jones, P. H. 1970. Geothermal resources of the northern Gulf of Mexico basin. *Geothermics* (Special Issue 2), 2(pt. 1): 14–26.

Lindal, B. 1973. Industrial and other applications of geothermal energy. In *Geothermal Energy*, H. C. H. Armstead, ed. Paris: UNESCO Earth Sciences Series, no. 12, pp. 135–48.

Mahon, W. A. J. 1966. Silica in hot water discharged from drillholes at Wairakei, New Zealand. *New Zealand J. Sci.* 9:135–44.

Mahon, W. A. J., and J. B. Finlayson. 1972. The chemistry of the Broadlands geothermal area. *Am. J. Sci.* 272:48–68.

Marshall, T., and W. R. Braithwaite. 1973. Corrosion control in geothermal systems. In *Geothermal Energy*, H. C. H. Armstead, ed. Paris: UNESCO Earth Sciences Series no. 12., pp. 151–60.

Mercado, S. 1970. High activity hydrothermal zones detected by Na/K, Cerro Prieto, Mexico. *Geothermics* (Special Issue 2), 2(pt. 2):1367–76.

Moskvicheva, V. N., and A. E. Popov. 1970. Geothermal power plant on the Paratunka River. *Geothermics* (Special Issue 2), 2(pt. 2):1567–71.

Ragnars, K., K. Saemundsson, S. Benediktsson, and S. S. Einarsson. 1970. Development of the Namafjall area, Northern Iceland. *Geothermics* (Special Issue 2) 2(pt. 1):925–35.

Rothbaum, H. P., and B. H. Anderton. In press. Removal of silica and arsenic from geothermal discharge waters by precipitation of useful calcium silicates. *Proceedings, Second U.N. Symposium on Development and Use of Geothermal Resources*, San Francisco, 1975.

Rothbaum, H. P., and P. T. Wilson. In press. The effects of temperature and concentration on the rate of polymerization of silica in geothermal waters. *Geochemistry 1975*, N.Z. Dept. Sci. and Ind. Research Bulletin.

Skinner, B. J., D. E. White, H. J. Rose, and R. E. Mays. 1967. Sulfides associated with the Salton Sea geothermal brine. *Econ. Geol.* 62:316–30.

Smith, J. H. 1958. Production and utilization of geothermal steam. *N.Z. Engineering* 13:354–75.

Smith, J. H. 1970. Geothermal development in New Zealand. *Geothermics* (Special Issue 2) 2(pt. 1):232–61.

Smith, M. C. 1973. The dry hot rock geothermal resource. In *Geothermal Energy*, Hearing before Subcommittee on Energy, Committee on Science and Astronautics, U.S. House of Representatives, 93rd Cong., 1st sess., H. Rept. 8628. Washington: U.S. Gov. Printing Office. pp. 110–28.

Tamrazyan, G. P. 1970. Continental drift and geothermal fields. *Geothermics* (Special Issue 2) 2(pt. 2):1212–25.

Weissberg, B. G., and M. G. R. Zobel. 1973. Geothermal mercury pollution in New Zealand. *Bull. Environmental Contamination and Toxicology* 9:148–55.

White, D. E. 1970. Geochemistry applied to the discovery, evaluation, and exploitation of geothermal energy resources. *Geothermics* (Special Issue 2). 1: sec. 5.

White, D. E. 1974. Diverse origins of hydrothermal ore fluids. *Econ. Geol.* 69: 954–73.

White, D. E., L. J. P. Muffler, and A. H. Truesdell. 1971. Vapor-dominated hydrothermal systems compared with hot-water systems. *Econ. Geol.* 66:75–97.

PART 5 *Mineral Resources*

A Second Iron Age Ahead?

Brian J. Skinner

The distribution of chemical elements in the earth's crust sets natural limits to man's supply of metals that are much more important to the future of society than limits on energy

Mention 1776 and most people think immediately of the historic Declaration of Independence by the American colonies. They would do well to enlarge their horizons and to include in their catalogue of the important events of 1776 a modest commercial transaction that took place in England. That transaction was a harbinger of changes in social structure even more far-reaching than the changes wrought by the political upheavals in the new United States; 1776 was the year that the newly formed engineering firm Boulton and Watt sold its first steam engine. Although this was not the first steam engine to be sold, it incorporated many of Watt's innovative ideas and thus was the first efficient example to be offered commercially.

Watt did not, of course, invent the steam engine, but prior to his time the

Brian J. Skinner is Eugene Higgins Professor of Geology and Geophysics at Yale University, a member of the Editorial Board of American Scientist, and Editor of the journal Economic Geology. Before joining the faculty at Yale in 1966, Dr. Skinner had been at various times a member of the U.S. Geological Survey, a faculty member at the University of Adelaide in his native country, Australia, and a geologist for several mining companies. He is a geochemist who focuses his research on the processes that form mineral deposits, and in recent years he has increasingly turned his attention to problems of resource evaluation and abundance. He recently served as chairman of the National Academy of Sciences' Committee on Mineral Resources and the Environment (COMRATE), a group that produced reports on copper, uranium, and fossil fuel abundances as well as other topics related to resource use and production. This paper has been extracted from material first presented by the author in his presidential address to the Geochemical Society in 1973. Address: Department of Geology and Geophysics, Yale University, New Haven, CT 06520.

old Newcomen engine was used exclusively as a steam pump; it was slow, cumbrous, and excessively wasteful of fuel. Watt transformed the old engine to a quick-working, powerful, and efficient device; then he adapted it to drive machinery of all kinds. A device made of metals dug from the ground could convert the chemical energy locked in wood and coal into useful mechanical energy. This was the key step that led to replacement of manpower by machine power. The door had been opened for the industrial revolution, and society was forever changed.

Following Watt's invention the use of mechanized devices proliferated, and demand for fuel to drive machines rose concomitantly with demand for materials to build more machines. Both demands were satisfied by a burgeoning mining industry that supplied first coal then oil and gas for fuel, together with a multitude of metals for ever more complex machines. Freed from the limitations imposed by puny human muscle power, able to travel and communicate widely and therefore no longer forced to subsist on food produced in the immediate environs of their habitations, populations grew rapidly and expanded into less and less hospitable parts of the world. Population densities rose, and as people were removed farther and farther from direct production of food, the infrastructure needed to supply and feed them became increasingly more complex. This trend continues apace, and because the infrastructure is built from and driven by mineral substances dug from the earth, every additional complexity brings demands for more fuels and more metals. Without those substances the structure must collapse and the population wither back

Figure 1. At El Tiro Pit, Silver Bell, Arizona—a modern, open-cut copper mine—huge quantities of rock are blasted out and hauled to nearby mills for processing. Unlike the rich veins of the underground mine in the cover photograph, the ore mined at El Tiro is of relatively low grade and the ore minerals are widely disseminated as small grains in a mass of valueless silicate minerals. The rock face at right has a greenish cast because the copper sulfide minerals in the ore have been oxidized after exposure. (Photograph by the author.)

Table 1. Comparison of the valuable metals in a cubic kilometer of average rock in earth's crust and in a cubic kilometer of seawater

Metal	Amount in crust (metric tons)	Amount in seawater (metric tons)
manganese	1,809,000	1.9
zinc	170,000	2.0
chromium	130,000	0.2
nickel	100,000	2.0
copper	86,000	2.0
cobalt	32,000	0.05
uranium	7,800	3.3
tin	5,700	0.8
silver	160	0.3
gold	5	0.01

Source: Skinner 1976

Table 2. Major chemical elements in the continental crust

Element	Amount (weight percent)
oxygen	45.20
silicon	27.20
aluminum	8.00
iron	5.80
calcium	5.06
magnesium	2.77
sodium	2.32
potassium	1.68
titanium	0.86
hydrogen	0.14
manganese	0.10
phosphorus	0.10
Total	99.23

to some small fraction of its present size.

The magnitude of the world's annual needs for newly mined substances has grown so large that it is difficult to view the volume in perspective. The annual per capita consumption of newly mined mineral products for all the peoples of the world now totals 3.75 metric tons. The total includes coal, oil, iron, copper, cement, and a myriad of substances used in countless different ways; the total is still rising, doubling approximately every decade, and there is no sign that it is likely to stop in the near future. If all 1975's newly mined mineral substances were somehow stacked on historic Boston Common, the result would be a column approximately six miles high. By 1985 the annual pile would be 12 miles high.

Annual per capita rates of consump-tion of mineral products are highest in the countries of the industrialized Western world (15 metric tons in the United States), but the rate of increase is highest in developing countries. We can anticipate that both aspirations and populations of the developing countries will continue to grow, and that consumption of mineral products will follow suit. But populations have now become so vast that we must question the adequacy of earth's mineral supplies to satisfy the aspirations. Ours is the first generation for which this question has had to be raised, and it is embarrass-ing because it means that we must examine life styles, too. As is commonly the case when difficult questions are first posed, initial responses are tentative and often conflicting.

Both the questions and their answers are many-sided. Most attention has naturally been focused on energy supplies, particularly on the fossil fuels oil, gas, and coal. The controversial reports concerning the magnitude of their supplies are typical of the confusion. However, despite doubts about the exact amounts of oil, gas, and coal that will ultimately be found, the attention focused on energy means that it is now widely appreciated that each of the fossil fuel sources is limited in size, and that mankind must ultimately develop and use other energy sources. But it is equally clear, fortunately, that of energy itself there is no shortage; provided we can learn to use energy safely from sources such as the sun, earth's internal heat, and naturally occurring radioactive elements, there need never be an energy shortage.

We fool ourselves, however, if we dwell on energy alone. The uses of all natural resources are intertwined. Oil is of little use without engines built of iron, copper, zinc, and other metals. Farmlands will yield maximum crops only if they are tilled by tractors and plows and fertilized with compounds of phosphorus, nitrogen, and potassium. A failure in the supply of one resource will inevitably influence the use of others. Viewing the panoply of natural resources, we see that one group, metals, occupies a unique position. Without metals we could not build machines to replace human muscle. Without metals we could use little of the available energy. Metals are, in effect, the enzymes of industry. If supplies of metals are limited, then society must ultimately be limited, too. It is my contention that the distribution of the chemical elements in nature means, inevitably, that there are natural limits to supplies of metals, and that these limits are much more important to the future of society than limits on energy. I also contend that, with sufficient work, the limits can be predicted. It is the purpose of this paper, therefore, to explore briefly the way metals occur and to attempt to place in perspective the limitations they may ultimately impose on us.

Where metals are found

With a few notable exceptions such as gold and platinum, metals do not occur in earth's crust in their elemental states. They are present instead as one of two or more elements combined into inorganic, crystalline compounds called minerals. To recover the metals from their entrapping compounds, it is natural to choose the least expensive options; we mine and process only those minerals from which the desired metals can be easily obtained. In practice this means that we choose those minerals that require the least expenditure of energy in order to effect the chemical disintegration and release of the metal.

The production of metals requires two separate and quite different kinds of steps. The first step is to mine and concentrate the desired mineral. Mining involves breaking and removing the rock containing the desired mineral from the ground (Fig. 1); concentration involves separating it from the valueless and unwanted minerals with which it occurs, then

Figure 2. A giant smelter is the focus of this Arizona landscape at Morenci, one of the world's largest copper-producing complexes. Ore from an open pit mine just out of view at left is brought by truck to the mill and smelter which make up the complex of buildings surrounding the two smokestacks. There the valuable copper minerals are separated from the containing rock, and the copper-rich concentrate is smelted down to copper metal. (Photograph by the author.)

gathering its grains into a pure aggregate. In concentration processes, the rock is crushed so that each fragment is a separate mineral. The mineral grains can then be separated on the basis of differing physical characteristics such as density, magnetism, or wetability in different liquids. By means of these processes, an ore that contains as little as 0.5% of the desired mineral can be upgraded to a concentrate containing more than 90% of the mineral.

By comparison with the second step in the metal recovery process—smelting and refining—mining and concentration are not very energy-intensive processes. During smelting and refining (Fig. 2) the concentrated mineral is broken down and the desired metal is in turn separated from the elements with which it is combined. This is an energy-intensive step because it involves disintegration of stable chemical compounds. To keep energy demands as low as possible, the kinds of minerals from which metals are preferentially sought are sulfides, oxides, hydroxides, and carbonates—not the highly refractory silicate minerals that make up the great bulk of earth's crust. The problem of assessing metal supplies therefore reduces to determining the frequency of the desired source minerals and evaluating the problems that will attend any change from preferred to less preferred minerals.

Crustal abundance of chemical elements. The only portion of the earth accessible to mining, and therefore the only portion we can consider as a source of metals, is the crust. It has sometimes been argued that seawater should be considered a potential source of metals, but the argument does not bear close examination because the concentration of most metals in seawater solution is vastly less than their concentration in the crust (Table 1). Seawater could be regarded as an ore only if a metal could be extracted at least one thousand times more efficiently from seawater solution than from solid rocks, and the likelihood of this being achieved is small. In terms of today's technological costs, for example, if a plant capable of handling a volume as great as 4.5 million liters a minute worked with a 100% extraction efficiency, the cost of the energy alone that would be needed to pump and process the water would far exceed the value of the extracted metals.

Earth's crust ranges in thickness from 10 to 50 km and contains at least trace amounts of 88 chemical elements. It can be subdivided into two distinctly different regimes: the oceanic crust that underlies the ocean basins and the continental crust. The two differ in composition—the oceanic crust being richer in iron, magnesium, and calcium, the continental crust being richer in silicon, aluminum, and alkali elements. The 88 natural elements are all present in both crusts, though in somewhat different concentrations. Nevertheless, only 12 elements, and the same 12 elements in each case, are present in amounts equal to or greater than 0.1% of the whole. Because our knowledge of the chemistry of the continental crust still far exceeds our understanding of the oceanic crust, most of the subsequent discussion will refer to the continental crust.

The 12 elements most abundant in the continental crust are listed in Table 2. These 12 account for 99.23% of the mass of the crust, and the remaining 76 elements account for a mere 0.77% of the mass. I shall refer to the 12 most common elements as being geochemically abundant, and those present at levels below 0.1% by weight as geochemically scarce. The geochemically abundant metals that are widely used in industry are aluminum, iron, magnesium, titanium, and manganese. All other metals, many of which have numerous uses, are geochemically scarce.

Geochemically abundant metals. Essentially every common type of rock is composed of an aggregate of minerals, one or more of which contains a geochemically abundant metal as an essential constituent. It is apparent that this must be the case by simply considering the abundance of elements, and it means that, starting with a common rock type, it is always possible to produce a mineral concentrate in which the desired abundant element is a major constituent.

The practical miner does not, of course, select an average crustal rock as a source of metal. Instead he seeks those rocks in which the desired metal has been concentrated by some natural process and in which the metal occurs in a desired mineral form such as an oxide or a carbonate. Nevertheless, because most of the minerals he seeks are widespread in common rocks, the miner can look forward confidently to a long future with respect to geochemically abundant metals. Because many of the present ore deposits grade slowly into common rocks, as the extreme local enrichments he now works become depleted, he will be able to continue

mining material that is less and less rich, and he will always be able to prepare concentrates of desirable minerals. The situation is as depicted in Figure 3. As reduction in the grade of ore mined (grade being a miner's term for the percentage of metal in an ore) declines arithmetically, the amount available increases geometrically.

One geochemically abundant metal that may seem at first glance to contradict this simple picture is aluminum. The common minerals that contain aluminum are all silicates, and the hydroxide compounds found in bauxite, which is the ore of aluminum mined at present, are not found in common rocks. Thus, once the world's resources of bauxite have all been mined out, we will have to learn to use silicate minerals such as anorthite ($CaAl_2Si_2O_6$), sillimanite (Al_2SiO_5), and kaolinite ($Al_4Si_4O_{10}(OH)_8$), as Russia, Sweden, Japan, and other countries have done at times of economic necessity. The principle of concentration of aluminum minerals will then be applicable and it will be possible to use common rocks as starting materials and to separate the anorthite, sillimanite, and kaolinite from other minerals in the rock. Carrying this line of reasoning to its conclusion, it is apparent that the entire crust can be considered a potential resource of geochemically abundant minerals and that no major technological barriers, beyond the ability to mine steadily declining grades of ore, will have to be overcome. The smelting practices available today will also work in the future.

Geochemically scarce metals. The geochemically scarce metals pose an entirely different situation and probably have a very different grade-distribution curve, as shown in Figure 4. Rarely do the scarce elements in common rocks form separate minerals. Instead, they are present as randomly distributed atoms trapped by isomorphous substitution in minerals of the geochemically abundant elements, an atom of a scarce element replacing an atom of an abundant element. For example, lead is found in most common rocks as an atomic substitute for potassium, while zinc appears as a substitute for magnesium. Thus, even though analysis of a common rock such as granite may reveal that it

contains many parts per million lead and zinc, no amount of searching will disclose either a lead or a zinc mineral (see Fig. 5a). The two metals are trapped in the atomic cages of common minerals such as orthoclase ($KAlSi_3O_8$) and biotite ($K_2(Mg,Fe)_6Si_6Al_2O_{20}(OH)_8$). A rock containing 10 ppm lead, in which all the lead is present in a feldspar, can yield a concentrate no richer than a pure feldspar concentrate. If the rock contains 10% feldspar, the maximum lead concentrate will contain only 100 ppm lead—hardly an encouraging result for a miner accustomed to producing concentrates with 70% lead or more. Furthermore, to release the lead from its silicate cage, the entire mineral must be broken down chemically and the lead atoms separated from all the other atoms. This is a complicated and very energy-intensive process, which at today's prices for energy could produce lead only at a cost of hundreds or thousands of dollars a pound. The present price of lead—less than a dollar a pound—is a bargain by comparison.

Miners never consider common rocks when they are seeking ores of geochemically scarce metals. Instead they seek those rare and geologically limited volumes of the crust where special circumstances have produced marked local concentrations of geochemically scarce elements, and have done so in a manner that leads to the scarce elements being present in compounds of their own, not as atomic substitutes in silicate minerals. These localized volumes (which we commonly call ore deposits) contain enrichments far above the average crustal abundance of an element—sometimes by as much as 100,000 times (see Fig. 5c). Lead, for example, has an average abundance in the continental crust of 0.0010%, but lead ores usually contain at least 2% lead, and some are known to be as rich as 20% lead. It is not surprising that such extraordinary local enrichments are limited both in number and in size. The vital question therefore is, What percentage of the scarce element content of the continental crust is concentrated into ore deposits?

Ore deposits are the result of unusual sets of circumstances. For example, most deposits of lead and zinc form as a result of reactions between brines circulating in the crust and the rocks

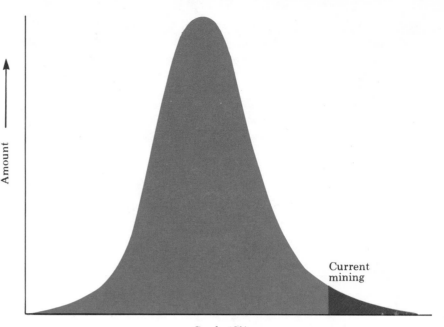

Figure 3. The curve represents the typical distribution of geochemically abundant metals such as iron, titanium, and aluminum in the earth's crust. The proportion of the total amount that has been mined up to the present is indicated at the right. The highest-grade ores have been mined first, but, because the same kinds of minerals occur regardless of grade, the concentrating techniques used in current mining processes to extract metal from high-grade ores can also be used in the future with less rich material and even common rock. As the grade, or percentage of metal in an ore, declines arithmetically, the curve indicates that the amount of metal available will increase geometrically down to a grade corresponding to the peak of the curve.

Figure 4. The bimodal curve represents the probable distribution of a geochemically scarce metal in the earth's crust. The large peak is the distribution in common rocks, where scarce metals occur not as separate minerals but as atomic substitutes for abundant metals; the small peak represents deposits produced by ore-forming processes such as those resulting from the circulation of brines in the crust. The relative positions of the mineral specimens labeled *a*, *b*, and *c* in Fig. 5 are shown. Current mining has already reached the point where the distribution curve for easily accessible scarce minerals turns downward. Further reductions in the grade of ore mined will produce declining tonnages of ore and will eventually bring us to a mineralogical barrier—which lies somewhere in the area between the two humps.

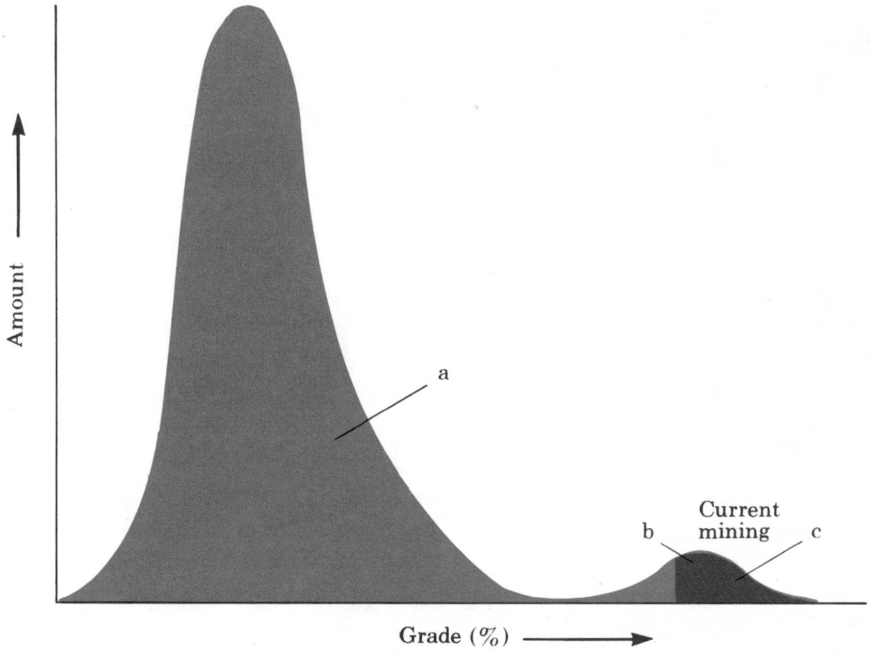

they encounter. The brines alter and cause recrystallization of the silicate minerals, and in the process lead and zinc trapped by atomic substitution in the silicate minerals pass into aqueous solution. The concentration of lead and zinc (and other scarce metals) in solutions is low—no more than a few hundred ppm—but it is sufficient, under suitable conditions, to form ore deposits. The heated brine and its dissolved load must flow in a restricted channel and, while doing so, must undergo a further set of reactions which cause precipitation of lead and zinc minerals—usually

the relatively insoluble sulfides PbS and ZnS. The rich vein shown in the cover photograph contains copper, lead, zinc, and silver minerals deposited by brines that flowed through and eventually sealed up a fracture in the barren volcanic rock enclosing the vein.

Thus, three steps—solution, restricted flow, and localized precipitation—are necessary, and although brines are common in the crust, the delicate timing of events and the chemistry needed to form lead and zinc ores coincide very rarely. Simi-

larly complex and therefore infrequent events lead to the formation of deposits of the other scarce metals. Just how the secondary concentration processes modify the distribution curve of geochemically scarce metals is not known with certainty, but the end result is, I suggest, the bimodal distribution depicted in Figure 4.

Figure 4 tells an important story. Ore deposits of scarce metals always contain local concentrations of minerals such as PbS and ZnS. It is possible to mine and concentrate the minerals in an ore deposit and thus

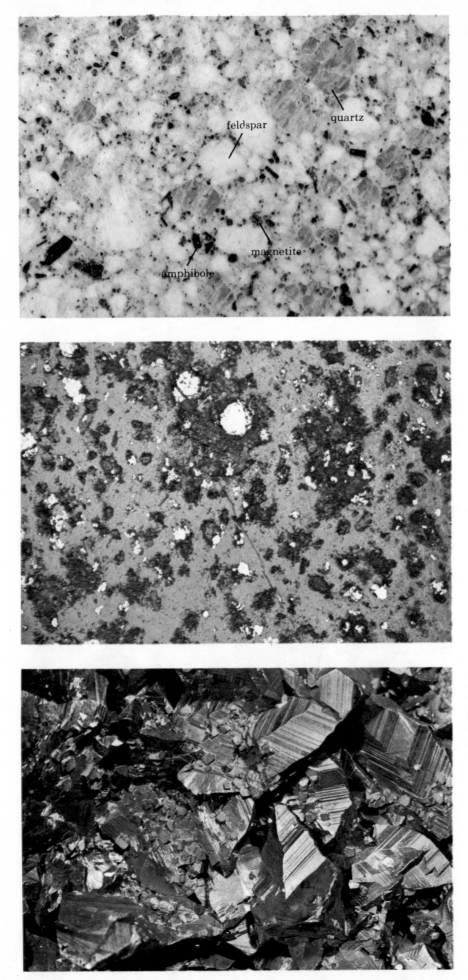

Figure 5. Geochemically scarce metals are found in two forms: as chemically dispersed atoms enclosed within the structures of common minerals in ordinary rocks, and as separate mineral grains in rare, localized volumes of rock called ore deposits. (*a*) Polished surface of granodiorite, a common igneous rock, reveals the presence of the minerals feldspar (Na + Ca alumino-silicate), quartz (SiO_2), amphibole (Fe + Mg + Ca silicate), and magnetite (Fe_3O_4), an ore mineral of iron. Magnetite is an example of an ore mineral of a geochemically abundant element that could, if necessary, be concentrated from ordinary rock. Although this rock sample also contains geochemically scarce elements—70 ppm of copper and 15 ppm of lead—they are present as atomic substitutes in the amphibole and feldspar, respectively. Field of view: 3 cm.

(*b*) Brilliant yellow grains of chalcopyrite ($CuFeS_2$), a common ore mineral of copper, were deposited in a porous volcanic rock that served as an aquifer for mineralizing solutions. The disseminated ore minerals can be readily separated from the valueless silicate minerals to yield a copper-rich concentrate; ore deposits containing as little as 0.4% copper are now being mined by large-scale mining operations like those at El Tiro Pit (Fig. 1). Field of view: 3 cm.

(*c*) Extraordinary, 2-cm-wide crystals of chalcopyrite are from a rich copper deposit at Ugo, Japan. (The smaller, cube-shaped crystals are galena, PbS.) Deposits like the one at Ugo and the similarly rich vein shown in the cover photograph can contain as much as 15 to 20% copper—but few such rich deposits have been found in the past 30 years. Field of view: 9 cm.

produce enriched aggregates ready for smelting. The area beneath the small hump in Figure 4 is therefore a measure of the percentage of a scarce metal in the crust that is freed from atomic substitution and is combined as separate minerals. The large hump, by contrast, is a measure of the percentage trapped by isomorphous substitution in abundant silicate minerals. Proof of the bimodal distribution of geochemically scarce elements is still lacking because analytical sampling of the crust is too crude. Considering the importance of the consequences, definitive sampling programs should stand high among the priorities of large government laboratories equipped to make the necessary investigations.

A bimodal distribution has two obvious consequences: first, as we mine the material under the small hump, starting with the richest grades first, there will be an initial period during which the declining grades will bring the reward of larger and larger tonnages of ore. This has been the situation faced by miners for most of this century. Eventually, however, the distribution curve turns down again, and further reductions in grade bring declining tonnages of ore. There are indications that this might already be happening in the mining of metals such as mercury, gold, and silver.

The second is less a consequence of the bimodal distribution than of mineral distribution in general: a point will eventually be reached where reduction in grade brings us to a mineralogical barrier—the point beyond which a scarce element occurs only as an isomorphous substitute and is therefore no longer amenable to concentration. The grade of the mineralogical barrier will vary for each element, each deposit, and each rock type. With a few exceptions, such as gold, uranium, and gallium, the barrier appears to lie at grades somewhere between 0.01 and 0.1%. That is, at grades below 0.01 to 0.1%, a scarce metal occurs solely by isomorphous substitution; above that grade, scarce metals form compounds of their own. Gold and uranium are exceptions on the low side, in that separate minerals are known to occur at grades well below 0.01%. Gallium is an exception on the high side. Indeed, so little gallium is concentrated into separate minerals that no large ore deposit has ever been found, even

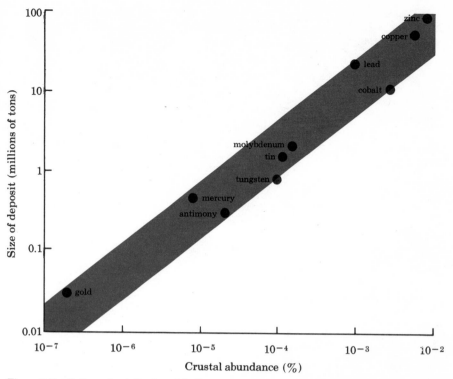

Figure 6. Predictions about the size of the "ore deposits hump" in Fig. 4 are based on certain mass properties of known ore deposits. For example, the graph shows that the largest known deposit of each scarce metal is approximately proportional to crustal abundance.

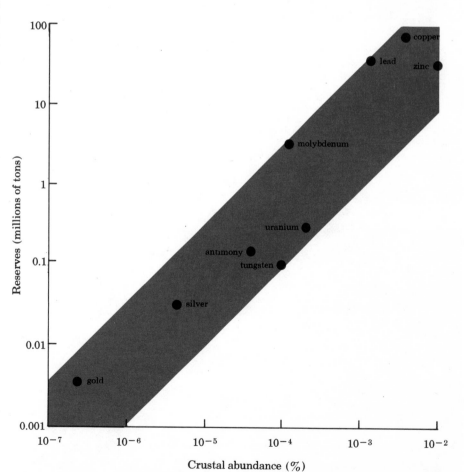

Figure 7. As shown in the graph, the known reserves of scarce metals in the U.S. are also found to be proportional to crustal abundance. Like the relationship illustrated in Fig. 6, this function suggests that the amount of scarce metal ultimately available to current mining techniques is directly proportional to the geochemical abundance of the element. (Data from Brobst and Pratt 1973.)

Table 3. Estimated maximum yield of geochemically scarce metals from ore deposits in the continental crust

Element	Average abundance in continental crust (%)	Maximum recoverable tonnage from ore deposits (millions of metric tons)
copper	0.0058	1,000
gold	0.0000002	0.034
lead	0.0010	170
mercury	0.0000002	0.34
molybdenum	0.00012	20
nickel	0.0072	1,200
niobium	0.0020	340
platinum	0.0000005	0.084
silver	0.000008	1.3
tantalum	0.00024	40
thorium	0.00058	100
tin	0.00015	25
tungsten	0.00010	17
uranium	0.00016	27

Note: The calculation assumes that mining will proceed no deeper than 10 km below the surface, and that 0.01% of all the metal in the continental crust is present in minerals available to mining and concentration. (The minerals may not be located in deposits rich enough to be considered ore by present standards.) The calculation includes that part of the continental crust that lies beneath the continental shelf.

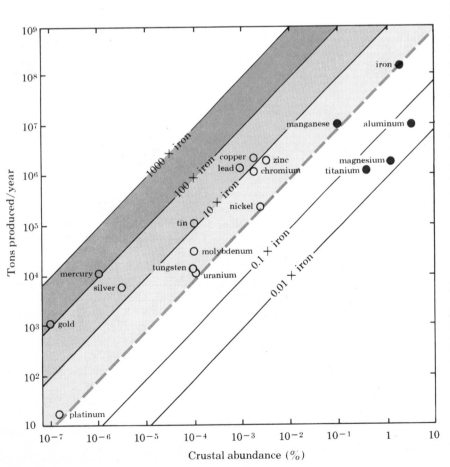

Figure 8. The graph shows the relation between annual world production of newly mined metals and their abundance in the continental crust. The dashed line drawn through iron, the most widely used geochemically abundant metal, may be considered a kind of baseline for use rates of metals: points lying on the line are produced at the same rate, relative to their crustal abundance, as iron. Metals below the line are mined proportionally slower; those above, proportionally faster. Metals farthest from the line—including many of the geochemically scarce metals—will be mined out first.

though gallium is twice as abundant as lead in the crust.

Can we predict how much of each metal is present in the small "ore deposit hump" in Figure 4? Until the last deposit is found and measured we cannot, of course, be sure that we are correct, but an indirect line of reasoning suggests that we can indeed estimate, at least to the correct order of magnitude, the size of the hump. A number of mass properties of ore deposits, such as the size of the largest known deposit for each scarce metal (Fig. 6) and the number of deposits that contain a million or more tons of a given metal (Skinner 1976) are proportional to the average content (or crustal abundance) of the element in the continental crust. As first pointed out by McKelvey (1960), even the discovered reserves of the scarce metals are proportional to the crustal abundances (Fig. 7). These relations suggest strongly that the size of the "ore deposit hump" is directly proportional to the geochemical abundance of an element. Other lines of evidence support this suggestion, but they are no more definitive than the arguments just presented. The contention is therefore still a suggestion, albeit one that has a high probability of being correct.

Accepting the conclusion that the size of the "ore deposit hump" is proportional to crustal abundance, we need only estimate the size of one "hump" to be able to estimate how much of the other scarce metals to expect. A recent report by the National Academy of Sciences' Committee on Mineral Resources and the Environment (COMRATE 1975) has shown how this might be done. The report estimates that the mineralogical barrier for copper is reached at a grade of 0.1%. It also estimates that no more than 0.01% of the total copper in the continental crust will be found concentrated in ore bodies with grades of 0.1% Cu or more. The committee's reasoning was based on the volume percentage of mineralized rock in the most intensely mineralized regions so far discovered and on the frequency of copper deposits in the crust. Their figure of 0.01% must therefore be taken as the maximum possible yield, but it is not likely to be too large by more than a factor of ten. Thus the size of the "ore deposit hump" will probably fall between 0.001 and 0.01% of the amount of any scarce metal in the crust. The maximum estimated yields of metals from concentrated ore deposits, calculated according to the assumptions given, are listed in Table 3.

Using this kind of reasoning, COMRATE estimated that present reserves (in the mining sense of material from which a profit can be made) plus past production of copper already amount to 3% of the world's ultimate yield. The equivalent figure for the United States is estimated to be 16%. It does not take much arithmetic to calculate that, with use rates growing as they now are, copper will change from its present position as a metal in apparently abundant supply to a position of strategic shortage by the end of the present century.

Relative use rates of metals

Because supplies of scarce metals are apparently proportional to crustal abundances, we should view the use rates of metals in the same terms. Using the most widely employed geochemically abundant metal, iron, as a basis of comparison, Figure 8 is an attempt to put relative use rates in a geochemical perspective. All metals falling on the dashed line are being used at the same rate, proportional to their crustal abundance, as iron. Metals that fall above the dashed line (and this includes most of the geochemically scarce metals) are being used at proportionally faster rates. For example, mercury and gold are being used at a rate about 110 times faster than iron, and lead about 40 times faster.

Assuming that we continue to use metals at rates that are not proportional to their abundances, we can read directly from Figure 8 those metals which are likely to be mined out first. The farther a metal plots from the dashed line, the earlier its demise. Metals apparently in trouble include such widely used commodities as mercury, gold, silver, copper, and lead. By contrast with geochemically scarce metals, all of the geochemically abundant metals seem to be underused. Their positions plot on or far below the dashed line. If my argument is correct, we have an unbalanced situation that cannot long continue. Clearly, we should be using abundant metals more and scarce metals less.

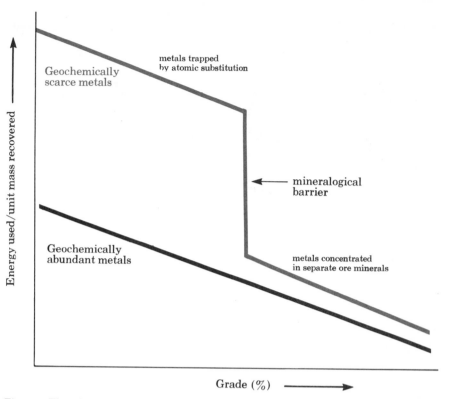

Figure 9. The relationship between the grade of an ore and the energy input per unit mass of metal recovered is shown for both scarce and abundant metals. A steadily rising amount of energy will be needed to produce even geochemically abundant metals from the leaner ores of the future, but the amount of energy needed to produce scarce metals will take a tremendous jump when the mineralogical barrier is reached. At that point, when ore deposits are worked out, mineral concentrating processes can no longer be applied, and the silicate minerals in common rocks must be broken down chemically to separate the atoms of scarce metals from all the other atoms.

What happens when ore deposits of scarce metals have all been found and mined—when the mineralogical barrier is finally reached? The situation is demonstrated in Figure 9. As grades decline in deposits of geochemically abundant metals, the energy input (and therefore the cost) per unit mass of metal recovered rises steadily. The smelting process remains the same because it is always possible to produce a concentrate. The steady rise in energy required is a result of the need to mine larger volumes as leaner and leaner ores are worked and to process these larger volumes by crushing and concentration.

The curve for geochemically scarce metals is very different. This curve parallels that for the abundant metals until the traditional ores have been worked out. Once the mineralogical barrier is reached, however, a tremendous jump in energy is needed, because mineral concentration processes can no longer be employed.

The host silicate mineral must be broken down in order to recover the trapped scarce metal. The magnitude of the energy increase will naturally vary with the kind of host mineral, but for most silicates the energy demand will jump by a factor of 100 to 1,000 times. It seems unlikely that we will choose to jump the mineralogical barrier. The relative costs of scarce and abundant metals, already widely separated, will become vastly more disparate. It will simply be cheaper to substitute iron and aluminum and put up with penalties, such as lower efficiencies in machines, that we do not now countenance.

Suppose, however, that abundant energy sources do become available and that it is feasible to overcome the mineralogical barrier. There is an obvious reason why a future technology built largely on geochemically abundant metals will pertain even in that event. Consider once again the relative abundance of elements. If we extract metals from silicate minerals,

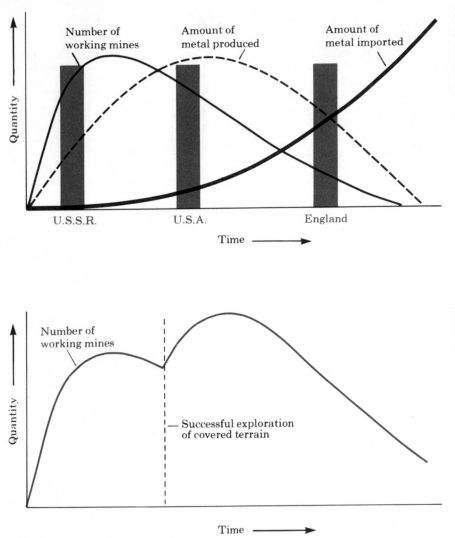

Figure 10. The historical development of metal production, the number of working mines, and the amount of metal imported are shown for three industrial countries. As time passes, the position of a country moves from left to right in the graph. The U.S. is today where England stood early in the last century; the U.S.S.R. is today in the same relative stage of development as the U.S. in about 1850.

Figure 11. When a successful method of prospecting for ore deposits beneath covered terrain is developed, the curve defining the changing number of working mines with time will move sharply up (compare the equivalent curve in Fig. 10). The situation is comparable to prospecting a new country.

and therefore succeed in using average rock as an ore, all metals would be produced in amounts approximately proportional to their crustal abundances. That is, a mining industry based on average rock would have production figures for *all* metals that would be along a line parallel to the dashed line in Figure 8. If we wished to use metals in the proportions in which we now use them, we would soon have vast surpluses of iron, aluminum, and other abundant metals. Instead of allowing huge unwanted stocks to accumulate, I have no doubt that we would soon find a way to reduce our demands for scarce metals and make do with the more abundant metals.

Hard times ahead

However one views the use of geochemically scarce metals in the future, it is clear that there are very real limits to the amounts available in traditional ore deposits of the conti-

nental crust. Efficient recycling, which surely must come as existing deposits are worked out and new ones become harder and harder to find, will guarantee that even the scarcest metals will always be available in at least small amounts. But recycling can at best sustain a declining use rate. While growth or even level use rates pertain, newly mined material must continue to be added. When the traditional deposits have all been found and mined, however, our responses will inevitably be governed by geochemical abundances, and little, if any, newly mined scarce metals will be available to be added to the pool in use.

This critical time in man's future technological development cannot be pinpointed, but it cannot be too far in the future. The date depends on future use rates of metals, and since some metals will be effectively used up before others, we are likely to see an extended decline rather than a

sudden cutoff. The decline has already started for gold, silver, and possibly a few other metals. The rest will follow during the next century, and by the year 2076, when the United States celebrates its tercentenary, mining of scarce metals will be increasingly a memory from the past.

The decline will be controlled, at least in part, by international politics. My estimate of the relative positions of three industrial countries, England, the United States, and the U.S.S.R., is illustrated in Figure 10, using relations first elucidated by Hewitt (1929). The curve defining the amount of metal produced starts at zero production when mining first commences in a country and ends again at zero when all ore deposits are worked out. The area under the curve is the total amount of metal produced in the country and corresponds to the "ore deposit hump" in Figure 4. The curve defining the number of working mines is a measure of the rate at which ore deposits are discovered, and the third curve indicates the rising amounts of metals imported to supply an industry that can no longer be maintained by internal supplies. A century and a half ago, England was a major metal producer, shipping its copper, lead, tin, and other scarce metals around the world. Since that time it has become, increasingly, an importer of raw materials. Now the United States is following the same route and, as each year passes, a higher percentage of essential raw materials must be imported (Morton 1973).

The future for production of scarce metals clearly lies with those continents where prospectors have not yet scoured every corner. These are, mainly, the less inhabited portions of Asia, such as Siberia, the central and northern portions of Africa, much of South America, Australia, and Antarctica. Each continent still has some unexplored ground of its own—that is, areas so deeply covered by soil and by young sediments that it is impossible to use any prospecting method to sense ore deposits below. For a country such as the United States, the amount of blind ground is close to 50% of the total land area. We await a breakthrough in research that will develop methods to carry out this deep prospecting. When it comes, the curve defining the number of working mines in Figure 10 will have to be modified as shown in Figure 11. No doubt hopes will rise when successful prospecting in covered terrain is achieved, and skepticism will then be expressed at the kind of predictions made in this paper. The end is entirely predictable, however. Figure 11 shows what it must be.

So far we have concentrated on the continental crust. The crust beneath the ocean floors also remains to be prospected, however, and we can consider it, in terms of Figure 10, as a huge new continent open for exploration. Unfortunately, the deep ocean floor, which is about two-thirds of earth's surface, seems to offer distressingly poor prospects for most scarce metals (Skinner and Turekian 1973). It is still much too early to draw final conclusions, but it looks as if considerably less than 0.01% of the metals in the oceanic crust are concentrated into ore deposits, even including such unusual deposits as ferromanganese nodules on the deep-sea

floor. It is not too surprising that this might be so, because the ocean floor is all less than 200 million years old—so young, geologically speaking, that concentrating processes have not had as much time to do their work as they have in the vastly older continental crust. Undoubtedly some deposits will be found—deposits of copper and nickel seem the most likely—but their recovery will offer a great many technological headaches.

Whichever way we turn, we are forced back to the realization that one day soon we will have to come to grips with the way in which earth offers us its riches. That day is less than a century away, perhaps less than half a century. When it dawns, we will have to learn to use iron and other abundant metals for all our needs. The dawn of the second iron age is much closer than most of us suspect.

References

Brobst, D. A., and W. P. Pratt. 1973. *United States Mineral Resources.* Geological Survey Professional Paper 820. Washington, DC: U.S. Govt. Printing Office.

COMRATE. 1975. *Mineral Resources and the Environment.* Report by the Committee on Mineral Resources and the Environment, National Academy of Sciences—National Research Council.

Hewett, D. F. 1929. Cycles in metal production. Transactions of Am. Inst. of Mining and Metall. Engineers, *Yearbook for 1929,* pp. 65–98.

McKelvey, V. E. 1960. Relation of reserves of the elements to their crustal abundances. *Am. J. Sci.* 258-A:234–41.

Morton, R. C. B. 1973. *Mining and Minerals Policy.* Second Annual Report of the Secy. of the Interior under the Mining and Minerals Policy Act of 1970. Vol. 1. Washington, DC: U.S. Govt. Printing Office.

Skinner, B. J. 1976. *Earth Resources.* 2nd ed. Englewood Cliffs, NJ: Prentice-Hall.

Skinner, B. J., and K. K. Turekian. 1973. *Man and the Ocean.* Englewood Cliffs, NJ: Prentice-Hall.

"Frankly, I'd be satisfied now if I could even turn gold into lead."

Donald A. Singer
A. Thomas Ovenshine

Assessing Metallic Resources in Alaska

Methods developed for estimating mineral resources in poorly explored regions of Alaska may have wide influence as land-use decisions become more deliberate and controversial

In the last two decades federal and state governments have become increasingly preoccupied with classifying public lands according to the uses that may be made of them. One outcome of the classifying can be a change in the land's legal status from one in which any use is tolerated to one in which only selected activities are allowed. Since such a change affects the economic and recreational opportunities of individuals and institutions, a variety of interest groups have developed to follow and influence the classifying process. Perhaps nowhere has the process been more closely scrutinized or more hotly debated than in Alaska, where land-use decisions for much of the state's 375 million acres are being made.

With heightening interest in the classifying process have come increasing requirements for documentation of the reasons for each classification decision. One response has been the development of resource inventories, or assessments of the different attributes of the land and

Donald A. Singer is a geologist with the Office of Resource Analysis, USGS. After receiving his Ph.D. from Pennsylvania State University in 1971, he worked in industry for several years as a systems analyst. His primary research interests have been the development of new mineral-resource assessment techniques and the improvement of mineral exploration through the use of statistical methods such as geometric probability. A. Thomas Ovenshine is chief of the Branch of Alaskan Geology, USGS. He joined the Survey in 1965 after completing degrees in geology at Yale, V.P.I., and U.C.L.A. His primary research interests and publications are in Paleozoic stratigraphy in southeastern Alaska, contemporary sedimentation in the intertidal zone, and regional geology in Alaska. Address: Mail Stop 84, USGS, 345 Middlefield Road, Menlo Park, CA 94025.

the life it supports. In theory, the assessments of different resources are weighed, and the qualities for which the land is most valuable determine the classification decision. Since synopses of resources are also a requirement of most environmental impact statements required by the National Environmental Policy Act of 1969, much of the contemporary workload of public scientific agencies involves making resource assessments.

Resource assessments for the purpose of land classification must be conducted differently for surface and subsurface resources. Surface resources can be assessed quickly and accurately by methods that are well proven if often quite complicated. Mammals and birds can be counted. Water can be fished, its volume measured, its quality tested. Timber can be appraised by cruising; recreational uses and game harvests can be tallied. Even scenic values can be photographed and described. For most assessment methods dealing with surface resources there is a generally accepted theoretical basis, decades of experience in application, and a common understanding of accuracy and precision.

In contrast, subsurface resources are usually concealed and are very difficult to inventory or assess. Particularly troublesome have been undiscovered deposits of oil or minerals, because it may require many years and many millions of dollars to find and prove new discoveries of metal or petroleum. Theories of mineral-resource assessment are still developing rather than accepted and established, and testing of the accuracy and precision of such assessments lags even further behind.

Assessing undiscovered mineral deposits is difficult enough in areas that have been geologically mapped in detail (at scales of 1 cm equal to 0.5 km or less) and partially explored by drilling. In Alaska the problem is compounded, because large regions have not been mapped even in reconnaissance fashion (typically at a scale of 1 cm equal to 2.5 km); geochemical and geophysical surveys are far from complete; and exploration by drilling is much less widespread than in western Canada or the western conterminous United States (DeYoung 1976).

This report describes how we and numerous colleagues in the United States Geological Survey assessed Alaska's deposits of valuable metals during the years 1974–78. Some of the methods and concepts used were new and may have applications to other problems.

Although Alaska had a turn-of-the-century gold rush, was a leading producer of copper during the 1910s and 1920s, and has accounted for much of the small United States platinum production, the state has not been a major supplier of minerals. In recent years it has had no large metal mines in production. Thus the question arises: Is Alaska as a whole so sparingly endowed with metallic mineral resources that it is not worth further investigation?

One way of answering this question is to apply something called the unit regional value concept (Griffiths 1978). This involves comparing the amount of a mineral already mined in a region such as Alaska with amounts of the mineral recovered from geologically similar regions. If the amount produced from the area in question is very much below that in the comparable regions, we infer that there may be a large untapped endowment of the mineral. Of course, production data should be expressed per unit area, since regions vary greatly in size.

Alaska, which has long been recognized as part of the Cordilleran belt of western North America, is geologically similar to western Canada and the western conterminous United States. Comparison of the total production of certain metals from the western states and the western provinces of Canada with Alaska's production demonstrates that, for all metals but platinum, Alaska has produced less than the analogous region (Fig. 1). Alaska has "overproduced" platinum by a factor of 30, but "underproduced" gold by a factor of 6, chromite by 20, mercury by 29, copper by 34, silver by 79, tungsten by 361, lead by 378, and zinc by 136,000.

These comparisons do not necessarily imply that Alaska is endowed with, or will produce, the exact amount yielded by the comparable region: production from western Canada and the U.S. still continues; some differences in resource endowment doubtless stem from differences in local geology, and mining costs may always remain so high in Alaska that low-grade deposits of low-priced metals may never be utilized. But the underproduction of most metals and the magnitude of the underproduction suggest that Alaska's mineral resource endowment is quite large and that further attempts to pinpoint its location are certainly warranted.

Areal resource assessment

Alaska's total metal endowment, as estimated by the unit regional value concept, is much larger than can be accounted for by the known deposits. This means that a resource assessment of Alaska requires careful consideration of undiscovered and incompletely explored mineral deposits. And for use in making land-classification decisions, resource assessments must be made for individual tracts of land within the country.

A mineral *resource* is a concentration of naturally occurring solid, liquid, or gaseous material in or on the earth's

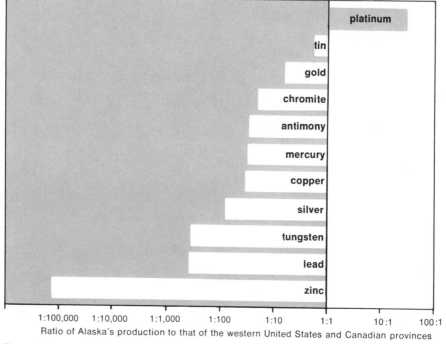

Figure 1. Comparison of metal production per unit area in Alaska with that in the geologically similar western United States and Canada suggests that Alaska has a significant endowment of untapped metallic resources. Even platinum, of which mining in Alaska has been extensive compared with that in nearby areas, is most likely still present in important quantities. Values were used for quantities produced through 1974 in Arizona, California, Colorado, Idaho, Montana, Nevada, New Mexico, Oregon, Utah, Washington, and Wyoming; and through 1975 in British Columbia and the Yukon. (Data compiled by Griffiths and Labovitz, pers. comm.)

crust in such a form that economic extraction of a commodity is either currently or potentially possible. *Reserves* are the part of resources that have been explored in detail and are at present economic to mine (U.S. Dept. of Interior 1974). Mineral resources should be considered conditional supplies of mineral raw materials. Estimation of the *reserves* entails what could be considered the simple task of adding together well-documented numbers; but estimation of the remaining portion of resources is much more complex: economic feasibility, uncertainty of future technologies and economics, and uncertainty of existence of acceptable raw materials should all be considered.

In principle, then, resource assessment must include estimates not only of quality and quantity of mineral endowment but also of all the economic and technological factors that can affect prices and costs, such as exploration intensity and effectiveness, future demand, development of other supplies, development of new refining methods, and government

actions that affect mining. Even if a generally accepted method of estimating mineral endowment is found, it is unlikely that consensus could be reached on assumptions and estimates of future economic and technological conditions. Moreover, such variables are inherently difficult to predict, and an assessment built upon specific values for those variables would probably become obsolete almost immediately.

One way to eliminate some of these difficulties is to estimate mineral endowment in a disaggregate manner (Singer 1975). This method requires the separate estimation of those variables related to the quality and quantity of mineralized areas that can affect possible economics and technologies of exploitation. These variables include grade and tonnage estimates; the physical, chemical, and mineralogical features of the rock that could affect its metallurgical treatment and recovery; the geographical location; the geologic structure and hydrologic conditions; and the spatial distribution of mineralization and overburden.

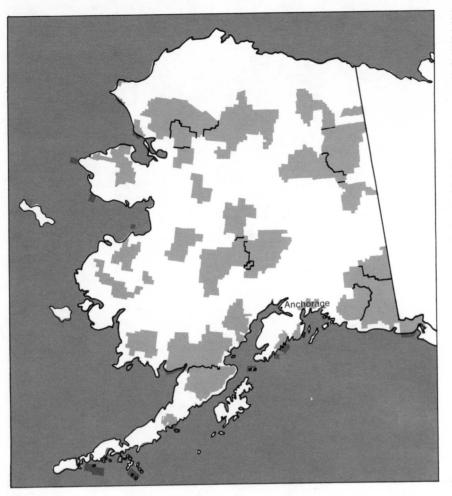

Anchorage

Figure 2. The areas proposed for withdrawal as National Interest Lands by the U.S. Department of the Interior in 1973 are delineated here in color. Such lands would be included in one of the four national preservation systems, and mining would not be permitted within them.

large regions may be quite regular and predictable. Thus, if the size of the tracts delineated is large enough, estimates will be stable. Just how large an area must be to produce stable estimates has not been determined; the size will vary with deposit type, the favorability of the tract for that deposit type, and the extent of exploration in the tract.

If statistical distributions derived from well-explored deposits serve as models for undiscovered or incompletely explored deposits, estimates based upon them must be tempered by evaluation of geological conditions that might be related to the number, quality, or quantity of the undiscovered mineral deposits. While variations within a deposit type are *comparatively* small, it has long been recognized that tracts of land which are geologically only slightly different may have different numbers of deposits. Recently, cases of significant regional differences in the grades and tonnages of deposits within types have been documented (Menzie and Singer, in press). For some deposit types, these differences may be predictable (Divi et al., in press); for others, differentiation factors have not yet been found. Where regional differences in grade or tonnage are suspected and a predictive model based on geology is not available, regional trends or tonnage-and-grade distributions from geologically similar areas near those for which the estimates are to be made are desirable.

Level of precision in estimates of number of deposits in a tract is indicated by statistical probabilities associated with the estimates. Level of precision in estimates of grade and tonnage of mineral deposits is represented by the proportion of deposits expected to have different combinations of grades and tonnages. The level of precision has important implications for policy-making. Moreover, specifying the degree of precision should help counter a common tendency to make conservative,

For an area such as Alaska, about which we have little detailed information, it makes more sense to focus on mineral deposit types—which may include more than one economic commodity—than to focus on the individual commodities themselves. Deposits are typed according to host and associated rock types, contained metals, ore minerals, and geologic setting. This method facilitates extrapolation from known deposits, because in many cases deposits of one type occur in similar kinds of rock and have similar physical, chemical, and mineralogical features (although they may not share a common age or origin). Thus, average grades and tonnages of well-explored deposits of each type can be used as models for the grades and tonnages of yet-to-be-discovered deposits in geologically similar settings; variation of these characteristics within deposit type is relatively small. Furthermore, the use of deposit types fits nicely with the need to break down assessments by relatively local tracts: by delineating tracts according to characteristic

geologic environment, one has already, to a certain extent, classified deposit types.

The use of deposit types allows resource assessments to be performed in three basic steps. First, areas are delineated according to the kinds of mineral deposits their known geologic character will permit. Second, the number of deposits within each tract is estimated. And third, the amount of metal and the characteristics of the ore in the deposits are estimated by means of models of grades and tonnages based on similar deposits. The relative economic importance of each tract can then be judged on the basis of these last two evaluations.

Obviously, the difficulty of estimating variables increases dramatically as one moves from well-explored deposits to lesser known prospects to undiscovered deposits. Although single rare events such as the occurrence of an undiscovered mineral deposit are unpredictable, statistical averages of the number of deposits in

Figure 3. Outlined and numbered tracts were delineated by the types of mineral deposits that local geology permitted. Individual assessments were then made for each tract; Table 1 shows a summary of the assessment for tract 10. (After MacKevett et al. 1978.)

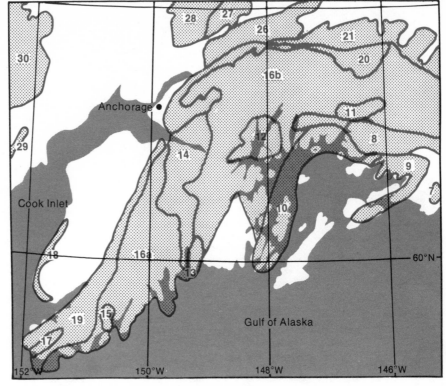

"safe" estimates—for example, by neglecting undiscovered deposits in resource assessments of incompletely explored areas. Neither conservative estimates nor exaggerated estimates, especially without an accompanying indication of degree and direction of uncertainty, are a good basis for policy decisions.

The assessments

In late 1978, President Carter withdrew 116.2 million acres in Alaska from commercial use in an "emergency" measure, designating 56 million acres of these lands as national monuments. But even before this action, the Alaska Statehood Act of 1958 entitled the state to select 102 million of the 375 million acres for whatever the state decides. The Alaska Native Claims Settlement Act of 1971 entitled aborigines to select 44 million acres for ownership by the local or regional native corporations that are chartered by the act. In addition, the 1971 act authorized the Secretary of the Interior to withdraw up to 80 million acres for possible inclusion in the four preservation systems (National Parks, Wildlife Refuges, Wild and Scenic Rivers, and National Forests). These 80 million acres are the "National Interest Lands," also frequently called "d-2" lands after the section of the act authorizing their withdrawal (see Fig. 2). Although the act empowered the secretary to withdraw lands and propose classifications, the final authority is the United States Congress, which, under the act, was required to complete classification by 18 December 1978.

In 1974 Congress requested that the United States Geological Survey assess the mineral potential of the d-2 lands of Alaska in time for the December 1978 decision. Initially, the USGS selected for study a group of 1:250,000-scale quadrangles that appeared to have significant mineralization and were being considered at least in part for withdrawal. A mul-

tidisciplined approach was to be used to produce folios including geologic, geochemical, and geophysical information; data on mineral occurrence; and earth-satellite maps.

By early 1976 it became evident that a systematic quadrangle-by-quadrangle approach at the 1:250,000 scale could not cover all of the d-2 lands by 1978; nor could it adjust to the many boundary and area changes that were being proposed by Congress and the Department of the Interior. The plan was therefore changed to provide for 1:1,000,000-scale assessments; these were completed for 80% of Alaska by February 1978. The 1:250,000-scale quadrangle assessments that had been completed—e.g. the assessment for Nabesna quadrangle by Richter et al. (1975)—or partially completed provided reconnaissance and in some cases detailed information for the 1:1,000,000-scale assessments.

In preparation for the regional mineral resource assessments, a team of geologists with an average of 20 years of field and laboratory experience in investigating the geology and mineral deposits of Alaska assembled information that became the foundation of the assessments. Geologic maps at 1:1,000,000-scale were prepared from

primary sources or revised from existing compilations in such a way as to emphasize rock units of metallogenic significance. New and comprehensive inventories of known mineral deposits were synthesized from existing government files, the mineral industry, and limited field studies. Experienced scientific judgment was required in discerning deposit types from data of variable age and authenticity. Gravity and aeromagnetic maps were compiled at the 1:1,000,000 scale, and geochemical data were assembled for local areas.

The state was divided into four regions: Brooks Range, Seward Peninsula, central Alaska, and southern Alaska. Excluded were the North Slope, the Aleutian Islands, islands of the Bering Sea, and the "panhandle" or southeastern region of Alaska, because none of the d-2 proposals included these areas. Assessments of the four regions shared a common philosophy, though methodology, format, and details of the individual assessments differed according to the amount and mixture of information available.

Tracts geologically permissive for the occurrence of various types of mineral deposits were delineated in each re-

Table 1. Metalliferous mineral resources of tract 10 shown in Figure 3

MAJOR TYPES OF KNOWN DEPOSITS	(a) Cu(Ag, Au, Zu)—submarine volcanogenic (b) Au—quartz lodes in Orca Group (c) Zn, (Ag, Au, Cu)—breccia cemented by sulfides
SUSPECTED OR SPECULATIVE TYPES OF MINERAL DEPOSITS (INCLUDES MINOR OCCURRENCES)	Cu—magmatic; occurrence of Tertiary gabbro that contains disseminated pyrrhotite and chalcopyrite
GEOLOGIC CONTROL(S) OF MINERAL RESOURCES	Contains the most important submarine volcanogenic deposits of the Prince William Sound area; area underlain by Orca Group (Tertiary) flysch and mafic volcanic rocks and scattered Tertiary felsic plutons (a) Consist of massive and disseminated sulfides, mainly pyrite, pyrrhotite, chalcopyrite, and sphalerite, in Orca Group; generally localized in or near shear zones; related to submarine volcanic processes (b) Small gold-bearing quartz veins, stringers, and veinlets in Orca Group near Tertiary felsic plutons (c) One small known deposit; on brecciated Tertiary pluton; breccia partly cemented by zinc and copper sulfides
PRODUCTION AND RESOURCE INFORMATION	Between 1900 and 1930 14 mines on the volcanogenic deposits produced about 97,000 tonnes (214 million pounds) of copper and subordinate amounts of gold, silver, and zinc; two mines, the Latouche and Ellemar, accounted for more than 96 percent of the production; the few gold mines in the area probably produced a total of not more than 31 kg (1,000 ounces) of gold; resource data are sketchy but the submarine volcanogenic deposits probably represent substantial copper resources; one prospect (Rua Cove) has estimated reserves of at least 1,020,000 tonnes (1,125,000 st) containing 1.25 percent copper
STATUS OF GEOLOGIC INFORMATION	Modern reconnaissance mapping accompanied by geochemical and geophysical studies by U.S. Geological Survey for that part of area within Deward quadrangle; U.S. Geological Survey sponsored mapping and some sampling for remainder of area; topical studies of some volcanogenic deposits by government agencies and industry; recent exploration of some volcanogenic deposits by industry
ADDITIONAL COMMENTS	The resource potential of the submarine volcanogenic deposits dwarfs that of other deposit types in the area; some of the volcanogenic deposits contain large amounts of pyrite, which have been investigated as a possible source of sulfuric acid; the main concentrations of known submarine volcanogenic deposits are in three areas: Knight and Latouche Islands and east of Valdez Arm; despite fairly thorough prospecting these largely vegetation-covered areas are favorable for new discoveries particularly of concealed deposits; the less prospected terrain west of Valdez Arm is also favorable
SUMMARY OF MINERAL RESOURCE POTENTIAL	(a) Over 50 mafic volcanogenic deposits are known; many have been incompletely explored and others probably remain to be found. Estimated number of deposits is only for deposits with tonnages comparable to those used in the grade-tonnage model. (b) Several small-tonnage gold-quartz veins are known; others possible. (c) One small breccia cemented by zinc and copper sulfides is known.
ESTIMATED NUMBER OF DEPOSITS (CHANCE THAT THERE ARE THE NUMBER PRESENTED OR MORE DEPOSITS)	(a) 90% 50% 10% chance that there are 2 4 8 deposits or more
GRADES AND TONNAGES FOR THIS DEPOSIT TYPE	(a) mafic volcanogenic model

SOURCE: Mackevett et al. 1978.

gion. These tracts were outlined on the basis of their known deposits and their potential for undiscovered deposits as inferred on the basis of occurrences in similar geologic settings elsewhere. Where the scale of the maps permitted, the outer limit of delineated tracts was not allowed to extend beyond geologic units that are permissive for the occurrence of a single deposit type, and tracts were further restricted in areal extent if closer examination demonstrated absence of mineralization. In many cases, however, the scale of the maps required that tracts be delineated on the basis of more general geologic units and therefore include more than one deposit type. Areas mantled in thick covers of unfavorable rocks, glaciers, or unconsolidated surficial deposits were not included in the tracts, since even though these areas may contain concealed deposits at depth, the chances for discovering and exploiting such deposits are minimal. Several delineated tracts in the Brooks Range have open ends, because not enough geologic information was available to close the boundaries. Figure 3 shows some delineated tracts in southern Alaska.

There are 144 delineated tracts in all; they are keyed numerically to tables giving succinct descriptions of deposit types; contained metals; geologic settings; geologic, geochemical, and geophysical indications of favorability; extent and adequacy of exploration and geologic knowledge; and, for some deposit types, estimates of the number of deposits and indications of tonnages and grades extrapolated from models based on better-known deposits. In most cases the basic data were insufficient to justify more than qualitative resource estimates. In some instances, however, the data were adequate to permit quantitative estimates of the number of deposits of a specific type that may be present in a given area and their possible grades and tonnages. For approximately 17% of the 501 listings of deposit types, quantitative estimates of the number of deposits were provided, and for about 23%, grade and tonnage models were used. If placer gold deposits and the various vein deposits that typically have low tonnages are excluded, the frequency with which the number of deposits was estimated rises to 32% and the usage of grade-tonnage models to 43%. Much of the information about the geologic factors that affect the economics of the resources thus was provided in tables. Table 1 shows part of one resource table for tract 10, shown in Figure 3.

In tract 10 the mafic volcanogenic deposits are of the greatest interest. This type of deposit occurs as massive and irregular sulfide bodies in or closely associated with mafic (dark, silica poor) volcanic rocks that originally formed in submarine troughs. The deposits are localized in or near these mafic volcanics, which are mixed with sediments and coarse-grained igneous rocks, all of which are mapped as the Orca Group. Two of these deposits in tract 10 produced over 90,000 tons of copper and lesser amounts of gold, silver, and zinc between 1900 and 1930. As indicated in Table 1, we have estimated that the chances are excellent that additional deposits exist in this area.

In a limited but important number of cases, grade-tonnage models were of use for estimating the resources contained in unexplored deposits. Models were based on well-explored deposits for which tonnages of mineralized rock could be calculated by adding past production to estimated reserves or resources. Deposits known to be only incompletely explored were excluded in constructing models. For most types, appropriate deposits from anywhere in the world were used; for two types—porphyry copper and island-arc porphyry copper—only known deposits in Alaska and western Canada were used, because the consistently low grades of partially explored Alaskan deposits of these types seem to fit most closely with these nearby counterparts. For some deposit types it was impossible to find adequate well-explored prototypes on which to base a model.

Grade-tonnage models were constructed for the deposit types

- porphyry copper
- island-arc porphyry copper
- porphyry molybdenum
- skarn copper
- mafic volcanogenic sulfide
- felsic and intermediate volcanogenic sulfide
- nickel and copper sulfides associated with small intrusions
- skarn tungsten

In some cases information on both grades and tonnages of mineralized

rock was not available, but information on the tonnage of contained metal within deposits was. Models based on metal content were developed for the deposit types

- podiform chromite
- mercury
- vein gold

The chromite model was based on deposits in California; the mercury model, on deposits in Nevada, Oregon, Idaho, and Washington; and the gold model, on deposits in British Columbia, Yukon, and Alaska.

Observed frequency distributions of

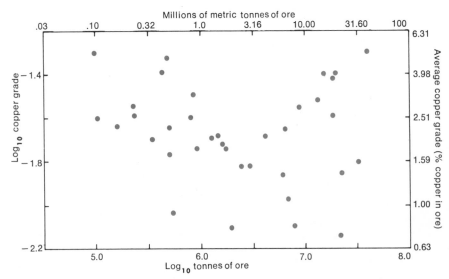

Figure 4. As part of a grade-tonnage model for the mafic volcanogenic sulfide deposit type, 37 well-explored and exploited copper deposits were plotted for tonnage and average grade. The model is used to predict grades and tonnage of copper in unexplored deposits, and correlation coefficients such as these permit calculation of the likelihood of any particular combination of grade and tonnage occurring. Each point represents one deposit.

average grade, tonnage, and, in some cases, contained metal were compared to theoretical lognormal distributions, which generally were found to represent adequately the shapes of the observed distributions. The statistical properties of the lognormal distribution are such that it is relatively easy to estimate proportions of the total number of deposits that should have different grade or tonnage characteristics. Thus, for a particular deposit type we could use the lognormal distribution to estimate that 90% of the deposits would have average grades at least as high as some rather low grade, that 50% of the deposits would have average grades above a somewhat higher grade,

that 10% would be expected to have average grades above a still higher value. This procedure was also used to estimate proportions of deposits that would have various tonnage and contained-metal characteristics. In Figure 4, average copper grades are plotted against their associated tonnages for those deposits used to construct the mafic volcanogenic grade-tonnage model.

Correlation coefficients among observed grades and tonnages were calculated in order to determine the degree of linear association with the logarithmic data. Correlation coefficients, along with the estimated probability of deposits having certain grade or tonnage characteristics, allow calculation of the likelihood that any combination of grades and tonnages occurs, given that such deposits exist (see, for example, Table 2).

Mafic volcanogenic deposits are mined for copper and frequently for zinc, gold, and silver. The range of possible tonnages is quite large; the variability of copper grades is considerably less. Because grades are independent of tonnages, the chance of a particular combination of grades and tonnages occurring is easily calculated as the product of the indi-

Table 2. Grade and tonnage model for mafic volcanogenic sulfide deposits (metric units)[†]

Variable (units)	Number of deposits used to develop model	Correlation coefficients (with tonnage)	90% of deposits have at least	50% of deposits have at least	10% of deposits have at least
Tonnage (millions of tonnes)	37		0.24	2.3	22.0
Average copper grade (%)	37	−0.13*	1.1	2.2	4.1
Average zinc grade excluding deposits without reported grades (%)	19	0.03*	0.3	1.3	5.5

SOURCE: Mackevett et al. 1978.

* not significant

[†] There are also significant local concentrations of gold, but average grade has not been determined.

vidual probabilities. Average copper grades are relatively high, and the presence of gold, silver, and zinc help to make many of these deposits economic to mine even in remote locations.

Probabilistic estimates of the number of deposits by type were made for each delineated area where such estimates were warranted by available information. Estimates were presented in a probabilistic form in order to show the degree of certainty concerning the number of deposits that might occur. The estimates were made subjectively; that is, rules whereby estimates were constructed cannot be exactly stated. A variety of pertinent considerations were integrated, including degrees of geologic, geochemical, and geophysical favorability; extent and adequacy of exploration and geologic knowledge; and extent of exploration of deposits already recognized.

Typically, the estimated number of deposits for which we can say there is a 90% chance of occurrence is closely related to the number of known deposits of that type in the delineated area. The reliability of the estimate depends largely on the degree to which the "known" deposits have been explored; that is, not all "known" deposits will necessarily be found to belong to a particular deposit type when the deposits are completely explored. Consistent rules for estimating the number of deposits with a 50% chance of occurrence are more difficult to define, because, in

some situations, the estimate was based on the number of prospects and completeness of exploration; in others it may have been based on the proportion of the area mantled by unconsolidated surficial deposits or ice. The estimates associated with a 10% chance can generally be considered speculative, in that they are limited by the number of deposits that could fit in an area, and may be based on a variety of indicators, such as the number of related deposits, altered areas, or geochemical anomalies. In a few cases, estimates of the number of deposits in tracts were directly based on the number of deposits in well-explored geologically similar environments. In general, estimates referred only to deposits with tonnages and grades comparable to those used in the grade-tonnage models, though a few were made for deposits that lack associated grade-tonnage models.

In some cases, explored deposits were known to be much lower in tonnage or grade than the deposits used in the grade-tonnage models and therefore were excluded from the estimated number of significant deposits. For example, in tract 10, described in Table 1, over 50 mafic volcanogenic deposits are known, but the majority of the well-explored deposits are so small in size that their tonnage is not estimated. Similar, unestimated small deposits are known near many of the deposits used to construct the mafic volcanogenic grade-tonnage model. To include the known very small deposits in the estimated number of deposits for tract 10 when it was not

possible to include similar deposits in the grade-tonnage model would be misleading.

Applications elsewhere

In order to provide information for decisions on Alaskan land, an attempt was made to present information concerning the geologic properties that can affect the economics of exploitation in a disaggregate manner. A variety of economic and technologic assumptions could be tested and the effects of the assumptions on the potential for exploitation of the resources could be evaluated. As far as possible the results were presented in a probabilistic manner, to make explicit the uncertainty and variability that exists in such estimates.

The direct usefulness of these assessments can only be determined by the effect they have on the interested parties in determining the boundaries of certain parcels of Alaskan land. These assessments may, however, also have a large indirect effect on future mineral-resource assessments elsewhere.

In adapting the general procedures used in our assessments for other regional assessment programs, the suitability of the final product—the form in which the accumulated information is presented—must be considered. In our assessments for Alaska, tracts of land were delineated by the mineral deposit types their geology allowed; the basis for determining the favorability of each tract was indicated in tables; and where possible, numbers of deposits were estimated and appropriate grade-tonnage models were provided. The disaggregated nature of the Alaskan assessments allows identification of specific problems that require further research, and it facilitates the updating of the assessments.

It might appear that one of the multivariate methods would be suited for such assessments, but we found none that could provide unbiased estimates combining the continuous, discrete, and spatial variables present in the Alaskan assessments. The use of subjective estimates for the number of deposits may be criticized, but as Harris (1977) has noted, subjective methods may have some advantages over multivariate methods in that "the information on basic geoscience to which the field geologist has been exposed may exceed by a large margin the information available to the public for the construction of a multivariate model." In the fields of medicine and meteorology, Hogarth (1975) reported that knowledgeable experts could increase the accuracy of predictions beyond that which could be achieved by the best available statistical model (no comparable study has been performed in mineral resources). We do, however, recognize the value of corroborative statistical models to ensure that the estimates are reproducible and to allow others to examine the methods critically.

In applying our method of mineral assessment in better-explored areas, it may be necessary to adjust the grade-tonnage models to reflect the possibility that the largest deposits might have been discovered first. In Alaska, it is believed, a complete range of deposits—from large-tonnage to small—remains to be discovered and explored; in better-explored areas, all that might remain are the smaller deposits. The possibility of regional differences in grades or tonnages within a deposit type must also be considered.

In some cases construction of the grade-tonnage models can be difficult because of the lack of agreement among economic geologists about the basis for classifying deposits. Deposits of more than one type can inadvertently be placed in the same model. The need for consistent methods of classifying deposits based on their geologic characteristics became evident as we constructed our grade-tonnage model.

Grade-tonnage models are based on two kinds of sources: those that represent the economic portion of deposits and therefore have relatively high cutoff grades, and those that represent the total endowment of mineralized rock in a deposit type and therefore have relatively low cutoff grades. Integrating these two kinds of sources means that tonnages and average grades in the models are associated with cutoff grades which are typically lower than those it is now economic to mine but higher than the endowment of the deposits. It is possible, for many deposits, to increase the average grade and decrease the tonnage by raising the cutoff grade; therefore some deposits that appear to have grades too low to mine economically, based on the grade-tonnage model, may in fact have parts that are economic.

Ideally, grade-tonnage models would be constructed for all the mineralized rock in deposits at a fixed low cutoff grade, and another model would be built to show how tonnage and average grade vary with cutoff grade. Unfortunately, information necessary to determine this relationship is closely held by individual mining companies and is not generally available. Such information would increase the value not only of the type of assessment considered here but also of most mineral-resource assessments to which more than one economic or technologic condition may be applied.

Many of the regional mineral-resource assessments being considered or conducted now are for tracts of land that are small in area. The direct application of the methods used in the Alaska assessments is of questionable value because of the problem of making probabilistic estimates of the number of deposits in small tracts. Subjectively estimating a single event that has a very low probability of occurrence is a treacherous process.

Many of the problems discussed in this paper have existed in previous mineral-resource assessments but have not been explicitly recognized. The Alaskan assessments may be considered a useful first step in the improvement of mineral-resource assessments.

References

DeYoung, J. H., Jr. 1976. Recent changes in Canadian tax laws affecting the mineral industries. In *Comparative Study of Canadian–United States Resource Programs*, prepared for the Committee on Appropriations, U.S. Senate, chap. C. Reston, VA: USGS.

Divi, S. R., R. I. Thorpe, and J. M. Franklin. In press. Application of discriminant analysis to evaluate compositional controls of stratiform massive sulphide deposits in Canada. *J. Math. Geol.*

Griffiths, J. C. 1978. Mineral resource assessment using the unit regional value concept. *J. Math. Geol.* 10:441–72.

Harris, D. P. 1977. *Mineral Endowment, Resources and Potential Supply: Theory, Methods for Appraisal, and Case Studies.* Tucson: Minresco.

Hogarth, R. M. 1975. Cognitive processes and the assessment of subjective probability distributions. *Am. Stat. Assoc. J.* 70:271–89.

MacKevett, E. M., Jr., D. A. Singer, and C. D. Holloway. 1978. *Maps and tables describing metalliferous mineral resource potential of southern Alaska.* USGS Open-File Rept. 78-1E.

Menzie, W. D., and D. A. Singer. In press. Some quantitative properties of mineral deposits. In *Small Scale Mining of the World*, ed. R. F. Meyer. Pergamon.

Richter, D. H., D. A. Singer, and D. P. Cox. 1975. Mineral resources map of the Nabesna quadrangle, Alaska. USGS Misc. Field Studies Map MF-655K.

Singer, D. A. 1975. Mineral resource models and the Alaskan Mineral Resource Assessment Program. In *Mineral Materials Modeling*, ed. W. A. Vogely, pp. 370–82. Baltimore: Resources for the Future.

U.S. Dept. of Interior. 1974. *New Mineral Resource Terminology Adopted* (news release, 15 April).

James H. Zumberge

Mineral Resources and Geopolitics in Antarctica

The physical obstacles to exploitation of mineral resources in Antarctica are currently prohibitive, but complex political issues will be raised if such exploitation becomes profitable

The continuous world search for mineral and fuel resources has led to the discovery of recoverable ores and fossil fuels on all inhabited continents and their continental shelves. The only continental platform on which mineral exploration has not occurred is Antarctica. This is not because geologists believe it to be barren of useful mineral resources. On the contrary: we have learned enough about the geology of the Antarctic Continent during the twentieth century to know that its rocks are no different from rocks on other continents. Moreover, the geologic history of Antarctica is closely related to the geological evolution of South America, Australia, and southern Africa, all of which contain mineral deposits of considerable value. Therefore, from the geological point of view, there is every reason to assume that Antarctica contains its share of mineral concentrations.

There are two basic sets of reasons why mineral exploration has not been attempted in Antarctica. The first includes the meager geophysical and geological data base, the technological problems, the extreme environmental hazards, and the long distances from

James H. Zumberge is the chairman of the Committee on International Polar Affairs of the Polar Research Board of NAS, and the U.S. delegate to the Scientific Committee on Antarctic Research of the International Council of Scientific Unions. Since 1974 he has been a member of the National Science Board. He is a geologist by training (Ph.D., U. Minn., 1950) and is the coauthor (with C. A. Nelson) of Elements of Geology (1973) and Elements of Physical Geology (1976). Currently he is Professor of Geology and President of Southern Methodist University. Address: Southern Methodist University, Dallas, TX 75275.

civilization, which make the search for mineral resources in Antarctica economically unattractive to the extractive industries (Fig. 1). The second is the political circumstances currently controlling the international scene on the continent, which are very complex. Even though territorial claims on Antarctic lands have been made by several countries, no sovereignty is exercised by these states because all of Antarctica and islands south of lat. 60° S are controlled by the terms of the Antarctic Treaty, which has been in force since 23 June 1961.

This paper will deal with these two major factors bearing on the mineral potential of Antarctica. First, I will consider the geographic setting of the continent and its surrounding ocean and identify the areas most likely to attract future exploration for mineral resources. Second, I will examine the unique international political circumstances that are critical in any assessment of the mineral potential of Antarctica.

A continental profile

The physical environment of Antarctica is described in a series of atlases published by the American Geographical Society (Craddock et al. 1970; Gordon and Goldberg 1970; Heezen and Tharp 1972; Goodell et al. 1973). The paragraphs that follow draw freely from these sources and from an unpublished report prepared by the Institute of Polar Studies at Ohio State University for the U.S. State Department (Elliot 1977).

Antarctica is a circumpolar continent of 13.5 million km² (Fig. 2), which is about 9% of the land surface of the earth. It is surrounded by the South-

ern Ocean, whose area is about 36 million km²—about 10% of the world's oceans. The Antarctic ice sheet covers 98% of the continent, leaving only about 260,000 km² of nonglacial terrain on which direct geological observations are possible. The ice sheet averages about 2,300 m in thickness, and its maximum known thickness is 4,000 m. The volume of glacier ice in Antarctica is sufficient that, if melted, the water released would raise sea level about 60 m. Shackleton and Kennett (1974) believe that the glacial history of Antarctica spans tens of millions of years.

The continental ice flows from the interior toward the coastal regions, where it discharges directly to the ocean in the form of icebergs. Most of the icebergs are calved from ice shelves: the Ross Ice Shelf and the Ronne Ice Shelf are the largest of these. The ice-free areas of Antarctica are concentrated in the Transantarctic Mountains, the Ellsworth Mountains, and the Antarctic Peninsula, although some are scattered in small tracts of relatively flat land along the coastal regions.

Geographically, the continent can be divided into East Antarctica, West Antarctica, and the Antarctic Peninsula. The Transantarctic Mountains of East Antarctica, rising above 4,000 m, are the highest range, although the highest single peak on the

Figure 1. Only a few rock outcroppings are visible above the ice sheet that covers 98% of Antarctica. Many are *nunataks*—peaks completely surrounded by glacial ice—such as these three in the Dufek Massif. The difficulties of exploration and sampling are obvious. (Photo courtesy of the U.S. Navy.)

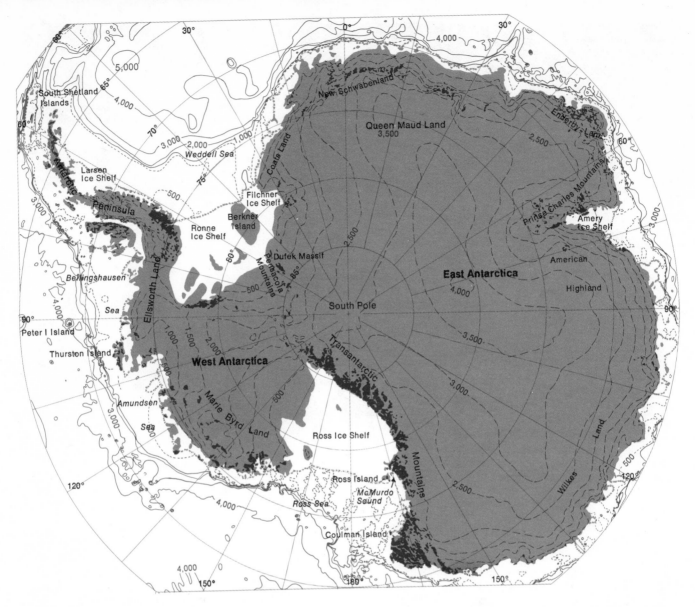

Bathymetric contours (meters)

Ice-sheet surface contours (meters)

Ice shelves

Areas of exposed rock

Figure 2. Antarctica is a circumpolar continent surrounded by the Southern Ocean. Its existence was inferred by the ancient Greeks, but it was not discovered until 1820. Its continental proportions were not fully known until the early 20th century, and not until satellite imagery became available were the details of its coastline and interior mountains determined accurately. (Adapted, by permission, from American Geographical Society 1970.)

continent, Vinson Massif (5,140 m), occurs in the Ellsworth Mountains of West Antarctica. Elsewhere in West Antarctica and on the Antarctic Peninsula, the mountains are rugged, but their summits generally range in elevation from 2,000 m to 3,500 m.

Temperatures on the Antarctic Continent are very severe. The mean annual temperature on the polar plateau of East Antarctica is −50°C and in the coastal regions is about −15°C. The temperatures of the coldest months on the plateau are not uncommonly below −70°C, while in the coastal areas they are in the −20°C to −30°C range. An exception is the north and west sides of the Antarctic Peninsula, where temperatures are about −10°C during the winter. In the warmest summer months the mean temperature is about −35°C on the plateau and around 0°C on the coasts and on the Antarctic Peninsula.

Antarctica is aptly described as the windiest continent on earth. Winds of high intensity and long duration are commonplace in Antarctic coastal regions. They are due to the strong katabatic—mountain—winds that flow from the interior toward the coasts. At high velocities, these winds produce the blizzard conditions that have plagued explorers in Antarctica since the earliest expeditions on the continent. On the Antarctic Peninsula's northern and western sides the winds are less intense and blizzard conditions less frequent than elsewhere around the continent's rim.

The Southern Ocean separates the Antarctic Continent from the civilized areas of the Southern Hemisphere, and consists of qualitatively distinguishable parts of the southern extremities of the Atlantic, Pacific, and Indian oceans (Dunbar 1977).

The nearest continent to Antarctica is South America, whose southern tip lies 880 km across the Drake Passage from the Antarctic Peninsula. The southern tip of New Zealand is 2,240 km from the Antarctic coast, and Tasmania and South Africa are more distant.

The continental shelf is relatively deep and narrow. Its seaward edge is about 900 m deep and its area is about 4 million km². Sea ice overlies all of the shelf for nine to ten months of the year. Much of it becomes broken into pack ice during the austral summer and moves northward, leaving open sea above the shelf. Sometimes, however, the presence of several years' accumulation of ice in the Amundsen, Bellingshausen, and Weddell seas makes ship passage impossible and ice-breaker movement difficult.

A clockwise (easterly) circumpolar current called the West Wind Drift flows around the Antarctic Continent (Gordon and Goldberg 1970; Gordon 1971). This is the major element in the surface circulation of the waters of the Southern Ocean. Between the West Wind Drift and the Antarctic coast is a counterclockwise (westerly) current called the East Wind Drift. This current is broken into gyres where it impinges against deep embayments such as the Weddell, Ross, and Bellingshausen seas.

In addition to these two major surface currents, considerable vertical movement of water masses is caused by differences in salinities and temperatures of the cold, low-saline coastal water and the warm, highly saline water of lower latitudes. The zone where these two masses meet is known as the Antarctic Convergence, or Polar Front Zone (Gordon 1971). The Antarctic Convergence generally lies between lats. 50° and 60° S and marks the surface boundary of the Southern Ocean (Gordon and Goldberg 1970).

Antarctic geology

Knowledge of Antarctic geology is rudimentary and unrefined in comparison with the geologic information available for the other continents. Serious and systematic geological exploration of Antarctica began only after the International Geophysical Year, 1957–58 (Adie 1964, 1972;

Craddock, in press). Geological field studies are hampered by the paucity of rock outcrops, the general lack of topographic base maps on which geological features are normally plotted, the hostile environment, and the extremely difficult logistical problems. In spite of this, geologic maps of Antarctica have been published in the United States (Craddock 1972a) and the Soviet Union (Ravich and Grikurov 1978).

The Antarctic continental platform consists of two major geological provinces, East Antarctica and West Antarctica. The former is a Precambrian shield of igneous and metamorphic rocks overlain in part by relatively undeformed sedimentary strata. The latter province is characterized by younger, highly deformed and metamorphosed strata that are intruded by igneous rocks. In addition, volcanic deposits are found, and some volcanic activity persists today. None of the rocks nor any of the geological relationships that occur in Antarctica is unique. Antarctica is a "normal" continent, geologically speaking, with one exception: 98% of its surface is covered by glacier ice. Otherwise, there is no reason to assume that the continent, because of its isolation, is peculiar with respect to its geologic history.

In 1937, Alexander Du Toit published *Our Wandering Continents*, a treatise based on the postulate of Wegener (1929) that continents have moved with respect to one another over geological time. Antarctica was regarded by Du Toit as the key to his hypothesis of continental drift. Subsequent geological studies—which led to the theory of plate tectonics—have proved him correct, because the major geologic components of Antarctica can be linked to other continents and landmasses of the Southern Hemisphere (for a detailed summary of these intercontinental relationships see Elliot 1975). Figure 3 is a reconstruction of the supercontinent of Gondwanaland prior to its breakup. The geological connections between Antarctica and South America, South Africa, and Australia are apparent. Future geologic studies in Antarctica will permit more precision in the reconstruction of Gondwanaland. However, the major elements in this crustal jigsaw puzzle have been established with reasonable confidence.

The economic geologist divides mineral resources—all the known and unknown mineral deposits on earth—into four categories (Brobst and Pratt 1973). The known or identified resources are of two kinds, *reserves* and *conditional resources*. Reserves are identified mineral deposits that are economically recoverable under current economic conditions and extraction technology. Conditional resources are those that have been identified but are not recoverable under current economic conditions and extraction technology. The unknown or undiscovered mineral resources are divided into *hypothetical resources*, undiscovered deposits in known mining districts, and *speculative resources*, undiscovered deposits outside the boundaries of known mining districts.

In Antarctica, one can speak only in terms of speculative mineral resources. The continent therefore remains an unknown quantity with respect to mineral resources; and that is precisely the reason why nations are interested in knowing more about its mineral potential.

Areas for exploration

The first step in attempting to assess the mineral-resource potential of Antarctica is to identify areas of highest mineral prospectivity. The paucity of detailed geological information for Antarctica makes this task difficult, but at the request of the Antarctic Treaty nations, a group of specialists of the Scientific Committee on Antarctic Research (SCAR) of the International Council of Scientific Unions identified several geographic areas on the Antarctic Continent and its surrounding continental shelf as possible areas for future exploration for mineral resources (Table 1) (Zumberge 1977).

Figure 3 shows the Antarctic Peninsula to be an extension of the Andean Orogen of South America. (An orogen is a belt of folded and metamorphosed sediments, commonly intruded by igneous rocks, that at one time in its geologic history was mountainous). The copper deposits of Chile and Peru have led some geologists to speculate about the prospectivity of the Antarctic Peninsula for mineral deposits of a similar geologic setting (Runnels 1970; Wright and Williams 1974). In fact, copper mineralization that has been studied in the Antarctic Peninsula is similar in many respects to typical copper deposits of the Andes and elsewhere (Rowley et al. 1975). These known mineralized zones are too low in grade to justify a classification higher than speculative, but they are indications that the Antarctic Peninsula could be a target area for future mineral exploration.

The Dufek Massif is a layered intrusion of basic igneous rock called gabbro. The geological name for a rock mass of this kind is a *stratiform igneous complex*. The Dufek Massif forms a major part of the Pensacola Mountains, with an areal extent of 34,000 km^2 and a thickness on the order of 7 km (Ford and Boyd 1968). Most of this formation lies beneath glacier ice, but parts are exposed in the many *nunataks* (mountain peaks surrounded by ice) in the northern part of the Pensacola Mountains of East Antarctica (see Fig. 1).

The Dufek Massif is particularly attractive as a target for mineral exploration because it is an igneous stratiform complex similar to others around the world from which significant amounts of copper, chromium, nickel, and platinum are mined. The Sudbury nickel deposits of Ontario and the minerals of the Bushveld Complex of South Africa (see Hunter 1978) occur in classic examples of igneous stratiform structures.

The base of an igneous stratiform body is the part of the rock mass where concentrated mineralization is most likely to occur, but the lower part of the Dufek Massif is not exposed in surface outcrops. Diamond-core drilling in selected parts of the Dufek Massif therefore seems to be the next logical step in determining the mineral potential of that body.

The rocks of the Transantarctic Mountains range from Precambrian crystalline masses to Paleozoic and Mesozoic sediments (Craddock 1972b). Among these are extensive coal beds of Permian age (Barrett et al. 1972). But even though some of these beds are of mineable thickness, the fact that coal is bulky and low grade means that their occurrence in mountainous regions at long distances from coastal areas limits the likelihood of their exploitation.

The main reason for assuming that other mineral resources might occur in the Transantarctic Mountains is that there are Gondwanaland relationships between them and the mineralized rocks of Tasmania and eastern Australia (see Fig. 3). The former contains deposits of commercial-grade lead, zinc, copper, tin, and tungsten; the latter contains economical ores of copper, lead, zinc, silver, and gold (Wright and Williams 1974). Transportation to the coast from these mountains would again be an expensive proposition, but if a high-grade ore deposit were discovered, the obstacles might be less forbidding. So far, however, only sporadic occurrences of minerals containing those elements are known from the presumed Antarctic extension of these mineralized zones in the Transantarctic Mountains (Stewart 1964).

The Precambrian rocks of East Antarctica contain several indications of iron deposits. The Prince Charles Mountains southwest of the Amery Ice Shelf contain a banded iron formation (jaspilite), and magnetite-bearing rocks in Queen Maud Land have been reported by Russian geologists (Ravich and Solov'ev 1969). Elsewhere along the coast from the east side of the Amery Ice Shelf to about long. 90° E, the occurrence of boulders of jaspilite in morainic deposits indicates a widespread occurrence of this iron-bearing rock beneath the ice farther inland (Wright

Table 1. Areas in Antarctica where commercially mineable resources might be found

	Possible mineral resource
Antarctic Peninsula	copper, molybdenum
Dufek Massif	chromium, platinum, copper, nickel
Transantarctic Mountains	copper, lead, zinc, silver, tin, gold
Prince Charles Mountains	iron
Continental shelf of West Antarctica and off Amery Ice Shelf in East Antarctica	petroleum

and Williams 1974). This Precambrian banded iron formation may be similar to rocks of the Lake Superior Iron District, but its inferred occurrence beneath a thick cover of glacier ice makes its potential for exploitation highly speculative, even in the long-term future.

The continental shelf is generally less than 200 km wide around East Antarctica, while in some stretches of West Antarctica it is 300–400 km wide. The average depth of the shelf is 500 m, which is more than twice as deep as continental shelves elsewhere in the world. This greater-than-normal depth is probably due to isostatic depressions of the crust by the Antarctic ice sheet (Kapitsa 1967; Denton et al. 1971).

Except for that part of the continental shelf beneath the Ross Sea and the western side of the Antarctic Peninsula, very little geophysical data regarding geological structures and thicknesses of sediments is known. Much of the geophysical data obtained from cruises of the *Eltanin* under the auspices of the National Science Foundation between 1962 and 1972 (Antarctic Journal 1973) are confined to the deeper waters of the Southern Ocean, where pack-ice conditions are less severe than in the shallower waters overlying the continental shelf. Hayes (1976, p. B-9) points out that the continental shelf of Antarctica is "less well surveyed than any portions of the world's ocean with the exception of the Arctic Ocean."

In spite of the paucity of geophysical information, the SCAR group of specialists suggested that the continental shelf on either side of the Antarctic Peninsula and beneath the Ross and Weddell Seas could attract exploration activities for hydrocarbons sometime in the future (Zumberge 1977). To be sure, the assignment of these areas of the shelf to a relatively high prospectivity category is based more on inference than fact. Nevertheless, the circumstantial evidence is fairly impressive.

Exploration for off-shore hydrocarbons is active off the coasts of southern Argentina (Zambrano 1975), South Africa (Biro 1976), Australia (Durkee 1976), and New Zealand (Katz 1976). Discoveries have resulted in producing oil fields off Ar-

Gondwanaland

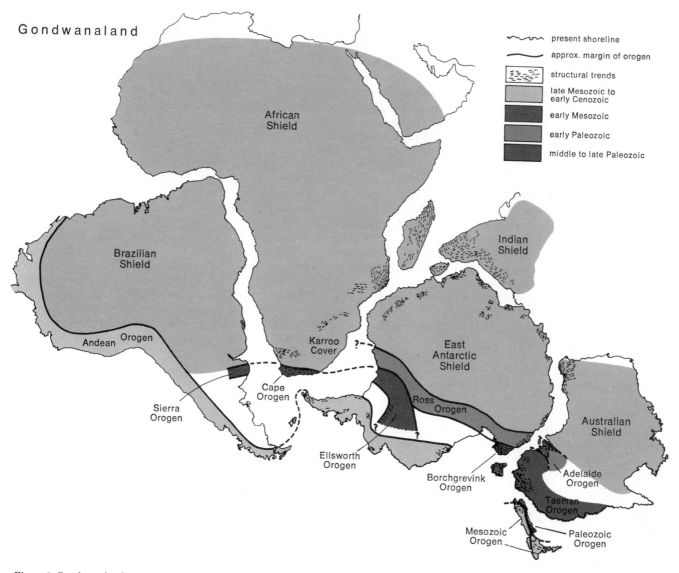

present shoreline

approx. margin of orogen

structural trends

late Mesozoic to
early Cenozoic

early Mesozoic

early Paleozoic

middle to late Paleozoic

African
Shield

Brazilian
Shield

Indian
Shield

Andean Orogen

Karroo
Cover

East
Antarctic
Shield

Sierra
Orogen

Cape
Orogen

Ross Orogen

Australian
Shield

Ellsworth
Orogen

Borchgrevink
Orogen

Adelaide
Orogen

Tasman
Orogen

Mesozoic
Orogen

Paleozoic
Orogen

Figure 3. Gondwanaland was a giant continent of the early Mesozoic, when what are now South America, Antarctica, Africa, India, and Australia were linked as one landmass. The assumed relationships between geological formations in Antarctica and those already known in the other continents of the Southern Hemisphere lead geologists to believe that there are mineral deposits in Antarctica comparable to those currently being exploited in its former neighbors. (Adapted, by permission, from Craddock et al. 1970.)

gentina, Australia, and New Zealand. Even though these off-shore deposits of hydrocarbons are far distant from Antarctica, the juxtaposition of the landmasses containing them during the existence of Gondwanaland at least suggests that the Antarctic continental shelf may not be barren of similar deposits.

The Ross Sea sector of the continental shelf is perhaps the most promising. It is by far the best known in terms of its gross geological structure and the thicknesses of its sedimentary strata (Hayes 1976). Moreover, the continental shelf beneath the Ross Sea was contiguous with the Kingfish, Barracouta, and Halibut oil fields (Franklin and Clifton 1971) between

Tasmania and Australia at the time of Gondwanaland.

Interest in the possible occurrence of off-shore hydrocarbons beneath the Ross Sea was stimulated by the 1972 discovery of traces of methane, ethane, and ethylene in three of four holes drilled by the *Glomar Challenger* in connection with the NSF-sponsored Deep Sea Drilling Project in the Southern Ocean (McIver 1975). Methane is a common occurrence in deep-sea cores and does not necessarily indicate the nearby presence of crude oil. Ethylene, on the other hand, often occurs with petroleum (Heirtzler 1975). It must be noted that the Ross Sea sediments in which the hydrocarbon gases were found are

of Miocene age and were deposited long after the separation of Antarctica from Australia. Though it is unlikely it is not impossible, however, that the gases migrated from older sediments at greater depth, which could be of the same geologic age as those from which hydrocarbons are now being produced off the coast of southern Australia.

The possibility of off-shore hydrocarbons being extracted from Antarctica caused a flurry of reports in the popular press and trade journals several years ago. The *Wall Street Journal* (Spivak 1974) attributed to the U.S. Geological Survey a report that 45 billion barrels of oil and 115 trillion cubic feet of natural gas could be recovered from the continental shelf of West Antarctica. There is some question as to the source of this estimate, because it was not included in U.S.G.S. Circular 705, in which the mineral resource potential of Antarctica was assessed in some detail (Wright and Williams 1974).

The United States delegation to the special preparatory meeting of the treaty nations was more conservative in its estimate. It said that "it appears that the Antarctic continental shelf could contain potentially recoverable oil in the order of magnitude of tens of billions of barrels" (U.S. Department of State, unpubl.). For the sake of comparison, the North Slope of Alaska is believed to contain 8 billion barrels of oil (some sources give 9.6 billion barrels) and 26 trillion cubic feet of gas (Panitch 1977).

Prospecting for petroleum on the Antarctic continental shelf presents unique problems. Much of the shelf is covered with sea ice for most of the year, and some parts always contain moving pack ice that makes the maneuvering of exploratory vessels difficult and the use of geophysical sensing arrays towed by such ships almost impossible. The Ross Sea is ice-free during part of the austral summer, but the Weddell Sea has heavy pack ice year-round.

Icebergs of enormous size and draft present another hazard to exploration, especially exploratory drilling. Icebergs with dimensions of up to 70 × 100 km have been documented (Swithinbank 1969). Smaller icebergs can be pushed aside by icebreakers and tugs, but the larger ones present obvious hazards to drill ships. Moreover, the larger icebergs are thick enough to scour the seabed in water depths up to 200 m. Production wells in iceberg-infested waters must therefore be constructed in a manner that leaves no structures protruding above the seafloor. Icebergs also present a hazard to tankers that would transport crude oil from production wells to refineries. Since the first sighting of icebergs in the Southern Ocean by Captain Cook in 1773, the occurrence of icebergs has been recorded as far north as lat. 35° S in the Atlantic Ocean, and lat. 40°–45° S in the Pacific and Indian oceans (Nazarov 1962).

Thus, no matter how geologically attractive the continental shelves of West Antarctica might appear, the combined hazards of severe pack ice, prolonged storms of high intensity, and bottom-scouring icebergs present technological difficulties of immense proportions for anyone contemplating petroleum exploration and, ultimately, extraction in the high latitudes of the Southern Ocean.

In the deep ocean around Antarctica are found manganese nodules and encrustations that have attracted some attention (Goodell 1973). There are, however, vast areas of the floor of the Pacific Ocean that are more amenable to exploitation than the hostile regions of the Southern Ocean. Moreover, there is some evidence that the copper, cobalt, and nickel content of the ferromanganese nodules of the Pacific Ocean is latitude-dependent (Horn et al., unpubl.) and that amounts of these minerals are significantly less in Antarctic concretions than in those closer to the equator. Nevertheless, the origin of ferromanganese nodules is related to seafloor volcanism associated with the mid-ocean ridges, and it is not impossible that further exploration in Antarctic waters near such features would uncover deposits of more desirable mineral content. On the basis of present knowledge, however, it seems very unlikely that manganese nodules in the Southern Ocean will be a prime target for exploration or exploitation in the foreseeable future.

The mineral resource of immediate economic interest in the Antarctic oceans is the ice in the icebergs themselves. The idea of transporting icebergs to water-deficient parts of the world was examined seriously in 1973 by Weeks and Campbell, and Hult and Ostrander, but in 1977 this once bizarre idea became popularly respectable when the First International Conference on Iceberg Utilization was held at Iowa State University (Husseiny 1978). The conference was sponsored by several organizations, including NSF and the King Faisal Foundation of Saudi Arabia. Papers presented at the conference covered subjects ranging from the selection of suitable icebergs by remote sensing techniques to their protection and transport across the open sea and ultimate utilization at their destinations.

A paper by Weeks and Mellor (1977) concluded that icebergs on the order of 2 km long, 0.5 km wide, and 200 m thick would be most suitable for towing. They also presented calculations to show that no unprotected iceberg, regardless of size, would survive a journey from Antarctica to low latitudes, because of excessive melting. They do believe that an unprotected tow at a speed of 1 knot could deliver a payload of ice to southern Australia. Hult and Ostrander (1973) are of the opinion that the costs connected with iceberg transit to southern California are competitive with desalinization.

Prince Mohammed Al-Faisal of Saudi Arabia is the impetus behind further research and development on the feasibility of using icebergs as a source of fresh water, and he has organized a company to accomplish this end. But until the first berg is delivered to a port in Australia or elsewhere, it will be impossible to assess the future potential of iceberg harvesting. Strangely enough, however, of all the potential mineral resources in Antarctica, icebergs hold the largest promise for successful exploitation in the near-term future.

The political scene

The foregoing account of Antarctica's mineral-resource potential deals only with the question of where additional geological data are needed before the next level of resource evaluation can be attempted. But even if a mineral resource of significant economic value should be discovered, the current political situation in Antarctica casts considerable doubt as to utilization of the resource.

All lands south of lat. 60° S are controlled by the Antarctic Treaty of 1961 (SCAR Manual 1972a). The treaty emerged shortly after the IGY because the twelve nations that conducted extensive research there desired to continue these activities under the same conditions of cooperation and freedom of access that prevailed in 1957–58. The key to maintaining the political status quo of the IGY was the agreement among the twelve contracting parties to suspend all further territorial claims and to hold in abeyance all existing claims. Seven nations—Argentina, Australia, Chile, France, New Zealand, Norway, and the United Kingdom—had filed claims prior to the

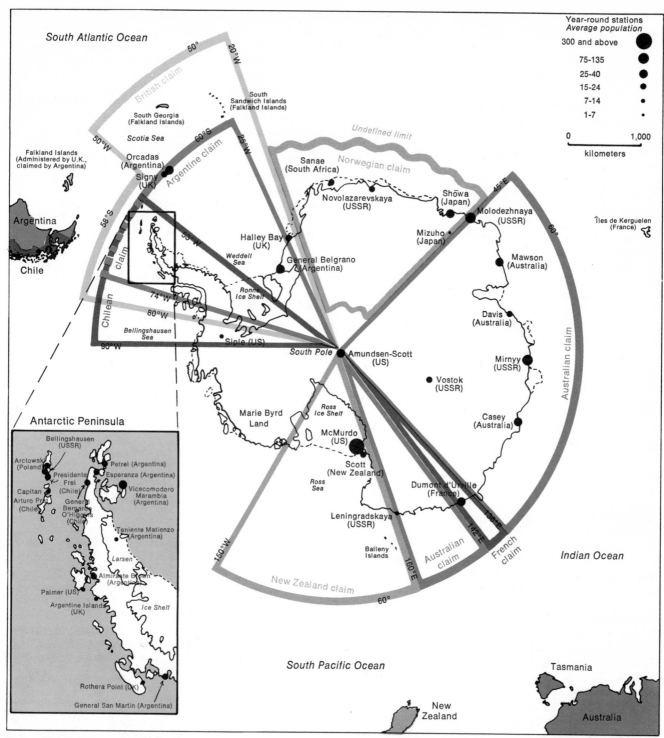

Figure 4. Territorial claims in Antarctica made prior to the Antarctic Treaty are held in abeyance by the treaty, which allows any nation to establish year-round scientific research stations anywhere on the continent. The unclaimed sector includes Marie Byrd Land, the principal area of U.S. exploration prior to the International Geophysical Year. The United States and the Soviet Union have made no territorial claims and they recognize no claims of other nations. (After CIA 1978 and SCAR Bulletin 1977.)

IGY, but the other five signators—Belgium, Japan, the Soviet Union, South Africa, and the United States—had made no claims, nor did they recognize the claims of the other seven. Three of the claimant nations—Argentina, Chile, and Great Britain—have overlapping claims that include all of the Antarctic Peninsula (Fig. 4). Poland acceded to the treaty in 1961 but did not become a full participant until 1977.

The treaty is intended to run indefinitely, but in 1991, any of the treaty nations may call for a general conference of the signatory nations to review the operation of the treaty. If amendments proposed at that conference and agreed to by a *majority* of the contracting parties are not ratified by one or more of the participating nations within two years of the conference date, those nations are free to withdraw from the treaty and will no longer be obliged to abide by its terms.

Actually, a treaty nation may withdraw from the treaty any time before 1991 if its government fails to ratify an amendment to the treaty within two years after an amendment has been agreed to by *unanimous* consent of the contracting parties. The requirement of unanimity for any amendment or changes in the terms of the treaty has worked effectively in keeping the treaty intact, but no such guarantee can be assumed after 1991, because any changes or additions proposed then will require only a majority vote, thereby leaving the door open for withdrawal by nations that are opposed to such changes or additions.

The treaty makes explicit reference to the status of territorial claims in Antarctica in section 2 of Article IV:

No acts or activities taking place while the present Treaty is in force shall constitute a basis for asserting, supporting or denying a claim to territorial sovereignty in Antarctica or create any rights of sovereignty in Antarctica. No new claims or enlargement of an existing claim to territorial sovereignty in Antarctica shall be asserted while the present Treaty is in force. [SCAR Manual 1972a]

Other provisions of the treaty provide for the free exchange of scientific personnel and information, prohibition of military activities other than logistic support of expeditions, a ban of nuclear explosions and nuclear waste disposal, complete freedom of access at any time to all parts of Antarctica, and the right of inspection by observers of the treaty nations of all stations, equipment, and installations in Antarctica.

The treaty provides for the preservation of living resources; and several lengthy recommendations on measures to preserve Antarctic flora and fauna, the designation of specially protected areas, and conservation of Antarctic seals have emerged from many of the consultative meetings (SCAR Manual 1972b) of the treaty nations. The treaty does not, however, speak to the question of mineral resources.

For the first eleven years of the life of the treaty the question of mineral resources never surfaced officially.

When it did, in 1972, the treaty nations took two years to get it on the agenda and another three years to agree to face the difficult legal and political aspects of the mineral-resource issue. This is not to criticize the treaty nations, for they truly are treading new ground on the international scene. The need to go slowly in resolving these complex issues is understandable. Antarctica is an entire continent where land ownership does not exist, where no government or private party can issue licenses, lease or sell mineral rights, or receive royalty payments. Until such matters are resolved, and an agreement is reached on the question of environmental protection, there is small likelihood that a rush to gain riches from Antarctic mineral resources will materialize from any quarter.

At the Ninth Consultative Meeting of the treaty nations, in 1977, the nations adopted a statement entitled Recommendation IX-1 Antarctic Mineral Resources (Anonymous 1977). In it the contracting parties endorsed the following principles:

(1) The Consultative Parties agree to play an active and responsible role in dealing with the question of mineral resources in Antarctica; (2) the Antarctic Treaty must be maintained in its entirety; (3) protection of the unique Antarctic environment and of its dependent ecosystems should be a basic consideration; and (4) the Consultative Parties, in dealing with the question of mineral resources in Antarctica, should not prejudice the interests of all mankind in Antarctica.

Recommendation IX-1 also urges the United States, which will be the host government of the Tenth Consultative Meeting, to convene a group to consider the legal and political aspects of the mineral-resource issue. In addition, nationals of the treaty and other states are urged to refrain from all exploration and exploitation of mineral resources during the time the treaty nations are attempting to make progress toward the adoption of an agreed regime concerning Antarctic mineral-resource activities. Finally, the Ninth Consultative Meeting agreed to place the subject of "Antarctic Resources: The Question of Mineral Exploration and Exploitation" on the agenda of the Tenth Consultative Meeting, scheduled for 1979. This meeting could lead to a workable regime for the responsible management of Antarctic mineral

resources by the time the Antarctic Treaty comes up for review in 1991.

It is quite possible that no mineral resources, not even icebergs, will ever be exploited in Antarctica. But since no one can predict with any degree of confidence how Antarctica will figure in any future demand for mineral commodities, the risk associated with a laissez-faire attitude on the part of the treaty nations is too high.

The final chapter obviously is yet to be written on this matter. Yet, the future is not without hope. The treaty nations have, for instance, managed to gain an interim agreement on the protection of the Antarctic marine living resources (Anonymous 1977, p. 15). Continued experience by the consultative nations in resolving these complex issues should lead to greater confidence in tackling the more difficult problem of an Antarctic mineral-resource regime.

References

Adie, R. J., ed. 1964. *Antarctic Geology: Proceedings of the First International Symposium on Antarctic Geology, Cape Town, 16–21 September 1963.* Amsterdam, The Netherlands: North-Holland.

————. 1972. *Antarctic Geology and Geophysics: Symposium on Antarctic Geology and Solid Earth Geophysics, Oslo, 6–15 August 1970.* Oslo, Norway: Universitetsforlaget.

American Geographical Society. 1970. *Antarctica,* 1:5,000,000. American Geographical Society.

Anonymous. 1977. Recommendations adopted at the Ninth Antarctic Treaty Consultative Meeting. *Antarctic Treaty: Report of the Ninth Consultative Meeting, London, 19 September–7 October,* pp. 11–13. London, England: Foreign and Commonwealth Office.

Antarctic Journal of the United States 8 (3). 1973. (Key papers related to cruises of *Eltanin*.)

Barrett, P. J., et al. 1972. The Beacon Supergroup of East Antarctica. In *Antarctic Geology and Geophysics,* ed. R. J. Adie, pp. 319–32. Oslo, Norway: Universitetsforlaget.

Biro, P. 1976. Petroleum development in central and southern Africa in 1975. *Am. Assoc. Petroleum Geol. Bull.* 60 (10): 1813–64.

Brobst, D. A., and W. E. Pratt, eds. 1973. *United States Mineral Resources.* USGS Professional Paper 820, p. 4.

Central Intelligence Agency. 1978. *Polar Regions Atlas,* CIA publication GC 78-10040.

Craddock, C. 1972a. *Geologic Map of Antarctica,* 1:5,000,000. American Geographical Society.

————. 1972b. Tectonic evolution of the Pacific margin of Gondwanaland. In *Gondwana Geology,* pp. 609–18. Canberra: Australian National University Press.

————, ed. In press. *Proceedings of the Third International Symposium on Antarctic Geology and Geophysics, Madison, Wisconsin, 1977.* Univ. of Wisconsin Press.

Craddock, C., et al. 1970. Geologic Maps of Antarctica. *Antarctic Map Folio Series,* Folio 12. American Geographical Society.

Denton, G. H., et al. 1971. The late Cenozoic glacial history of Antarctica. In *Late Cenozoic Ice Ages,* ed. K. Turekian. Yale Univ. Press.

Dunbar, M. F., ed. 1977. *Polar Oceans.* Calgary, Canada: Arctic Institute of North America.

Durkee, E. 1976. Petroleum developments in Australia in 1975. *Am. Assoc. Petroleum Geol. Bull.* 60 (10): 1957–64.

Du Toit, A. 1937. *Our Wandering Continents: An Hypothesis of Continental Drifting.* Edinburg, England: Oliver and Boyd.

Elliot, D. H. 1975. Tectonics of Antarctica: A review. *Am. J. Sci.* 275-A:45–106.

———. 1977. *A Framework for Assessing Environmental Impacts of Possible Antarctic Mineral Development,* part 1. Institute of Polar Studies, Ohio State Univ. (U.S. Dept. Commerce, National Technical Information Service, PB-262 750.)

Ford, A. B., and W. W. Boyd, Jr. 1968. The Dufek intrusion, a major stratiform gabbroic body in the Pensacola Mountains, Antarctica. *Proceedings of the 23rd Congress on International Geology, Prague, 1968,* sec. 2, pp. 213–28.

Franklin, D. H., and B. Clifton. 1971. Halibut Field, Southeastern Australia. *Am. Assoc. Petroleum Geol. Bull.* 55 (8): 1262–79.

Goodell, H. G., et al. 1973. Marine sediments of the Southern Oceans. *Antarctic Map Folio Series,* Folio 17. American Geographical Society.

Gordon, A. L. 1971. Recent physical oceanographic studies of Antarctic waters. In *Research in the Antarctic,* ed. L. O. Quam, pp. 609–29. AAAS Publication, no. 93.

Gordon, A. L., and R. D. Goldberg. 1970. Circumpolar characteristics of Antarctic waters. *Antarctic Map Folio Series,* Folio 13. American Geographical Society.

Hayes, D. 1976. The circum-Antarctic seafloor and subsea floor. In *A Framework for Assessing Impacts of Possible Antarctic Mineral Development,* ed. D. Elliott, part 2, appendix, section B. Institute of Polar Studies, Ohio State Univ.

Heezen, B. C., and M. Tharp. 1972. Morphology of the earth in the Antarctic and Subantarctic. *Antarctic Map Folio Series,* Folio 16. American Geographical Society.

Heirtzler. J. R. 1975. The Southern Ocean floor. *Oceanus* 18, no. 4: pp. 28–31.

Horn, D. R., et al. Metal content of ferromanganese deposits of the ocean. *Technical Report No. 3.* NSF Office of International Decade of Ocean Exploration (1973, unpubl.).

Hult, J. L., and N. C. Ostrander. 1973. *Antarctic Icebergs as a Global Fresh Water Resource.* The Rand Corporation, R-1255-NSF.

Hunter, D. 1978. The Bushveld complex and its remarkable rocks. *Am. Sci.* 66:551–59.

Husseiny, A. A., ed. 1978. *Iceberg Utilization: Proceedings of the First International Conference, Ames, Iowa, 2–6 October 1977.* Pergamon Press.

Kapitsa, A. P. 1967. Antarctic glacial and subglacial topography. In *Proceedings of the Symposium on Pacific-Antarctic Sciences, University of Tokyo, 1966,* ed. I. Nagata, pp. 82–91. Tokyo, Japan: Department of Polar Research, National Science Museum.

Katz, H. R. 1976. Petroleum development in New Zealand in 1975. *Am. Assoc. Petroleum Geol. Bull.* 60 (10): 1947–56.

McIver, R. D. 1975. Hydrocarbon gases in canned core samples from Leg 28 Sites 271, 272, and 273, Ross Sea. In *Initial Reports of the Deep Sea Drilling Project, Volume 28,* ed. D. E. Hayes et al., pp. 815–17. U.S. Govt. Printing Office.

Nazarov, V. S. 1962. L'dy antarkticheskikh vod, Results of the IGY. *Okeanologiya,* no. 6. Academy of Sciences, Soviet Geophysics Committee.

Panitch, M. 1977. Alaskan gas: Impact of pipeline on Canadian north stirs debate. *Science* 195:1308–9.

Ravich, M. G., and G. E. Grikurov, eds. 1978. *Geologic Map of Antarctica.* Moscow: Ministry of Geology of the USSR.

Ravich, M. G., and D. S. Solov'ev. 1969. *Geology and Petrology of the Mountains of Central Queen Maud Land (Eastern Antarctica).* Jerusalem: Israel Program Science Translations.

Rowley, P. D., et al. 1975. Copper mineralization along the Lassiter Coast of the Antarctic Peninsula. *Econ. Geol.* 70:982–87.

Runnells, D. R. 1970. Continental drift and economic minerals in Antarctica. *Earth and Planetary Science Letters* 8:400–02.

SCAR Bulletin. 1977. Antarctic resources—effects of mineral exploration (initial response by SCAR, dated May 1976, to Antarctic Treaty Recommendation V111–14). *Polar Record* 18 (117): 631–36.

SCAR Manual. 1972a. Antarctic Treaty, 1959, pp. 81–86. Cambridge, England: International Council of Scientific Unions, Scientific Committee on Antarctic Research.

SCAR Manual. 1972b. Selected recommendations for consultative meetings I–VI of the Antarctic Treaty Nations, 1961–70, pp. 87–125. Cambridge, England: International Council of Scientific Unions, Scientific Committee on Antarctic Research.

Shackleton, N. J., and J. P. Kennett. 1974. Paleotemperature history of the Cenozoic and the initiation of Antarctic glaciation: Oxygen and carbon isotope analyses in DSDP sites 277, 279, and 278. In *Initial Reports of the Deep Sea Drilling Project,* ed. J. P. Kennett et al., vol. 29. U.S. Govt. Printing Office.

Spivak, J. 1974. Now, the energy crisis spurs idea of seeking oil at the South Pole. *Wall Street Journal,* 21 February 1974, p. 1.

Stewart, D. 1964. Antarctic mineralogy. In *Antarctic Geology: Proceedings of the First International Symposium on Antarctic Geology, Cape Town, 16–21 September 1963,* ed. R. J. Adie, pp. 395–401. Amsterdam, The Netherlands: North-Holland.

Swithinbank, C. W. M. 1969. Giant icebergs in the Weddell Sea, 1967–68. *The Polar Record* 14 (91): 477–78.

U.S. Department of State. Antarctic Mineral Resources, Annex C, Document presented by the U.S. Delegation to the Special Preparatory Meeting in Paris, 28 June 1978 (1976, unpubl.).

Weeks, W. F., and W. J. Campbell. 1973. Icebergs as a freshwater source: An appraisal. *J. Glaciology* 12:207–33.

Weeks, W. F., and M. Mellor. 1977. Some elements of iceberg technology. In *Iceberg Utilization,* ed. A. A. Husseiny, pp. 45–98. Pergamon Press.

Wegener, A. 1929. *The Origin of Continents and Oceans,* 4th rev. ed., trans. John Biram (1962). Dover.

Wright, N. A., and P. L. Williams. 1974. *Mineral Resources of Antarctica,* Geological Circular 705. Reston, Virginia: USGS National Center.

Zambrano, J. 1975. Perspectivas petroliferas de la platforma continental Argentina. *Revista Petrotecnia,* vol. 15. Buenos Aires, Argentina: Instituto Argentino del Petroleo.

Zumberge, J. H., convenor. 1977. A Preliminary Assessment of the Environmental Impact of Mineral Exploration/Exploitation in Antarctica. (Available from SCAR Secretariat, Scott Polar Research Institute, Lensfield Road, Cambridge, England.)

"Our plan is to extract sulphates, bromides, copper, silver, and gold from sea water. All we've managed to get so far, however, is salt."

Sam H. Patterson

Aluminum from Bauxite: Are There Alternatives?

With the cost of bauxite escalating, alternate raw materials and processes for manufacturing aluminum are under close investigation

Less than a century ago, aluminum was considered a rare metal. In 1885, at the dedication of the Washington Monument in Washington, D.C., an aluminum casting weighing 100 ounces was installed on the tip of the monument. The casting is thought to have been the largest single mass of aluminum made up to that date. It was valued at $1.10 an ounce and was considered such a notable achievement that it had been displayed at Tiffany's in New York City. Only a few years ago, a pound of aluminum was priced at 23¢ and is now 44¢ to 48¢.

Because of the technological advances that have made the production of aluminum from bauxite in large quantities possible, our dependency on the metal has grown tremendously during the last few decades. Countries supplying bauxite are demanding greater and greater returns for their exports, and thus alternate materials and processes are being considered. The U.S. Geological Survey is compiling information on domestic alunite resources, investigating kaolin and bauxitic clay in Georgia, and updating information on both do-

mestic and worldwide resources of bauxite and other potential sources of aluminum. This article is a byproduct of that effort.

Aluminum is a silvery white, ductile, lightweight (specific gravity of 2.7) metal with efficient electrical conductivity and resistance to oxidation. Because of its properties, abundance, and low cost, aluminum has become one of the most important metals in industry; only iron surpasses it in tonnages used. Major uses of the metal are in the construction industry and in the manufacture of aircraft, motor vehicles, electrical equipment and supplies, fabricated metal products, machinery, beverage cans, containers, and a wide variety of home-consumer products. The availability of aluminum plays an important role in maintaining our standard of living, and it is a strategic commodity for military defense.

Plastics, wood, and metals including copper, steel, lead, zinc, magnesium, and titanium could be substituted for some of the uses for aluminum. However, none of these materials has all the properties of aluminum, except for a few specialized uses. Both alumina (the oxide of aluminum, Al_2O_3) and bauxite (the major ore of aluminum) are used in the manufacture of refractories, abrasives, and chemicals.

The breakthrough that made possible the growth in use of aluminum came in two steps. The first was improvements by European chemists in methods of processing bauxite and preparing alumina, the intermediate product required for the production of aluminum. This work led to the flexible Bayer process in which bauxite, under pressure and moder-

ately high temperature, is subjected to an alkaline leach. The second and more significant advancement came in 1886, when C. M. Hall, in the United States, and Paul L. T. Héroult, working independently in France, both discovered a commercially successful electrolytic process for reducing alumina to metal. The Hall-Héroult process opened the way to a new industry, and it is used today, basically unchanged, by most aluminum manufacturers.

Bauxite, defined as aggregates of aluminous minerals, more or less impure, in which aluminum is present as hydrated oxides, has been the source of all but minor quantities of the aluminum metal produced to date. Though the Earth's crust is 8% aluminum and there are literally hundreds of aluminum-bearing minerals and vast quantities of aluminous rocks, bauxite best fulfills the requirements of an ore of this metal. It is richest in aluminum of all rocks occurring in large quantities. Natural alumina—the mineral corundum (including the gem varieties ruby and sapphire)—does not occur in large deposits of high purity and therefore is an impractical source for making the metal. Most of the many abundant nonbauxite aluminous minerals are silicates, and, like all silicate minerals, they are refractory, resistant to chemical dissolution, and extremely difficult to process. The aluminum silicates are therefore generally not competitive with bauxite because considerably more energy is required to extract alumina from them.

The fact that alumina is an essential intermediate product in the preparation of aluminum should be noted because it has been overlooked even

Dr. Patterson is the Commodity Geologist for aluminum and clays for the U.S. Geological Survey in Reston, Virginia. He is a member of six professional societies and the editorial board of Economic Geology and is the author or coauthor of numerous reports on bentonite, fuller's earth, kaolin, refractory clay, and U.S. and world bauxite. The author wishes to acknowledge the helpful suggestions of D. P. Cox and H. Klemic of the U.S. Geological Survey, W. D. Mitchell of Reynolds Metals Co., J. W. Shaffer and C. P. Bingham of Kaiser Exploration and Mining Co., and J. A. Barclay and H. F. Kurtz of the U.S. Bureau of Mines. Address: U.S. Geological Survey, Reston, VA 22092.

in some of the most authoritative chemistry texts. Alumina is made from bauxite commonly at great distances from electrolytic plants, and large quantities of this oxide are shipped across oceans.

The use of aluminum, after the invention of the Hall-Héroult process, progressed slowly at first but accelerated rapidly in the last two decades (Fig. 1). World production in 1918, the peak World War I year, was only 144,000 short tons, and in 1943, the year of maximum production during World War II, only 2.1 million tons were produced. The maximum annual production to date was in 1974, when the total world output was 14.5 million short tons. For comparison, it is interesting to note that, in recent years, world production of aluminum has been nearly twice that of copper but only a fraction of the total for iron, which is the only metal used more by man.

World bauxite resources are far greater than imagined a generation ago, and they are more than adequate to sustain the growing demands for aluminum well into the next century. Most of the world bauxite reserves, estimated to be more than 17 billion tons, plus subeconomic resources that are two or more times this amount, have become known mainly through discoveries since World War II. The largest reserves are in laterite deposits in tropical regions of Australia and in the Republic of Guinea, and in recently discovered deposits in the Amazon Basin, India, Indonesia, Cameroon, and elsewhere.

The fundamental reasons for the exploration for and the discovery of the very large lower-grade reserves were the rapid exhaustion of high-grade deposits, which spurred the continued and diligent efforts by industry and government agencies, and the aluminum industry's ability to use lower-grade bauxite. At the beginning of World War II, the U.S. alumina plants were designed to process bauxite containing 55% Al_2O_3 or more, no more than 8% SiO_2 (silica), and a low percentage of iron. During that war, a combination Bayer process was developed and applied to bauxite having a lower alumina content and containing 8 to 15% SiO_2. By 1950, the U.S. aluminum industry had learned to process high-iron Jamaican bauxite. Elsewhere, ways were found to process very low-grade, high-silica bauxite. For example, deposits now mined on a large scale in the Darling Range, Australia, contain available alumina in the 30–45% range and much SiO_2 that is nonreactive in the Bayer process.

Politics and bauxite

Because of recent demands by the major bauxite-producing countries for greater returns for exporting their natural resources, bauxite is now considered a critical commodity in the U.S. Political activities by the major bauxite-producing countries have taken the form of nationalization of the industry, sharply increased taxes and royalties, government operation of new mines and plants and joint ownership with industry, and efforts for cartel-type joint price control.

The first nationalization step in the present cycle came in 1971 when Guyana took over the holdings of Alcan International (1975) Ltd., in that country. Reynolds Metals Co. holdings in Guyana were nationalized in January 1975. Purchase of company-owned bauxite lands by the Jamaican Government, beginning in 1974, also is a form of nationalization, and other countries, including Surinam, have bought aluminum-industry investments.

The pace in new revenues was set by Jamaica in the summer of 1974, when the government enacted legislation imposing a bauxite-production levy, retroactive to January of that year. This levy is based on a percentage of the price of primary aluminum; the percentage has varied from 7.5 to 8.5% in 1976. At a price of 48¢ per pound for primary aluminum, the levy ranges from $16.75 to $19.00 per long dry ton of bauxite. The rate of the royalty paid by the companies on Jamaican bauxite production was also increased to U.S. 55¢ per ton. Prior to these increases, taxes and royalties on bauxite came to about $2 per ton. Haiti, the Dominican Republic, and Surinam have exacted similar new bauxite-production taxes. These four countries supply nearly 80% of the bauxite requirement of the U.S. aluminum industry, and the new levies have caused an increase in the cost of primary aluminum of 2¢ or more per pound.

Government ownership of mines and plants and joint ownership with industry either exist or are planned in several countries. According to one plan for an integrated aluminum-producing complex, which has been announced by the governments of Jamaica and Mexico, Jamaica will mine 1.3 million metric tonnes of bauxite per year and build a caustic soda plant and an alumina plant, and Mexico will build an aluminum

Figure 1. The world production of aluminum has increased greatly since the 1885 development of the Hall-Héroult electrolytic process.

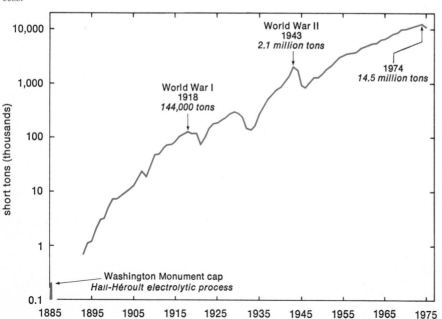

smelter of 132,000 short tons capacity. Another suggested joint venture is for an alumina plant on Tobago to be operated by the governments of Jamaica, Guyana, and Trinidad. The government of Guinea, a country whose bauxite reserves rank with Australia's as the world's largest, operates bauxite mines at Boké and Kimbo-Fria, jointly with a consortia of U.S. and European companies, and at Kindia with assistance from the Soviet Union. Another major joint government-industry-owned bauxite district scheduled to be a producer soon is in the Trombetas region in the Amazon Basin. Cia. Vale do Rio Doce, a company representing the Brazilian government, owns 41% of this mining operation; the consortium owning the remainder includes corporations in Brazil, Canada, The Netherlands, Norway, Spain, the United Kingdom, and the United States.

The International Bauxite Association (IBA) was organized in 1974 as part of a broader effort to expand the participation of the developing countries in management of the world's resources and in the distribution of the world's wealth. Some of things IBA is reported to be looking into are employment of the host countries' citizens by foreign corporations, particularly in high executive positions, possibilities for bauxite shipping fleets, and the sharing of aluminum technology that would help member countries build their own plants. In 1976, eleven countries—Australia, the Dominican Republic, Ghana, Guinea, Guyana, Haiti, Indonesia, Jamaica, Sierra Leone, Surinam, and Yugoslavia—belonged to the IBA. These countries produce more than 75% of the Western world's bauxite.

The search for nonbauxite sources

Though world bauxite resources are very large, a vigorous upsurge in research and evaluation of nonbauxite sources of aluminum is underway in response to economic and political pressures. Nonbauxite sources of aluminum have long been sought in the industrialized countries, but the 1973 energy crisis and subsequent inflation motivated much of the renewed activity. The principal concern

Figure 2. The manufacture of one pound of aluminum from bauxite requires many materials and much energy.

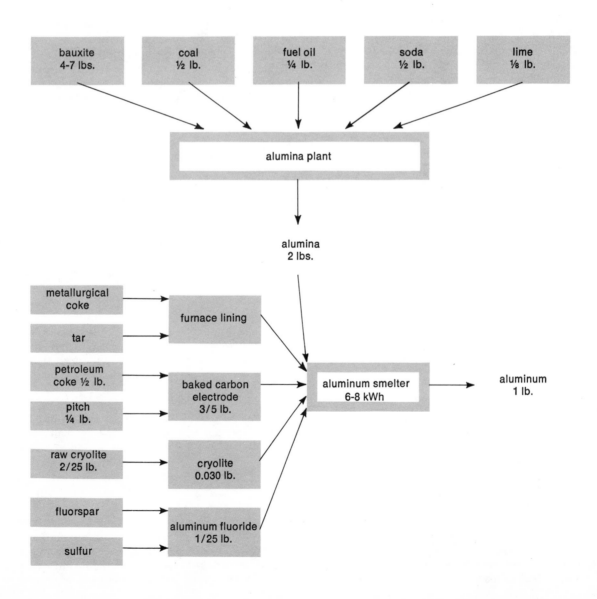

in producing aluminum is transportation and electricity costs, as bauxite or alumina must be shipped great distances, and 6 to 8 kilowatt hours of electricity are required to reduce 2 pounds of alumina to 1 pound of metal (Fig. 2). Though the major share of the energy needed for aluminum is for power for the electrolytic processing, significant quantities are needed to extract alumina from the basic raw material. Because of the inflation in the cost of making alumina from bauxite, interest in a cheap process for recovering it from nonbauxite sources is growing.

The upsurge in research on nonbauxite sources of aluminum is in both industry and government, particularly in the U.S., Canada, France, Scandinavia, the USSR, and East European countries. These industrialized countries in the Northern Hemisphere lack adequate bauxite reserves, but most have very large resources of other aluminous rocks and minerals.

Nonbauxite sources of aluminum are not a new challenge. During World War II, several countries had to devise processes for such materials, but in most cases the need did not last beyond the war. During the war, a small part of the German supply of aluminum was made from domestic clay (Table 1) and andalusite, Al_2SiO_5, concentrated from mine wastes in Sweden. Part of Japan's aluminum was made from Korean and domestic alunite, domestic clay, and Chinese, Manchurian, and Korean shale. In the U.S., four experimental plants were built by the government to test the feasibility of extracting alumina from nonbauxite materials: (1) Laramie Range anorthosite in a plant in Laramie, Wyoming; (2) South Carolina kaolin, in Harleyville, South Carolina; (3) alunite from the Marysvale district in Utah, in Salt Lake City; and (4) clay from Oregon and Washington, in Salem, Oregon. None of these plants produced commercial alumina.

Although minor quantities of aluminum were made from nonbauxite rocks in several countries during the World War II emergency, the USSR has the only viable peacetime aluminum industry based partly on such materials. The nonbauxite ores of aluminum in the USSR are a nepheline syenite that contains apatite—

$Ca_5(F, Cl)(PO_4)_3$—on the Kola peninsula, a related igneous rock in central Siberia, and alunite in the Zaglik district in Azerbaijan.

The Kola nepheline operation, active for more than a decade, was possible because of the large production of apatite for fertilizer, the availability of nepheline concentrate reported to contain 28 to 30% Al_2O_3, and the production of portland cement as a coproduct. In the process used to recover alumina sufficiently pure for electrolytic reduction, the nepheline concentrate is mixed with limestone and ground and sintered at 1,300°C, producing dicalcium silicate and aluminates of sodium and potassium. The sinter is ground and the alkali aluminates leached with caustic soda solution; aluminum hydroxide is

precipitated from the solution by treatment with carbon dioxide. Nepheline syenite masses in Egypt have been investigated with Soviet assistance as possible sources of alumina for an aluminum plant now powered by electricity generated at the high Aswan Dam. The central Siberian syenite supplies an alumina plant of recent vintage reported to have a capacity of 500,000 tons per year. Alunite apparently supplies a small alumina plant that has had difficulty fulfilling production quotas.

The U.S. situation

Aluminum was recognized as one of the materials shortages by the U.S. Congress in 1974, and it is considered one of the critical shortages by the

Table 1. Mineralogy and alumina content of nonbauxite rocks and materials that are possible sources of aluminum

Rock or material	Major aluminous minerals and structural formulas	Theoretical percent of Al_2O_3 in mineral	Percent of mineral in whole rock or material	Percent of Al_2O_3 in whole rock or material
Aluminum phosphate rock*	wavellite $Al_3(PO_4)_2(OH)_3 \cdot 5H_2O$	37.1	—	—
	millisite $(NaK)CaAl_6(PO_4)_4(OH)_9 \cdot 3H_2O$	37.4	variable	5–15
	kaolinite $(OH)_8Si_4Al_4O_{10}$	39.5	—	—
Aluminous shale	illite $(OH)_4K_2(Si_6Al_2)(Mg \cdot Fe)_6O_{20}$	38.5	variable	15–30?
	kaolinite	39.5	—	—
Alunite rock	alunite $KAl_3(SO_4)_2(OH)_6$	37.0	15–40	5.5–14.8
Coal washings	illite and kaolinite	38.5 } 39.5 }	variable	25–30?
Coal ash	mullite (?) $Al_6Si_2O_{13}$	84.0	minor?	5–35
	glass	?	variable	—
Dawsonite in oil shale	dawsonite $NaAl(OH)_2CO_3$	35.4	8±	3
High-alumina clay	kaolinite	39.5	75–95	>30–35
Igneous rocks, anorthosite	plagioclase feldspar $NaAlSi_3O_8$ $CaAl_2Si_2O_8$	19.4 36.6	65–95	22–30
Nepheline syenite	nepheline $(Na,K)(Al,Si)_2O_4$ alkalic feldspar $KAlSi_3O_8$	35.9 18.4	— 80±	— 20–25†

*Several related aluminum phosphate minerals are not listed.
†Nepheline concentrate from Kola Peninsula used for alumina in USSR is reported to be 28–30 percent Al_2O_3.

Interior Department. The Bureau of Mines and other government agencies foresaw this shortage earlier and in 1970 supported a National Materials Advisory Board panel of industry experts that evaluated processes for extracting alumina from nonbauxite resources. In order to investigate several processes for using nonbauxite materials, this panel recommended that two 5-ton-per-day pilot plants be built by the Bureau of Mines, with financial help and cooperation from aluminum producers, for the purpose of process and cost evaluations. The first raw material investigated was kaolin, because of its high Al_2O_3 content (30 to 35%) and its availability in large tonnages. One pilot plant was to test a hydrochloric acid process and the other, a nitric acid process for extracting alumina from clay. A recommendation was also made to build a pilot plant with a capacity of 50 to 100 tons per day, so that data necessary for scaling up to commercial-plant size could be obtained if the results from the first plant were successful.

As an outgrowth of these recommendations, the Bureau of Mines began operating a miniplant at its Boulder City, Nevada, facility in July 1973, and subsequently ten companies joined in on a cooperative basis. The nitric-acid processing of Georgia kaolin was tested first. A miniplant for processing clay by using hydrochloric acid is now being operated, and one for the lime-sinter processing of anorthosite is under construction. A contract is being negotiated with an engineering construction firm to select the process having the greatest potential for commercial success and to prepare the preliminary design of a 10- to 50-ton-per-day pilot plant based on that process.

In addition to providing advice and financial support for Bureau of Mines testing of extraction processes for high-alumina clays, aluminum companies have carried on their own research. Industry's view of high-alumina clays as a promising alternate source of alumina is indicated by its renewed interest in acquiring clay lands in Georgia.

The largest domestic resources of the better grades of high-alumina clay (30 to 35% Al_2O_3) are in the Georgia-South Carolina kaolin belt, the Andersonville district of Georgia, and in

deposits associated with bauxite in Arkansas. Other extensive sources are the underclays below coal beds in Appalachia and other regions. Large tonnages of these clays are lost each year, because they are not recovered when the coal is strip-mined and the land reclaimed.

In many parts of the world, varieties of igneous rocks containing low silica, iron, and magnesium have attracted considerable interest as a possible source of alumina because they occur in very large masses and because of the success of the Soviets in processing nepheline-bearing rocks. In most countries, calcium-rich anorthosite, an igneous rock very rich in plagioclase, is considered more promising than nepheline syenite because of its higher alumina content.

In the U.S., Alcoa purchased 8,000 acres (3,238 hectares) of anorthosite land in the Laramie Range, Wyoming, in 1972. This mountain range is estimated to contain as much as 30 billion tons of anorthosite to depths of no more than 33 m below the surface, and the mass probably extends to much greater depths. The alumina content of this rock is 25 to 30%; therefore, the holdings of this one company, extending over nearly half the anorthosite outcrop, contain more aluminum than our nation can use for many decades. Alcoa has announced that research on the recovery of alumina from this rock is under way, and it is their hope to reduce the costs of the process used by the government during and following World War II; the results to date are privileged information.

Several of the anorthosite masses in the U.S. are in scenic mountain ranges where the environmental impact of mining would have to be considered. The Laramie Range is in an unpopulated region, in a state having many mountains, and large-scale mining would have a minimum effect on the environment. Other large masses of anorthosite occur where environmental restrictions might rule out mining, such as in parklands in the Adirondack Mountains in New York.

Another raw material for the production of alumina being studied in the U.S. is alunite, a mineral now used in the USSR. Investigations in the U.S. center on deposits in the Wah

Wah Mountains in southwestern Utah, which contain an estimated 680 million tons of alunitized rock; much of it is 40% alunite, or 14.8% Al_2O_3. This alunite is controlled by a joint venture of Earth Sciences, Inc., National Steel, Inc., and National Southwire, Inc. In a pilot plant in Golden, Colorado, Earth Sciences is evaluating its own process together with modifications in technology obtained from the USSR. The process involves a reduction roast to drive off SO_2, a water leach to remove K_2SO_4, and a caustic digestion to take Al_2O_3 into solution and to permit separation of it from SiO_2, an impurity in the alunite rock. If the pilot-plant tests are successful, a large plant is to be built near the deposits in Utah. Plans call for the SO_2 from the process to be used for making superphosphate from ore shipped from Idaho; K_2SO_4 will be produced as another fertilizer ingredient.

Coal-mine washings, waste, and ash are considered potential sources of alumina. Alcoa has announced that pilot-plant tests of the recovery of alumina in coal washings have been made with favorable results. Little is known about the amount of coal washings available or their mineralogical and alumina content. However, impurities in coal are known to be quite variable, and probably this is true of alumina concentrations in washings. Tested washings reportedly contain about 28% Al_2O_3. Most aluminous minerals in washings are fine particles, and kaolinite is abundant in the coal-bed underclays, which are also considered an aluminum resource. The little research that has been published on the subject suggests that the alumina in washings is much more soluble than that in coal ash. Solubility increases with calcining to about 500°C and drops sharply at the higher temperatures reached in burning. Coal washings are attractive because coal companies bear the mining costs and because the washings contain significant energy potential in the form of fine-grained coal, which may provide some of the fuel necessary for calcining the washings. The future of coal washings as a source of alumina is uncertain, but large quantities are likely to become available because of increased demands for coal resulting from the energy crisis.

Coal waste was investigated as an

alumina source in the early 1960s in a pilot plant operated by the North American Coal Co. at Powhatan Point, Ohio. The material tested was roofstone that had to be removed in mining Pittsburgh coal to provide headroom. The tests were unsuccessful, but this type of waste may be worth further consideration, because of the large quantities available and because it might be practical to mix the waste with coal washings and recover alumina from both simultaneously.

Coal ash is generally considered to have little potential as a source of alumina in the U.S., probably because of the low solubility of the alumina it contains and because sufficiently large quantities are not available at any one place. However, the amount of coal ash is likely to increase sharply as the demands for energy and a clean environment force the concentration of large power-generating centers in coal fields, where atmospheric emission can be controlled more effectively than at many scattered plants.

Wastes from coal mining and coal ash probably have attracted more attention overseas than in the U.S. An experimental plant for extracting alumina from ash was active in Czechoslovakia during World War II and apparently for a few years afterward. Coal ash, aluminous rocks associated with coal, and limestone, reportedly, are to be the raw materials for a large alumina plant planned in Poland. Research on extraction of alumina from coal ash is being carried on in the USSR; one of the possible sources of ash is the large coal-fired generating plant at Alma Ata, Kazakh S.S.R.

A potentially important process that may be suitable for recovery of alumina from high-alumina clay, coal ash, aluminous shale, and other aluminous rocks with low calcium content is the PUK, or Pechiney H+, process. Developed by Pechiney-Ugine-Kuhlmann in France, this process is reported to involve a sulfuric-acid treatment, filtering, crystallization of aluminum sulfate, filtering, a gaseous hydrochloric-acid treatment, precipitation of aluminum chloride, and reduction to alumina by calcination. Alcan has joined Pechiney in the construction, now under way, of a large pilot plant to evaluate

the process. This venture by the Canadian and French companies, which are two of the six leading aluminum producers in the Western world, ranks with the principal efforts to find sources of nonbauxite aluminum.

Alumina sources that are receiving less intense study in this country include aluminous shale, aluminum phosphate rock, and dawsonite. In the U.S., only the aluminous shale in coal-mine wastes has been tested in the last few years. However, in Europe, aluminous shale itself has been investigated as a source of aluminum. A Scandinavian shale is of interest mainly because of its uranium content; although it is reported to contain less than 20% Al_2O_3, plans to process 6 million tons a year for uranium make its alumina attractive. The uranium processing may leach out about 55,000 tonnes of alumina per year.

In the 1950s, uranium-bearing aluminum phosphate rock in a weathered zone overlying the productive deposits in the Florida phosphate field was investigated as a source of aluminum as well as phosphate and uranium. Several patents for processes were issued, but no production has been attempted. A recent news item in a mining magazine, however, indicated that a French company has started a thorough study of methods for refining aluminum from phosphate rock in Senegal. The recovery of aluminum and phosphate from Senegal's very large resources of phosphate rock has long been an attractive prospect.

Another potential source is dawsonite, a colorless or white soluble mineral that is a minor constituent of oil shale in the Piceance basin in Colorado. Because of its solubility and abundance—billions of tons of it are present in oil shale—it has been studied as a source of aluminum by the Denver Research Institute, Bureau of Mines, and several oil companies. Patent applications have been made for processes for recovery of alumina from spent oil shale and for recovery of alumina underground should any of the schemes for in situ extraction of oil from shale be successful. Dawsonite is found in several countries around the world, but only the USSR is reported to have investigated its deposits—in the Kusnetsk

and Pripyat basins—as a source of aluminum.

The outlook

Escalation of the costs of bauxite, because of increases in taxes in producing countries, and inflation in energy costs are opposing factors bearing on the development of nonbauxite sources of alumina. Unquestionably, the tax increases have greatly narrowed the cost gap between bauxite and alternate sources. Industry representatives claim that the cost of alumina from nonbauxite materials is only 20% higher than from bauxite, but they want further research into the "economics." Probably, nonbauxite-alumina processing would be feasible were it not for inflation in energy prices and uncertainties about future supplies, because processing nonbauxite materials requires about twice as much energy as Bayer processing of bauxite.

For the foreseeable future, bauxite will continue to be the major source of aluminum in the U.S. and the rest of the noncommunist world, for several reasons. The industry is based on Bayer-processed bauxite, and a shift to another process and material would involve major changes in technology. Plant construction and debugging necessary for significant production would require more than five years, barring an all-out emergency effort. Another reason is that the industry has major overseas investments in mining, processing, and shipping installations (approximately 600 million dollars in Jamaica alone). Under the free-enterprise system, less loss may be suffered in the long range by paying taxes than by writing off investments. A third reason is that the present industry-government joint ownership of mines and other installations in some countries is based on long-term agreements, and more such agreements seem certain. Some countries with bauxite resources need industry technology and capital, and, with joint agreements, the returns to their governments are in the form of profits rather than taxes—for example, in the Amazon Basin, where very large bauxite deposits have been discovered recently. Thus supplanting bauxite by other aluminous materials seems premature, unless circumstances change drastically in the near future.

References

Barclay, J. A., and F. A. Peters. 1976. New sources of alumina. *Mining Cong. J.* 62(6): 29–32.

Guda, Henri A. M. 1975. Organization and priorities of the International Bauxite Association. In Proceedings of Bauxite Symposium No. 3, Kingston, May 17, 1975. *Geol. Soc. Jamaica J.* pp. 7–21.

Miller, G. S. 1976. Foreign-supply fears spur search in industry for alternate materials, aluminum research pushed as bauxite levies jump; economic, political risks, plastic outshines chromium. *Wall Street J.* 187(65):1, 27.

National Materials Advisory Board. 1970. Processes for extracting alumina from non-bauxite ores. Natl. Materials Advisory Board Rept. 278.

Parkinson, Gerald. 1974. Golden pilot plant points way to 500,000-tpy alumina-from-alunite mine and plant in Utah. *Eng. Mining J.* 175(8):75–78.

Shabad, Theodore. 1976. Soviet experiment in aluminum fails. *New York Times,* May 8, sec. 3, p. 5.

Shaikh, N. A. 1976. Non-bauxite sources of alumina in Scandinavia and Finland. Industrial Minerals Internatl. Cong., 2d, Munich, Germany, May 17–19, preprint.

U.S. Congress, House Committee on Interior and Insular Affairs, Subcommittee on Mines and Mining. 1974. Oversight hearings on mineral scarcity. U.S. 93d Cong., 2d Sess. Comm. Print, Serial No. 93-48.

U.S. Congress, Senate Committee on Government Operations, Permanent Subcommittee on Investigations. 1974. Materials shortages—aluminum. U.S. 93d Cong., 2d. Sess. Comm. Print.

U.S. Department of the Interior, Office of Minerals Policy Development. 1975. Critical materials: Commodity action analyses aluminum, chromium, platinum. U.S. Dept. Interior, Office Minerals Policy Develop., App. 1–6.

Walker, W. W., and D. N. Stevens. 1974. The Earth Sciences-National-Southwire alunite to alumina project. AIME Metallurgical Soc. TMS Paper Selection, Paper No. A74-64.

Index